The Electronic Packaging Series

Series Editor: Michael Pecht, University of Maryland

Published

Long-Term Non-Operating Reliability of Electronic Products
Judy Pecht and Michael Pecht

Advanced Routing of Electronic Modules
Michael Pecht and Yeun Tsun Wong

High Temperature Electronics
F. Patrick McCluskey, Richard Grzybowski, and Thomas Podlesak

Influence of Temperature on Microelectronics and System Reliability
Pradeep Lall, Michael Pecht, and Edward Hakim

Influence of Temperature on Microelectronics and System Reliability

Pradeep Lall
Advanced Manufacturing Technology Center
Motorola, Inc.
Plantation, Florida

Michael G. Pecht
CALCE Electronic Packaging Center
University of Maryland
College Park, Maryland

Edward B. Hakim
U.S. Army Research Laboratory
Fort Monmouth, New Jersey

CRC Press
Taylor & Francis Group
Boca Raton London New York

CRC Press is an imprint of the
Taylor & Francis Group, an **informa** business

CRC Press
Taylor & Francis Group
6000 Broken Sound Parkway NW, Suite 300
Boca Raton, FL 33487-2742

First issued in paperback 2019

ISBN-13: 978-0-8493-9450-8 (hbk)
ISBN-13: 978-0-367-40097-2 (pbk)
Library of Congress Card Number 96-39038

Library of Congress Cataloging-in-Publication Data

Lall, Pradeep.
Influence of temperature on microelectronics and system reliability / Pradeep Lall, Michael G. Pecht, Edward B. Hakim.
p. cm.
Includes bibliographical references and index.
ISBN 0-8493-9450-3 (alk. paper)
1. Microelectronics--Materials--Thermal properties. 2. Electronic packaging. 3. Electronic apparatus and appliances--Reliability.
I. Pecht, Michael. II. Hakim, Edward B. III. Title.
TK7870.25.L35 1997
621.381'046--dc20 96-39038
CIP

Visit the Taylor & Francis Web site at
http://www.taylorandfrancis.com

and the CRC Press Web site at
http://www.crcpress.com

Foreword

On the Arrhenius equation

It's but a walking shadow, a poor player,
That struts and frets his hour upon the stage,
And then is heard no more; it is a tale,
Told by an idiot, full of harm and danger,
Signifying nothing.

(After Macbeth)

This (somewhat adapted) quote crosses my mind every time I come across a paper on some aspect of Thermal Management of Electronic Parts or Systems in which the author refers to "a 10 ^{0}C decrease in temperature increases reliability by a factor of two," in order to motivate the reasons for his or her research. By now, there is no excuse to repeat the conclusions of doubtful experimental work that dates back to the Germanium age. The open literature is abundant with evidence that the statement has no relationship whatsoever with field failure data. Steady state temperatures are in most practical temperature ranges only weakly related to the physical failure mechanisms that rule reliability behavior. Clearly, some exceptions do exist but even these are not solved by launching "the statement".

The Arrhenius philosophy is still the basis for most specification protocols; unfortunately, this detracts the attention from the fact that the mechanisms that *are* responsible for reliability problems are not specified at all. As a result, a tremendous amount of money is lost in the search to meet useless specs, while far too little money is spent in the search for specs that make sense. Therefore, this book is most welcome, especially for people who are working in the area of micro-electronic devices, as it summarizes the state-of-the-art in reliability physics, points out the flaws underlying current specs, and provides physically sound alternatives.

Clemence Lasance
Sr. Scientist Thermal Management
Phillips Research Center

Preface

There has been a common belief that reliable electronics can be achieved by lowering temperature. Slogans such as "lower the temperature by 10 and double the reliability" have become the rules of thumb for some designers, without regard to the cost-effectiveness, and actual reliability benefits and auditability. The belief in the harmful effects of temperature has woven itself into today's screening and thermal management processes.

The often inappropriate importance given to the effects of steady state temperature is not without reason! Often the device failure signature, obtained from a "quick" failure analysis, appears to support the conclusion that steady state high temperature is the cause of failure. As an example, often design or misapplication errors result in failures characterized by melting or thermal breakdown. The common conclusion is that the temperature was too high. The question which arises and the one which is often left unanswered is — Was the failure a result of high temperature or was the failure a result of a more serious design problem which subsequently resulted in a temperature rise? While a high junction temperature could be a cause of failure, the high temperature could also have been produced from electrical over stress or accidental spikes of energy caused by lightening electrostatic discharge. Even in the case where a high junction temperature is the cause of failure, the amount of reliability improvement achieved by continually lowering the operating temperature is still a question which is seldom answered substantially.

Is steady state temperature alone, a sufficient icon for the effects of temperature on microelectronic device reliability? For example, temperature profile during temperature cycling is characterized by large spatial temperature gradients in the package resulting from thermal inertia and difference in thermal diffusivity of the package elements. The temperature gradients, often result in slow degradation of the die-package interface due to thermo-mechanical fatigue — an effect which is presently not even accounted for in temperature acceleration model for microelectronic devices.

An inescapable conclusion is that system reliability has been penalized unjustifiably by steady state temperature dependent models, and other temperature effects associated with temperature cycling, temperature gradient and time dependent temperature changes must receive more importance.

What this book is about

The purpose of this book is to raise the level of understanding of thermal design criteria. The goal is to provide the design team with sufficient knowledge to help them evaluate device architecture trade-offs and the effects of operating temperature, by answering the following questions.

- Is there a need for lowering temperature?
- If there is need for lowering temperature, what is the value of the lower temperature as a function of materials, manufacturing processes and device architecture and manufacture?
- How does the maximum operating temperature vary with microelectronic device design?
- What device design and manufacturing modifications are necessary to enhance the maximum allowable operating temperature for a desired mission life?
- How do manufacturing defect magnitudes affect the time to failure under temperature stress conditions?
- Is it possible to escape the penalty of the added cost and weight of a cooling system?

Who the book is for

This book is directed to the reader interested in the damage mechanisms associated with various forms of temperature stress in microelectronic packages. The reader should have prior knowledge of bipolar and MOSFET device fundamentals, semiconductor device physics, fracture mechanics, and elasticity dynamics.

This book will assist the reader with evaluating the effects of certain classes of manufacturing defects on operating life for high-temperature operation in order to tailor existing screens for maximum defect detection and overall device quality. Further, the book will assist the reader in evaluating the feasibility of escaping the penalty of added cost and weight of cooling system to achieve reliable design at high temperature.

What this book contains

The microelectronic package considered in the book is assumed to consist of a bipolar or MOSFET (silicon) semiconductor device; first-level interconnects that may be wirebonds, flip-chip, or tape automated bonds; die attach; substrate; substrate attach; case; lid; lid seal; and lead seal. Failure mechanisms actuated under various forms of temperature stress, including steady-state temperature, temperature cycling, temperature gradients, and time-dependent temperature change, have been identified for each of the package elements. This investigation covers damage mechanisms, approximately in the temperature range of -55°C to 125°C. At temperatures much higher or lower than this temperature range, the damage mechanisms in the microelectronic package can change considerably with respect to their stress dependencies. The temperature effects on electrical parameters of both bipolar and MOSFET devices have been investigated, and models quantifying the temperature effects on package elements have been identified. This book does not address damage mechanisms in device technologies other than bipolar and MOS, or assemblies such as circuit cards, printed wiring boards, sub-assemblies, and assemblies. Temperature- related models have been used to derive derating criteria for determining the maximum and minimum allowable temperature stresses for a given microelectronic package architecture.

Chapter 1 presents the motivation for this book, in the context of microelectronics reliability. Problems with some of the modeling strategies are outlined.

Chapters 2 and 3 expose the reader to microelectronic device failure mechanisms in terms of their dependence on steady state temperature, temperature cycle, temperature gradient, and rate of change of temperature at the chip and package level. The microelectronic package is considered to be an assembly of package elements, including the chip, chip metalization, operating devices on the chip, die attach, substrate, substrate attach, first-level interconnects (wirebond interconnects, tape automated bonds, flip-chip bonds), leads, lid, lead seal and lid seal. Physics-of-failure based models used to characterize these failure mechanisms are identified and the variabilities in temperature dependence of each of the failure mechanisms, are characterized. We also discuss the existence of temperature thresholds below which various failure mechanisms are not activated.

Chapters 4 and 5 expose the reader to the effects of temperature on the performance characteristics of MOS and bipolar devices. The parameters investigated for bipolar devices include the current gain, I-V characteristics, collector-emitter saturation voltage, and voltage transfer characteristics. The parameters investigated for MOS devices include threshold voltage, mobility, drain current, time delay, leakage currents, chip availability, dc voltage transfer characteristics and noise margins.

In Chapter 6, the reader is presented with the applicability of using high-temperature stress screens, including burn-in, for high-reliability applications is discussed. The burn-in conditions used by some manufacturers are examined, and a physics-of-failure approach is discussed. The physics-of-failure approach addresses the dominant failure mechanisms in the device architecture and tailors the screening stresses to effectively remove defective devices.

In Chapter 7, we overview, existing guidelines for thermal derating of microelectronic devices, which presently involve lowering the junction temperature. Then, the reader learns how to use physics-of-failure models presented in Chapters 2, 3, 4, 5 for various failure processes, to evaluate the sensitivity of device life to variations in manufacturing defects, device architecture, temperature, and non-temperature stresses. Stress margin curves for device life are derived for mechanisms with complex dependencies on stresses and defects. The cumulative effect of competing failure processes on device life is used to determine the values of operating temperature and non-temperature related stresses.

Acknowledgments and confessions

We acknowledge the help of **Joseph Kopanski** from NIST; **Fausto Fantini** from Telettra (Italy); **Sorin Witzmann** from BNR; Canada; **D.S.Campbell** from Loughborough University of Technology (U.K); **Pat O'Connor** from British Telecom (U.K); **J.R.Lloyd** from Digital Equipment Corp.; **Walter Winterbottom** from Ford Motor Company; **Michael Cushing** from AMSAA; **George L. Schnable** from David Sarnoff Research Center; **George A. Swartz** from RCA Lab.; **Steve Martell** from SONOSCAN; **Dennis Karr** from Airpax Corporation; **Charles T. Leonard** from Boeing Commercial Airplanes; **Vinod Maudgil** from CETAR Ltd.; **Anthony Gallo** from Dexter Corp.; **Dave Urech** from Allied Signal Aerospace Co., **Werner Engelmaier** from Engelmaier Associates; **Thomas Kwok** from IBM Research Center; **Robert Keyes** from IBM Research Division; **Lloyd Condra** from Eldec. Corporation; **Nick Lycoudes** from Motorola Semiconductor Products; **Imants Golts** formerly from Tektronix. Photos of Failure Analysis accumulated over years of root-cause analysis have been contributed by **Robert J. Mulligan** (worked for the National Research Corp. In Newton Massachusetts, prior to joining Motorola. He has worked at Motorola for 8 years with experience in various analytical equipment and is recognized as a sector expert in Acoustical Microscopy. He is currently a Senior Engineer at Motorola within the Electronic Materials Research Laboratory under the Advanced Manufacturing Technology Center at the Land Mobile Products Sector Facility in Plantation Florida) and **Russell Faryniak** (worked within the Aerospace and Computer Industries prior to joining Motorola. He has more than forty years of experience in design, development, manufacturing, quality and reliability related to electronic components. During his twenty-five year career with Motorola, he has defined detection techniques for a wide range of failure modes by applying analytical techniques including optical, fluorescence and scanning electron microscopy, radiology and chemical decapsulation).

Permissions are listed on page 299.

List of Symbols

α	-	coefficient of thermal expansion
α_h	-	hole-generation coefficient
α_r	-	temperature coefficient of resistance of the multilayer Metalization
$\beta_{an,ox}$	-	fraction of anode surface (that is susceptible to oxidation)
γ	-	field acceleration parameter
$\gamma_{ef,acc}$	-	electric field acceleration parameter
$\gamma_{p,att}$	-	plastic strain amplitude
$\Gamma_{sf,ar}$	-	surface energy per unit area
$\Delta\gamma_{p,att}$	-	plastic straining amplitude in the die attach
$\Delta g_{gb,ox}$	-	Gibbs free energy for oxide breakdown processes
Δh_{ox}	-	change in enthalpy required to activate the polyfilament
ΔT_0	-	temperature rise due to joule heating
ϵ_{cf}	-	fatigue ductility coefficient
ϵ^T	-	thermal strain
η_h	-	hole-trapping efficiency
λ_R	-	device failure rate
μ_{gb}	-	mobility of metal ions (along Metalization)
μ_{met}	-	shear modulus (of the Metalization)
$\mu_{v,met}$	-	vacancy mobility (in the Metalization)
ν	-	Poisson's ratio
π_i	-	functional factors for device technology, complexity, package type
π_T	-	temperature acceleration factor
ρ	-	resistivity (of the Metalization)
ρ_{chem}/Z_{chem}	-	resistivity of the electrolyte
$\sigma_{0,met}$	-	Metalization stress due to a temperature change
σ_{da}	-	static tensile stress perpendicular to crack
$\sigma_{psv,met}$	-	passivation-induced stress (in the Metalization)
$\sigma_{ult,att}$	-	tensile strength of the attachment material
σ_y	-	yield strength of the Metalization
τ_c	-	capture time constant
τ_{da}	-	shear stress at the die-attachment interface
τ_{hf}	-	shear strength of the attachment material

$\tau_{0,x0}$	-	room temperature value of the pre-exponential
τ_{rc}	-	time constant
$\tau_{v,met}$	-	average lifetime of vacancy in Metalization
Φ	-	back diffusion due to concentration gradient
ψ_{wf}	-	wave function of the electron
$\omega_{0,met}$	-	normalization width of the Metalization
ω_{b}	-	grain boundary thickness
ω_{met}	-	Metalization width
a	-	crack size
A	-	area
$A_{g,Black}$	-	Black's electromigration coefficient
A_{gt}	-	gate area
a_{i}	-	initial crack size
$A_{hc,cf}$	-	hot electrons coefficient
$A_{ox,}$	-	oxide dielectric breakdown
A_{paris}	-	Paris's coefficient
$A_{sddv,ok}$	-	Okabayashi's coefficient for stress-driven diffusive voiding
$B_{gm,fct}$	-	geometric factor for grain boundary migration
$B_{gt,fct}$	-	geometric factor for matrix diffusion
$B_{hc,ex}$	-	hot electrons exponent
b_{vt}	-	Burger's vector
C	-	capacitance
C_{by}	-	body capacitance
C_{cf}	-	fatigue ductility exponent
C_{hf}	-	heat transfer coefficient
$C_{i,bulk}$	-	ionic contamination (in the bulk)
$C_{mt,fac}$	-	material-related factor
$C^{0}{}_{v,met}$	-	vacancy concentration at thermal equilibrium
c_{ox}	-	oxide capacitance
$c_{p,vl}$	-	Venables and Lye's constant of proportionality
$C_{p,chp}$	-	density of the chip
$C_{pr,hm}$	-	Hickmott's constant of proportionality
$C_{rr,ox}$	-	reaction rate constant for oxide breakdown
$C_{st,fac}$	-	structure-related factor
$c_{th,em}$	-	thermal emission coefficient
$C_{v,met}$	-	vacancy concentration in the Metalization
$C'_{vc,f}$	-	critical value of vacancy concentration at which failure occurs
D_{0}	-	diffusion coefficient (of the Metalization)
$d_{ave,met}$	-	average grain size
d_{chp}	-	density of the chip
d_{dndt}	-	density of the dendrite
d_{sm}	-	mass density of the semiconducting material
$d_{v,met}$	-	density of the vacancies in the Metalization
$D_{v,met}$	-	vacancy diffusivity in the grain boundary
e	-	electron charge
E	-	modulus of elasticity
E_{a}	-	activation energy
E_{b}	-	activation energy of the pre-exponential
E_{c}	-	conduction bend energy level
E_{f}	-	Fermi-level energy
$E_{f,ch}$	-	electric field across dielectric/oxide
$E_{int,ox}$	-	internal energy of the dielectric

E_{ox}	-	electric field across the oxide
E_s	-	oxide permittivity
$E_{th,apt}$	-	apparent activation energy for time-dependent dielectric
F	-	Faraday's constant
F_{diff}	-	diffusion force
G	-	shear modulus
h	-	Plank's constant
$h_{0,met}$	-	normalization thickness (of Metalization)
h_{cond}	-	thickness of the conductor
$H_{f,chp}$	-	heat of fusion of chip per unit volume
h_{met}	-	convective heat transfer coefficient (of the Metalization)
H_{ox}	-	enthalpy of the dielectric
I_B	-	base current
I_c	-	collector current
I_e	-	emittor current
I_{fn}	-	Fowler-Nordheim current in oxide
I_p	-	peak current of discharge waveform
j_0	-	current density in a pore-free stripe
$J_{a,met}$	-	atomic flux (in Metalization)
J_{ce}	-	current density near the crack in the Metalization
j_{ctip}	-	current density at the crack tip
J_{inj}	-	density of the injected current
j_{met}	-	current density (in the Metalization)
j_p	-	current density (in a Metalization stripe) with pores
$J_{v,c0}$	-	atomic flux of vacancies in the bulk
$J_{v,met}$	-	vacancy flux (in the Metalization)
K_1	-	physical chemical properties
K_2	-	coating integrity factor
K_3	-	mission profile corrosion factor
K_4	-	environmental stress correction factor
$K_{1,2,3,4,wb}$	-	Wunsch-Bell constants of proportionality
K_B	-	Boltzmann's constant
$K_{f,v\text{-}die}$	-	stress intensity factors for vertical die cracks
$K_{f,h\text{-}die}$	-	stress intensity factors for horizontal die cracks
K_{met}	-	Metalization's thermal conductivity
l_{crkt}	-	crack length (in Metalization)
$l_{d,e}$	-	distance between electrodes material
$L_{f,s}$	-	latent heat for fusion of the semiconducting material
M_{at}	-	atomic weight
m_i	-	weight assigned to the activation energy of each failure mechanism
M_{met}	-	atomic weight of a metal conductor
M_{mf}	-	molten filament mass
m_{ok}	-	Okabayashi's Metalization width exponent
MTF	-	mean time to failure
N_a	-	atomic density (of the conductor Metalization)
n_{ce}	-	density of the states in the conduction band
n_{chem}	-	chemical valency of the Metalization
n_{corr}	-	corrosion exponent
N_d	-	concentration of donor
N_{da}	-	donor/acceptor atomic density
$N_{dp,ox}$	-	number of dipoles induced and/or oriented
$n_{e,Black}$	-	Black's electromigration exponent
N_f	-	number of cycles to failure

$n_{gb,met}$	-	grain boundary node density (in the Metalization)
$N_{i,met}$	-	density of the ions (in the Metalization)
$N_{inj,e}$	-	density of injected electrons
n_{ok}	-	Okabayashi's exponent
$N_{ot,eff}$	-	effective oxide trap density
n_{paris}	-	Paris's exponent
n_{sddv}	-	stress driven diffusive voiding exponent
N_{st}	-	density of states per unit volume
N_{trp}	-	spatial density of the interfacial traps
$N_{v0,met}$	-	vacancy density in the Metalization
$P_{av,rpn}$	-	average power in a biased junction
p_{da}	-	peeling stress
P_{dl}	-	electrical power dissipated in the filament
P_{gt}	-	power in the gate area
p_{ok}	-	Okabayashi's Metalization thickness exponent
P_{ox}	-	dielectric polarization
p_p	-	porosity (of the Metalization)
$P_{pr,occ}$	-	probabillity of occupancy
p_{sat}	-	saturated vapor pressure (above the meniscus or above a flat
$P_{sh,2}$	-	power dissipated due to changed resistance
$P_{sh,1}$	-	power dissipated due to a change in the resistivity of silicon
Q	-	mass transport rate
q_e	-	absolute value of the electronic charge
$Q_{cr,ox}$	-	critical charge density in the oxide
Q_d	-	power dissipated by the devices
$Q_{m,ox}$	-	mobile charger per unit area in oxide
Q_{ox}	-	density of the oxide trapped charge integrated over the thickness of quality, temperature and voltage
$R_{av,t}$	-	total average resistance
R_{by}	-	body resistance
R_{ct}	-	constant resistance
r_{dndt}	-	radius of curvature of the dendrite
R_{dv}	-	device resistance
RH	-	relative humidity
R_i	-	resistance of the multilayer Metalization at initial temperature
R_{im}	-	resistance of the refractory (in a multilayered Metalization)
r_{met}	-	aspect ratio (of the Metalization)
$r_{r,refct}$	-	temperature coefficient of resistance of the refractory
r_r	-	reaction rate
r_{st}	-	surface tension
R_T	-	resistance of the multilayer Metalization at temperature T
R_{th}	-	resistance as a function of temperature
S_{ox}	-	entropy of dielectric system
S_r	-	matrix of threshold stress value for device reliability
T	-	steady state temperature
t_{ann}	-	annealing time
t_b	-	time under bias
t_{bd}	-	time to oxide breakdown
TF	-	time to failure
Tf_{sddv}	-	time to failure due to stress-driven diffusion voiding

t_{inj}	-	length of the injection phase
$T_{m,chp}$	-	melting point of the chip
T_m	-	melting temperature
$T_{psv,dp}$	-	passivation deposition temperature
T_{ref}	-	reference temperature
t_{rlx}	-	relaxation time
V_{app}	-	applied voltage
$v_{bd,ox}$	-	breakdown voltage of the oxide dielectrics
V_{by}	-	discharge voltage on the human body
V_{ce}	-	collector-emitter voltage
v_{dft}	-	drift velocity
V_{dv}	-	device voltage
V_{fb}	-	flatband voltage shift
V_{met}	-	voltage applied
V_{mol}	-	molar volume
$V_{ov,pt}$	-	over potential
v_{th}	-	thermal velocity
$V_{vd,vl}$	-	void volume
$V'_{ox,vl}$	-	dielectric volume
w_{cond}	-	conductor width
x	-	distance (along the Metalization)
Z^*	-	ionic charge
Z_{elec}	-	ionic voltage of the electrolyte
Z_{met}	-	valence of metal ions

Subscripts

a	-	attachment
allow	-	allowable
amb	-	ambient
av	-	average
base	-	base
bg	-	bandgap
corr	-	corrosion
dev	-	device
df	-	diffusion
die	-	die
eff	-	effective
elec	-	electromigration
epi	-	epitaxial
f	-	final
flt		filament
flat	-	flat surface
gb	-	grain boundary
GS	-	gate-source
i	-	initial
in	-	input
int	-	interdiffusion
junc	-	junction
max	-	maximum
men	-	meniscus
met	-	metalization
min	-	minimum

mmig	-	metal migration
mult	-	multilayer
out	-	output
ox	-	oxide
ref	-	reference
sat	-	saturation
sub	-	substrate
th	-	threshold

Contents

Dedicated to the battles fought in search of truth and wisdom so the generations to come don't wander in the dark

Chapter 1

TEMPERATURE AS A RELIABILITY FACTOR

"We have a headache with Arrhenius"[1]

Many reliability engineers and system designers consider temperature to be a major factor affecting the reliability of electronic equipment. Unfortunately, in an effort to improve reliability, design teams have often lowered temperature without fully understanding the impact on cooling system reliability, in dollars, weight, and size, and the extent of any actual reliability improvement.

In this chapter, various modeling methodologies for temperature acceleration of microelectronic device failures are discussed, as are situations in which some current methodologies give misleading results. The aim is to raise the level of understanding of the impact of temperature on reliability and to define the objectives of a new temperature modeling and design methodology.

1. BACKGROUND

Reliability, defined as the ability of a device to fulfill its intended function, is often expressed in terms of number of years of useful life. Reliability-related failures render the device non-operational due to damage caused by a failure mechanism, actuated generally by external and internal stresses. Failure mechanisms determine device reliability; most often, some failure mechanisms will dominate and cause device failure before others.

A device may also fail to fulfill its intended function when its application to operating and environmental conditions lies outside the specification limits. Performance malfunctions may commonly arise due to a threshold voltage drift, a large leakage current, an unacceptable propagation delay, or noise margins, although normal operation is often resumed once the operating and environmental conditions return within specifications. Performance problems generally indicate either the need for a system design change or the unsuitability of the device technology for a beyond-specification, high-temperature application. For example, Figure 1.1 shows how the minimum output voltage for an output high (1) changes with respect to temperature. Clearly, different devices have different values; in this case, the lower the temperature, the smaller the margin for safe operation. The results are similar for

---

[1] Takehisa Okada, Senior General Manager of Sony Corporation, when asked about Sony's perspective on reliability prediction methods during a U.S. Japanese Technology Evaluation Center visit [Kelly et al. 1993].

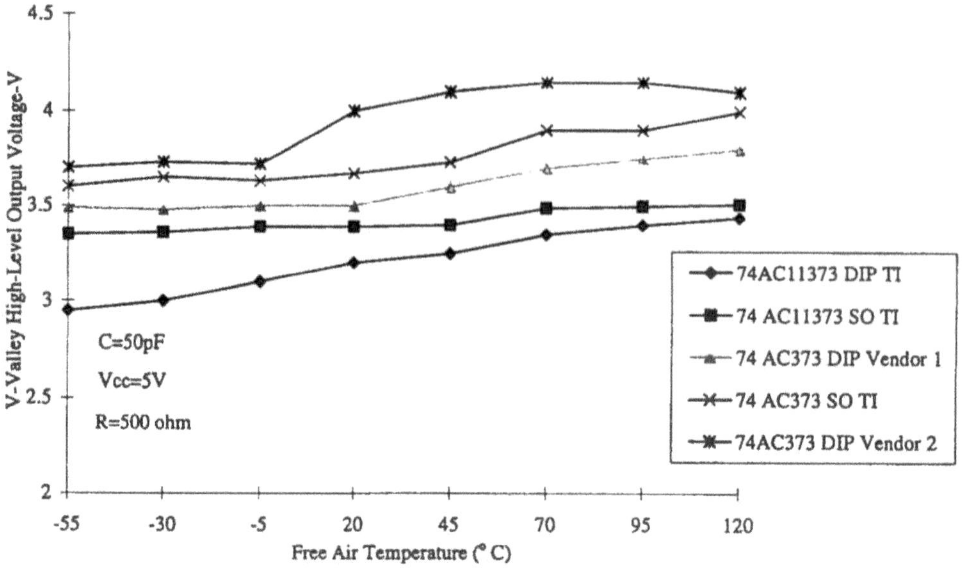

Figure 1.1 Valley high-level output voltage vs. free-air temperature (74AC11373 compared to end-pin product) [Texas Instruments Databook]

the maximum output voltage for an output low (0) (see Figure 1.2). Of course, other electrical device technologies may work the other way, with noise margins becoming worse at high temperature.

It is important that product engineers be knowledgeable of the effect of temperature on system performance requirements (as specified in the device catalog). This discussion will focus only on the influence of temperature on reliability.

2. ACTIVATION ENERGY-BASED MODELS

Steady-state temperature, temperature cycles, temperature gradients, and time-dependent temperature changes all have the potential to affect the reliability of modern electronic devices and equipment. However, because of the required use of reliability prediction methods such as Mil-Hdbk-217 [1991] and progeny [HRD5, 1995; CNET, 1983; Siemens, 1986], steady-state temperature has often been considered the only stress parameter affecting reliability.

Current methodologies are based on the work of Savante Arrhenius, a Nobel prize winner in chemistry in 1889. He published the results of an experimental study of inversion of sucrose, in which the steady-state temperature dependence of a chemical rate reaction was fit to the form

$$r_r = r_{r-ref} e^{-E_{a-chrt}/K_B T} \tag{1}$$

where r_r is the reaction rate (moles/meter^{2}second), r_{r-ref} is the reaction rate at a reference temperature (moles/meter^{2}second), E_{a-chrt} is the activation energy of the chemical reaction (eV), K_B is Boltzmann's constant (8.617 x 10^{-5} eV/K), and T is steady-state temperature (Kelvin).

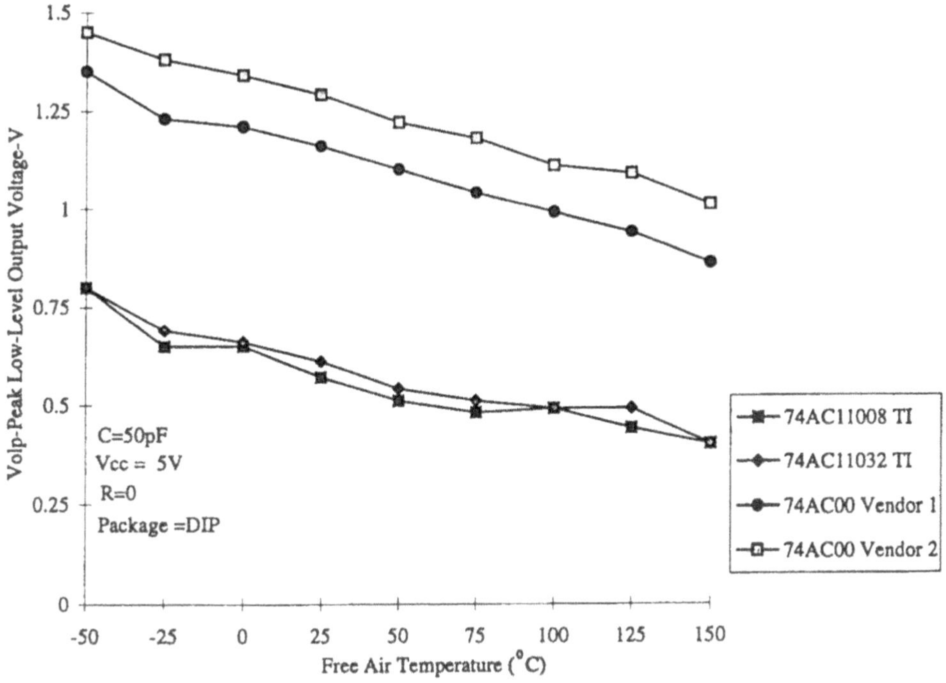

Figure 1.2 Evaluation of temperature effects (peak low-level output voltage vs. free-air temperature) [Texas Instruments Databook].

Equation (**1**), now called the Arrhenius equation, has been used to assess the temperature dependence of a wide variety of reaction rate constants and diffusion coefficients - often crudely, but sometimes quite accurately [Wong, 1990; Blanks, 1990; Witzmann, 1991, Klinger, 1991; Berry, 1980; Clark, 1979][2].

The Arrhenius-based models have been reformulated to predict the influence of steady-state temperature on electronic device reliability. In this case, the mean time to failure, *MTF* (hours), for a given steady-state temperature is represented as

$$MTF = MTF_{ref}\, e^{E_{a\text{-}dev}/K_B T} \tag{2}$$

where MTF_{ref} is the mean time to failure at a specified reference temperature and $E_{a\text{-}dev}$ is the device activation energy (eV). Figure 1.3 shows an Arrhenius plot in which the activation energy is obtained from curve-fitting and extrapolating experimental data to an Arrhenius equation.

---

[2] Theoretical work in kinetic theory, thermodynamics, and statistical mechanical treatments has developed forms that contain exponentials similar to the Arrhenius form [Eyring, 1980; Wigner, 1938; Evans, 1938]. At their core is the assumption that a state of equilibrium exists between the reactants and the products of a reaction, which are separated by a finite energy difference [Reif, 1965].

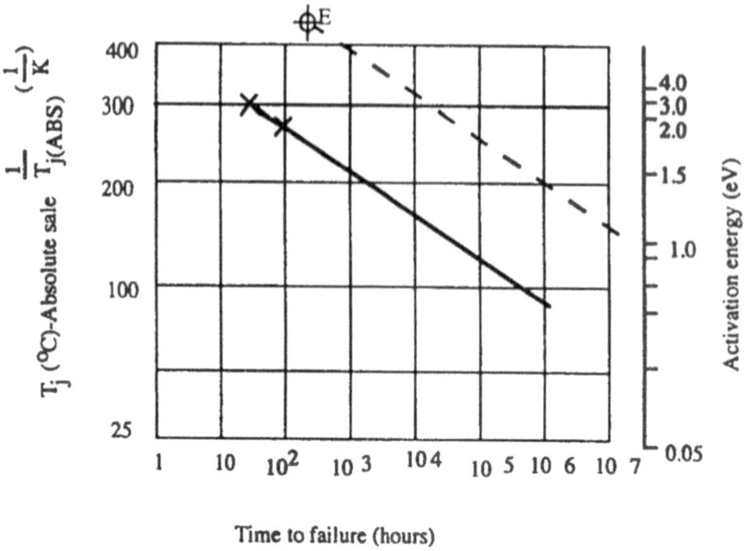

Arrhenius plot: inverse of junction temperature vs time to failure. To determine the activation energy, E_a, for the failure mechanism draw a line from the reference point E parallel to the arrhenius plot. The intercept of this line (dashed line in the figure) on the right hand side scale gives the required activation energy. [Boccaletti, Borri, D'Espinosa, Fioravanti and Ghio, Accelerated Tests, Microelectronic Reliability, vol.2]

Figure 1.3 The Arrhenius plot depicts the inverse of junction temperature vs. time to failure.

Table 1 Dominant VLSI failure mechanisms based on survey response [IITRI, 15 January 1988, RADC]. Clearly, no single activation energy can be assigned to the device because the failure depends on manufacturing processes and the actual failure mechanism experienced by the device can vary. Furthermore, few failure mechanisms remain dominant, or even significant, for very long [Witzmann, 1991]

Failure model/mechanism	Survey response					
	1	2	3	4	5	6
Electromigration						13%
Dielectric breakdown	X	50%	< 0.1%	98%		2%
Soft errors						
Parametric drift	X		1%			38%
Hot electrons	X					
Latch-up	X	10%	0.1%		X	
Electrical overstress		20%		2%	X	
Package related		20%	< 0.1%		X	28%
Other					X	19%

In the modeling formulation, activation energies of the various failure mechanisms that arise in the device are lumped together to generate a weighted average activation energy for the device [LSI Logic, 1991; Setliff, 1991]. That is,

$$E_{a-dev} = \sum_{i=1}^{n} m_i E_{a-i} \qquad 0<m_i<1 \tag{3}$$

where m_i is the weight assigned to the activation energy of each failure mechanism (dimensionless), and E_{a-i} is the activation energy of the I-th failure mechanism.

The use of an activation energy to describe a device failure rate is extremely complex and often misleading [O'Connor, 1990; Hakim, 1990]. First, the weighted activation energy approach given in Equation 3 is so sensitive to the variability in the relative dominance of the failure mechanisms (i.e., the assigned weight) that useful conclusions cannot generally be drawn[3]. For example, the effect of even a 0.1 eV variation in the value of activation energy on the *MTF* predicted by the Arrhenius model at a temperature of 70°C is:

$$F_{acc-MTF} = \frac{e^{\frac{E_{dev}+0.1}{K_B T}}}{e^{\frac{E_{dev}}{K_B T}}} \approx 22 \tag{4}$$

This means that a variation of 0.1 eV at 70°C results in an error of 22 times. This error can be orders of magnitude larger at lower temperatures. Figure 1.4 shows the sensitivity of a change in activation energy on the mean time between failures (MTBF) as a function of temperature [LSI Databook].

The mean time to failure predicted by the Arrhenius model is also very sensitive to the value of the activation energy even when the specific device failure mechanism is known. In other words, activation energies for any one failure mechanism also vary over a wide range (Table 2) and depend on the materials, geometries, manufacturing processes, and quality control methods. For example, the activation energy for electromigration varies from 0.35 to 0.85 eV, even for the same metallization system. Predicted reliability using this approach will have little useful meaning.

Another problem with the use of an activation energy is indicated by studies such as those observing failure rate and steady-state junction temperature for various semiconductor devices (See Figure 1.5). Finally, the effects of temperature on electronic devices are often assessed by accelerated tests carried out at extremely high temperatures. For example, electromigration

---

[3] Table 1 demonstrates the extreme variability of the dominant device failure mechanisms for different manufacturers of VLSI devices. Considering that activation energies for different failure mechanisms can have values ranging from - 0.06 eV (for hot electrons) to 2 eV (for intermetallic growth), this approach is highly sensitive to the manufacturers, and thus to the weighing factors, and therefore is not recommended.

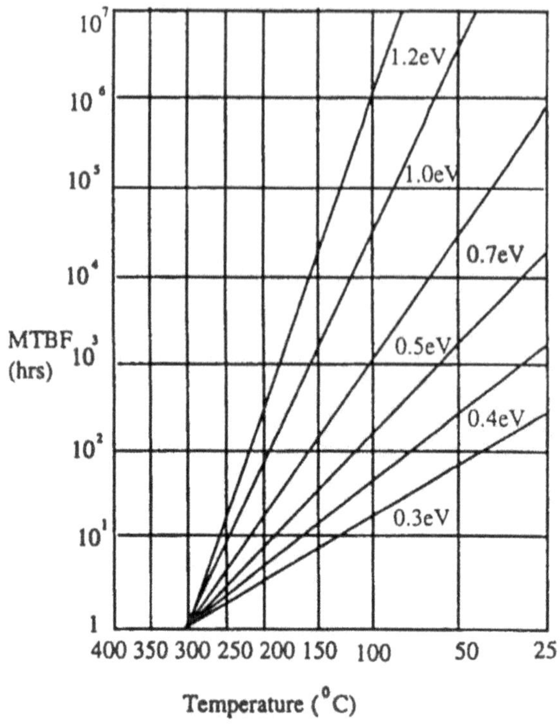

Figure 1.4 The sensitivity of a change in activation energy on the mean time between failures as a function of temperature.

tests are generally conducted at temperatures above 250 °C and at current densities ten times those applied in actual operation; the test results are then extrapolated to operating conditions to obtain a value for the thermal acceleration of device failures.

To determine the activation energy, E_{dev}, for a failure mechanism, draw a straight line through the experimental data points, then draw a parallel line from the reference point E. The intercept of this line (dashed line in the figure) on the right-hand scale gives the activation energy [Pollino, 1987].

Often implicit in the test strategy is the assumption that the failure mechanisms active at higher temperatures are also active in the equipment operating range, and that the Arrhenius relationship holds. Problems arise when the failure mechanisms precipitated at accelerated stress levels are not activated in the equipment operating range[4].

In particular, many failure mechanisms have temperature thresholds below which failure will not occur. In some other cases, high temperature can actually inhibit or de-accelerate a failure mechanism that will occur at a lower temperature i.e. hot carrier generation. Often, threshold information provides a more effective way to design and test a device and to provide stress management.

---

[4] A NIST study noted that, "there is ample evidence that a straight forward application of the Arrhenius equation, with activation energies determined from high temperature accelerated stress testing, is not strictly valid for predicting real device lifetime" [Kopanschi, et al. 1991].

Table 2 Activation energies for common failure mechanisms

Failure mechanisms	Activation energy	Reference
Die metallization failure mechanisms		
Metal corrosion	0.3 to 0.6 eV	[Hakim, 1989; Jensen, 1982; Amerasekera, 1987]
	0.77 to 0.81 eV	[Peck, 1986]
Electromigration	0.5 eV (small-grain Al)	[Black, 1982]
	0.43 eV (Al)	[Ghate, 1981; Towner, 1983]
	0.35 to 0.85 eV (Al)	[Lloyd, 1987]
	1.0 eV (large-grain glassivated Al)	[Nanda, 1978; Jensen, 1982]
	0.24 to 0.57 eV (Al)	
	0.7 eV (Al)	[Reimer, 1984]
	1.67 to 2.56 eV (Al-1%Si)	[Saito, 1974]
	0.58 eV (Al-1%Si)	[Suehle, 1989]
	0.96 eV (Al-1%Si)	[Schafft, 1985]
		[Fantini, 1989]
Metallization migration	1 eV	[Abbott, 1976]
	2.3 eV	[Jensen, 1982]
Stress-driven diffusive voiding	0.4 eV	[McPherson, 1987]
	1.0 to 1.4 eV	[Tezaki, 1990]
Device and device oxide failure mechanisms		
Ionic contamination (surface, bulk)	0.6 to 1.4 eV	[Amerasekera, 1987]
	1.4 eV	[Jensen, 1982]
Hot carrier	-0.06 eV	[Hakim, 1989]
Slow trapping	1.3 to 1.4 eV	[Jensen, 1982]
Gate-oxide breakdown		
ESD	0.3 to 0.4 eV	[Baglee, 1984]
	0.3 eV	[Crook, 1979]
TDDB	1 eV	[Hokari, 1982]
	0.3 eV	[Crook, 1979]
	2.1 eV	[Anolick, 1979]
	0.3 to 1.0 eV	[McPherson, 1985]
EOS	2 eV	[Anolick, 1979]
Surface-charge spreading	1.0 eV	[Hakim, 1989]
	0.5 to 1.0 eV	[Jensen, 1982; Amerasekera, 1987]
First-level interconnection failure mechanisms		
Au-Al intermetallic growth	0.5 eV	[Irvin, 1978]
	1.0 eV	[Hakim, 1989; Jensen, 1982]
	1.1 eV	[Mizugashira, 1985]
	2.0 eV	[White, 1978]

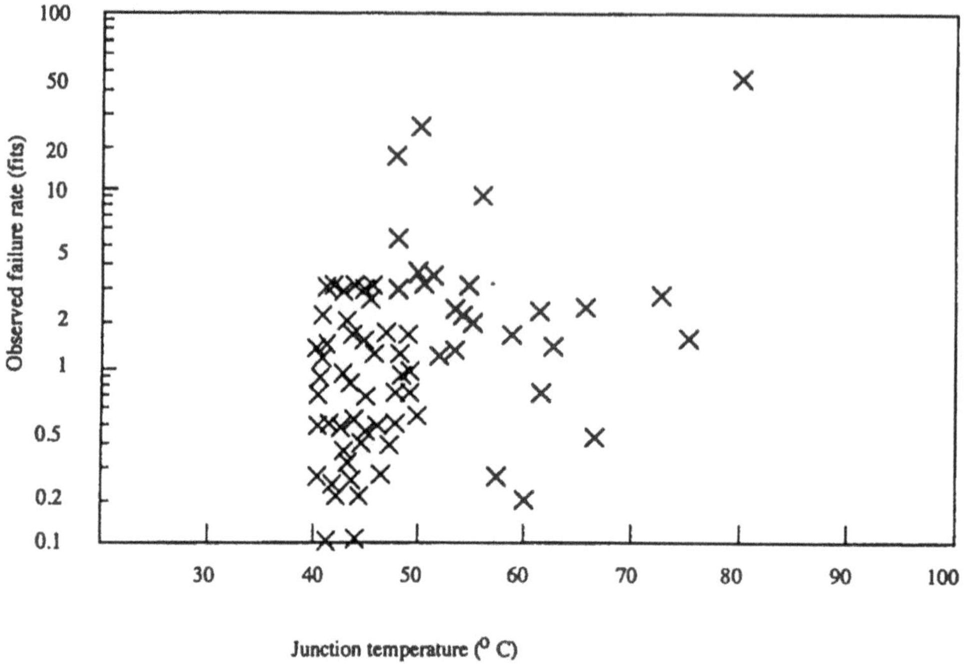

Figure 1.5 Scatter diagram showing the lack of correlation between observed failure rate and junction temperature of different types of bipolar logic ICs [Hallberg, 1994]

3. RELIABILITY PREDICTION METHODS

As noted previously, Arrhenius-based models have been incorporated into some reliability prediction methods. This section reviews these methods and the impact of temperature-dependent models on system effectiveness.

The reliability of electronic devices has often been represented by an idealized plot called a bathtub curve (Figure 1. 6), which consists of three regions. Region 1, in which the failure rate decreases with time, is called the infant mortality or early-life failure region. Region 2, in which the failure rate reaches a constant level, is called the constant failure rate or useful life region. Region 3, in which the failure rate increases, is called the wear-out region.

Modern semiconductor designs, manufacturing processes, and process controls have improved to the point where the infant mortality and useful life regions of the semiconductor devices have failure rates so near zero that the bathtub curve "no longer holds water" [Wong, 1990; Beasley, 1990]. Furthermore, for a device operated within specification limits, the wear-out portion of the curve has been delayed well beyond the useful life of most products [Hakim, 1990, McLinn, 1990, O'Connor, 1990]. In fact, Pecht and Ramappan [1992] found that the majority of electronic hardware failures over the past decade were not component failures, but were attributable to interconnects and connectors, system design, excessive environments, and improper user handling [Table 3]. Nevertheless, various non-scientific attempts to predict the failure rate of devices [HRD5, 1995; CNET, 1983; Siemens, 1986; Mil-Hdbk-217, 1991] are being used, even though they have been proven inaccurate, misleading, and damaging to cost-effective and reliable design, manufacture, testing, and support [Cushing et al. 1993, 1994]. An overview of these reliability prediction models can be found in Bowles [1992]. The models are typically of the form:

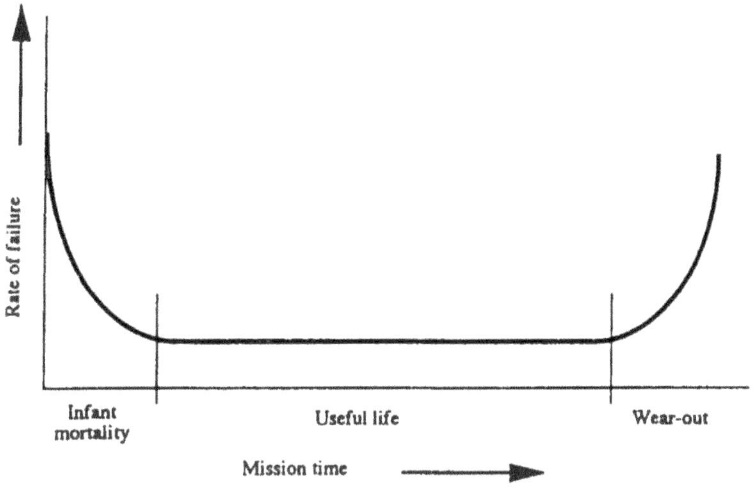

Figure 1.6 The reliability of electronic devices has often been represented by an idealized graphic called bathtub curve.

$$\lambda_{R-dev} = \lambda_{R-base} \prod \pi_i \quad (5)$$

where λ_{R-dev} is the device failure rate (failures/10^6hours), λ_{R-base} is the base failure rate (failures/10^6hours), and π_i are various functional factors for device technology, complexity, package type quality, temperature, and voltage (dimensionless). The temperature acceleration factor, π_T, generally has the form of an Arrhenius equation (see Table 4). Because steady-state temperature is the only temperature factor, or for that matter, generally the only stress parameter (i.e., temperature cycling, vibration, moisture, voltage, and current are usually not applicable), system designers often use temperature reduction as the primary means to improve reliability, often without understanding the actual reliability or the hidden costs associated with the temperature reduction.

As an example of the damage on product design and tradeoffs caused by current reliability prediction methods and the steady-state temperature relation, Figure 1.7 shows a plot of mean time between failures (MTBF) as a function of package case temperature for a Boeing E-3A multiplexer hybrid. The Mil-Hdbk-217 prediction is an order of magnitude pessimistic which could easily lead to the non-use of this electronic device.

As another example, Figure 1.8 comes from the U.S. Joint InterAgency Working Group (JIAWG), which developed reliability requirements for such new military systems as the F-22 and the Comanche (light helicopter). Figure 1.6 was developed, using the values and temperature relations from Mil-Hdbk-217, to provide guidance for reliability allocations of the new systems. To meet system reliability requirements, the maximum allowable component junction temperature was determined to be 65°C. For the Comanche, this dictated the development of a super-cooling system pumping air at -60°C in order to lower the temperature outside the sealed electronic boxes enough to get component temperatures to 65°C. Initially there was no consideration of the total reliability impact. In particular, on a hot day with

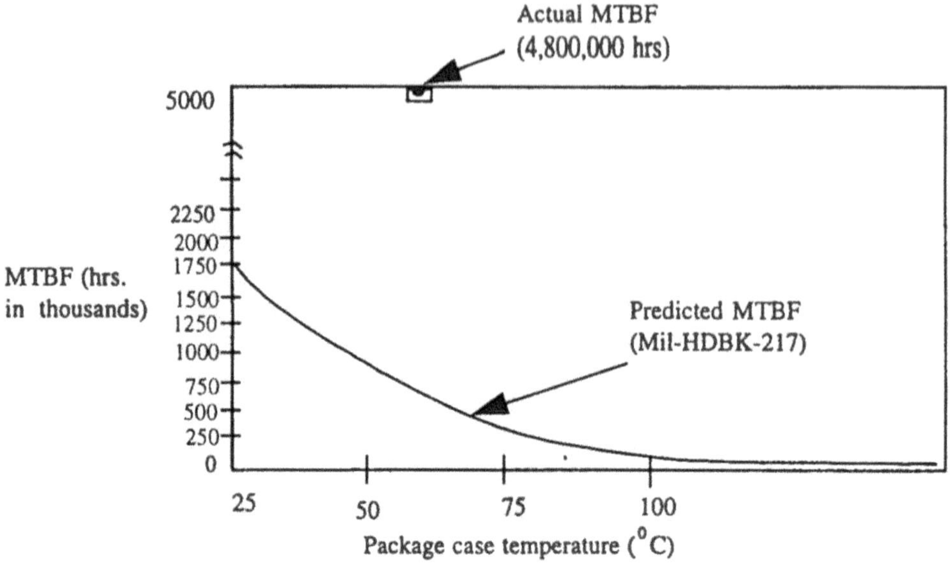

Figure 1.7 Plot of mean time between failures (MTBF) as a function of package case temperature for a Boeing E-3A multiplexer hybrid.

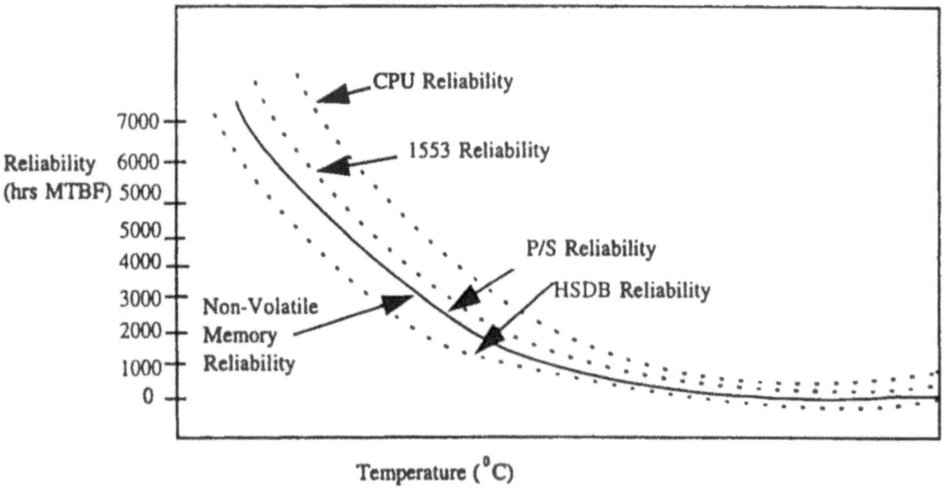

Figure 1.8 Guidance for reliability allocations of a system.

Table 3 Historical perspective of dominant failures in microelectronic devices [Pecht, 1992]

Source of data	Year	The dominant causes of failure
Failure analysis for failure rate prediction methodology [Manno, 1983]	1983	Metallization (52.8%); oxide/dielectric (16.7%)
Westinghouse failure analysis memos [Westinghouse, 1989]	1984-1987	Electrical overstress (40.3%)
Failure analysis based on failures experienced by end-user [Bloomer, 1989]	1984-1988	Electrical overstress and electrostatic discharge (59%); wirebonds (15%)
Failure analysis based on Delco data [Delco, 1988]	1988	Wirebonds (40.7%)
Failure analysis by power products division	1988-1989	Electrical overstress damage (30.2%)
Failure analysis on CMOS [Private correspondence]	1990	Package defects (22%)
Failure in vendor parts screened per MIL-STD-883	1990	Wire bonds (28%); test errors (19%)
Pareto ranking of failure causes per Texas Instruments study [Leonard, 1989]	1991	Electrical overstress and electrostatic discharge (20%)

Table 4 Temperature - acceleration factors [Bowles, 1992]

Source of procedure	Temperature acceleration factor
HRD5 1997	$\pi_T = 1, \quad for\ T_{junc} \leq 70°C$ $= 2.6\times10^4 e^{-3500/T_{junc}} + 1.8\times10^{13} e^{-11600/T_{junc}} \quad for\ T_{junc} > 70°C$
CNET 1983	$\pi_T = A_1 e^{-3500/T_{junc}} + A_2 e^{-11600/T_{junc}}$
Mil-Hdbk-217	$\pi_T = 0.1 e^{-A(1/T_{junc} - 1/298)}$
Siemens 1986	$\pi_T = A e^{11605 E_{siem,1} (1/T_{junc,1} - 1/T_{junc,2})} + (1-A) e^{11605 E_{siem,2}(1/T_{junc,1} - 1/T_{junc,2})}$

43°C outside ambient, cooling is started first; the electronic box will cool to around -40°C, then rise to around 60°C when the electronics are turned on. This extreme temperature cycling would occur every time the helicopter is started and stopped. In addition to fatigue damage, Boeing engineers estimated significant standby water in the bottom of the electronic assemblies due to condensation. When further reviewed by the Army, junction temperatures were raised and the use of Mil-Hdbk-217 in general, and the temperature functions specifically, was dropped. The final statement from Boeing was that "the validity of the *steady state temperature* relationship to reliability is constantly in question and under attack as it lacks solid foundational data."

The questionable validity of Mil-Hdbk-217 has not discouraged its use by the U.S. military. In fact, the military standards body has declined to evaluate the technological merit of the document and is waiting for a replacement. Unfortunately, the military electronics industry is somewhat relieved to shed the burden of technical validity.

4. HOW SHOULD DESIGN, THERMAL MANAGEMENT, AND RELIABILITY ENGINEERS WORK TOGETHER?

To address the actual impact of temperature, design, thermal management, and reliability engineers should work together, utilizing a physics-of-failure methodology. There are six steps to this method:

- Develop a thorough knowledge and understanding of the environment in which the equipment will operate. Usually, the customer will specify the operating environment in terms of absolute physical parameters, such as temperature ranges, or will quote the relevant chapter in some handbook or specification. While this may be a useful starting point for the designer, it rarely identifies the actual range of environments experienced by the equipment. It may be better, and from the customer's point of view, more contractually sound, to state where and how the equipment will be used. As a point of interest, consumer goods manufacturers, such as the automobile industry, have never had the benefit of a detailed environmental specification supplied by their customers (the public), but have been able to effectively ascertain the environment for themselves.
- Develop an understanding of the material properties and architectures used in the design. This involves tailoring the product design to requirements by modifying materials geometry, allowable manufacturing defects, and operating stresses.
- Learn how products fail under various degrading influences. This involves assessing the potential failure mechanisms and determining the role of stresses, including steady-state temperature, temperature cycling, temperature gradients, and time-dependent temperature changes, on the failure mechanisms.
- Examine field failure data providing information on how failures occur.
- Control manufacturing to reduce those variabilities that cause failure.
- Design the product to account for temperature-related performance degradation. Steady-state temperature has an influence on many electrical functional parameters, including propagation delays and noise margins.

5. SUMMARY

There are alternatives to the Arrhenius relation and the Mil-Hdbk-217 approach to reliability. In Japan, Taiwan, Singapore, and Malaysia, a physics-of-failure approach is used by most companies [Kelly et al. 1995] and in the U.S., the CADMP Alliance has developed methods and software to conduct reliability assessments [Evans et al. 1995].

Chapter 2

TEMPERATURE DEPENDENCE OF MICROELECTRONIC PACKAGE FAILURE MECHANISMS

The temperature dependence of a device is assessed by the effect of temperature stresses on dominant device failure mechanisms. The dominant failure mechanisms are a function of the package geometry, architecture and materials, and of accelerating environmental and operational stresses. The accelerating environmental stresses include temperature, relative humidity, pressure, and static charge, and their cycles, gradients, and transients; accelerating operational stresses include voltage and current. While high magnitudes of these stresses will overstress the device to failure, smaller values of the same stresses may not result in any failure during useful life of the device. To examine the thresholds of the accelerating stresses for each and every failure mechanism is beyond the scope of this book, which concentrates only on the effect of various temperature stresses and identifies the thresholds beyond which observable changes in device life can be measured. Both bipolar and MOS devices have been covered here. While some failure mechanisms are not specific to a device technology, others are specific to just bipolar or field-effect devices. The failure mechanisms discussed in this chapter are a superset of all the mechanisms that will occur in various package elements in both bipolar and field-effect devices.

Package elements include the chip (die) and the device packaging. The chip has been further subdivided into die metallization, device oxide, device, and device-oxide interface. Device packaging has been subdivided into first-level interconnects, package case, leads, lead seals, die(s), and substrate attaches. First-level interconnects include wirebonded interconnects, TAB, flip-TAB, and flip-chip. The emphasis is on the physics of the failure process, rather than any particular device type. Figure 2.1 illustrates the failure mechanisms for various package elements. These have been discussed in terms of their dependence on steady-state temperature, temperature cycle, temperature gradient, and time-dependent temperature change. For each mechanism, typical stress thresholds beyond which observable changes in device life will be obtained have also been identified.

Failure mechanisms occurring predominantly at the die level include slow trapping, hot electrons, electrical overstress, electrostatic discharge, dielectric failure, oxide breakdown, and electromigration. First-level package failure mechanisms generally arise from corrosion, differential thermal expansion between bonded materials, large time-dependent temperature changes, and large spatial temperature gradients, all of which can cause tensile, compressive, bending, fatigue, and fracture failures. Corrosion-induced failure mechanisms include electrolyte formation and galvanic and ionic corrosion. Corrosion failures are complex functions of contamination, temperature, humidity, and bias. All failure mechanisms have

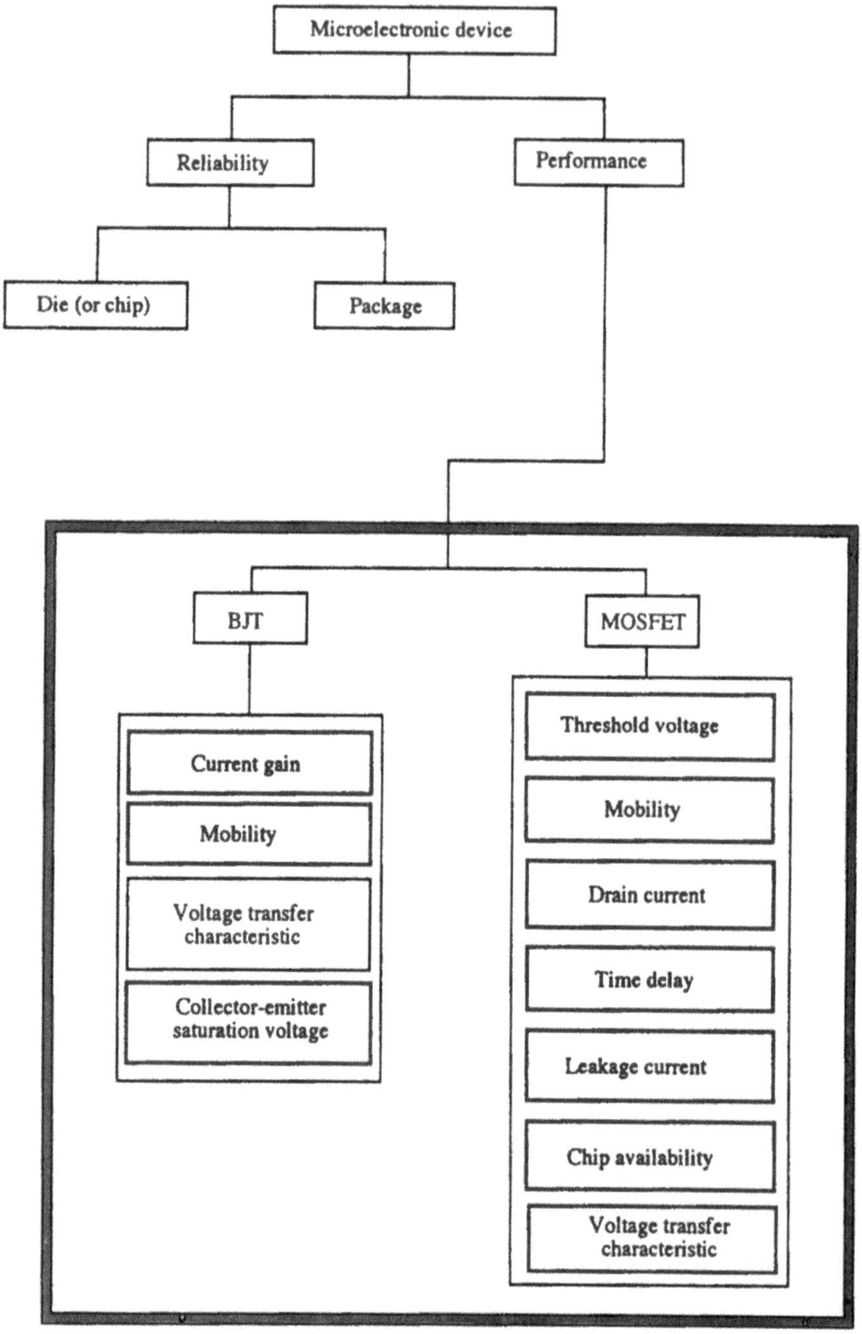

Figure 2.1 Failure mechanisms in microelectronic devices have been classified according to failure sites at die (or chip) level and first level package.

been discussed here in terms of their dominant temperature dependencies. Failure mechanisms with a dominant temperature dependence in the range of -55 to 125°C have also been identified. The reader is encouraged to focus on the logic and relevance of existing reliability modeling methodologies (discussed in chapter 1), while reading through the discussions on mechanisms and performance parameters, in the context of the Arrhenius dependence of device reliability on steady-state temperature.

1. TEMPERATURE DEPENDENCIES OF FAILURE MECHANISMS IN THE DIE METALLIZATION

This section discusses the temperature dependencies of failure mechanisms in the metallization of the die or the substrate. The failure mechanisms discussed in this section include corrosion, electromigration, hillock formation, metallization migration, and stress driven diffusive voiding. While the mechanics of each failure process are complex enough that books have been written on individual failure mechanisms, this section provides a brief overview of the failure signature of each mechanism, focusing on its temperature dependencies.

The acceleration by various processes inducing temperature-stress failure may vary, depending on the material and geometry - even for the same mechanism. For example, in thin films, electromigration is characterized by an open, while in multilayered metallizations, electromigration failures manifest themselves as resistance increases. Most known processes of failure have been discussed in cases having closed-form relations characterizing the effects of temperature, material properties, and geometry on the time to failure. For some mechanisms, more than one model has been proposed to characterize the temperature dependence of the failure process. Most of the commonly accepted models have been discussed in such cases, along with their implicit assumptions.

While most of the mechanisms discussed here have multiple temperature dependencies, some are steady-state temperature dependent. A typical case occurs when the mechanism does not assume a dominant dependence on steady-state temperature below a threshold temperature because the time to failure due to temperature stress is well beyond the mission life of the device. For such mechanisms, the typical temperature thresholds above which the failure mechanism becomes dominant have been identified.

1.1 Corrosion of Metallization and Bond Pads

Corrosion is typically defined as the chemical or electrochemical reaction of a metal with the surrounding environment. Corrosion can occur during manufacturing, storage, shipping, or service of the device. In a microelectronic package, the time to failure due to corrosion is dependent on the package type, corroding material, fabrication and assembly processes, and environmental conditions. The package and environmental conditions control the moisture ingress into the package. Corrosion of aluminum metallization has been a major concern in microelectronic devices; common types include atmospheric and galvanic corrosion.

Atmospheric corrosion. Atmospheric corrosion can be divided into two types: dry — such as oxidation of aluminum in air, and wet — in which the reaction occurs in the presence of an electrolyte, a moist environment, and an electromotive force.

Dry corrosion. Dry corrosion is of minor importance in semiconductor devices, since the corrosion process is self-passivating, forming a thin oxide film that prevents further oxidation. Dry corrosion at ambient temperatures occurs on metals that have a negative free energy of oxidation and, therefore, form a stable oxide film in the presence of oxygen. Oxide films are usually desirable because they are defect-free, non-porous, and self-healing and prevent further corrosive attack on the base metal. If the corrosion rate is completely governed by the elementary process of metal oxidation, the corrosion rate increases exponentially with temperature and is represented by an Arrhenius relationship.

Wet corrosion. Wet corrosion occurs in the presence of an ionic contaminant and moisture; the latter provides a conductive path for electrical leakage between adjacent conductors, resulting in dendritic growth, or corrosion of the device metallization or bond pad (see Figure 2.2). The ions most commonly found on the die surface that, in the presence of water, give rise to the electrolytic solution required to trigger the corrosion process include:

- halogens (especially Cl^-), arising both from inadequate removal of fabrication process residue and from plastic containers;
- alkalines (especially Na^+), derived from diffusion ovens, glass containers, and the hands of operators; and
- phosphorus, which, unlike contaminating ions, is specially incorporated in the surface passivation glass to improve its mechanical characteristics and restrict the effects of Na^+.

Temperature can affect wet corrosion in a number of ways. If an important corrosion constituent in solution has limited solubility, a temperature change can alter the concentration of that constituent, affecting the rate of corrosion. A classic example is the corrosion of iron in the presence of oxygen. The corrosion rate of iron in a system closed to the atmosphere has been shown to increase almost linearly with temperature from about 40°C to 160°C. However, in a system open to atmosphere, the corrosion rate increases up to about 80°C and then decreases. The change in the corrosion rate results from the decrease in oxygen solubility, which releases oxygen from the liquid. In a closed system, the oxygen cannot leave the vapor space above the liquid. As the temperature increases, the water vapor pressure increases until it maintains the oxygen concentration in the liquid. The corrosion rate tends to increase with temperature because of its effects on viscosity and diffusivity. In open systems, oxygen can escape from the immediate vicinity of the liquid. The vapor pressure remains constant. Above a certain temperature, the liquid-phase oxygen concentration in equilibrium with oxygen in the atmosphere decreases to the extent that the corrosion rate decreases. Further, the ionization constant of water increases with temperature. Pure water with a pH of 7 at one temperature will have a lower pH at a higher temperature. Thus, an increase in temperature affects corrosion by moving the pH from a neutral to an acidic value. As surface temperature increases, moreover, the corrosion rate will rise sharply to a point at which the electrolyte evaporates. At this temperature, the corrosion rate will decrease quickly.

Comizzoli [1980] studied the effect of passivation defects on the metallization layer by monitoring corrosion currents on test structures. The test structures consisted of four pairs of identical combs, each with 32 interdigitated stripes 12 μm wide, 770 μm long, and with 12 μm spacing. The passivation on the test structures had various magnitudes of known defects, including holes and slots. The test structures were subjected to various temperatures and relative humidities, a voltage was applied, and the corrosion current was measured. The corrosion current was linearly dependent on steady-state temperature between 60°C and 100°C. The corrosion current at 90% relative humidity using passivated chips decreased with a decrease in temperature between 60°C and 100°C. The corrosion current for various temperatures between 60°C and 100°C was greater in unpassivated stripes than in passivated stripes with pinholes.

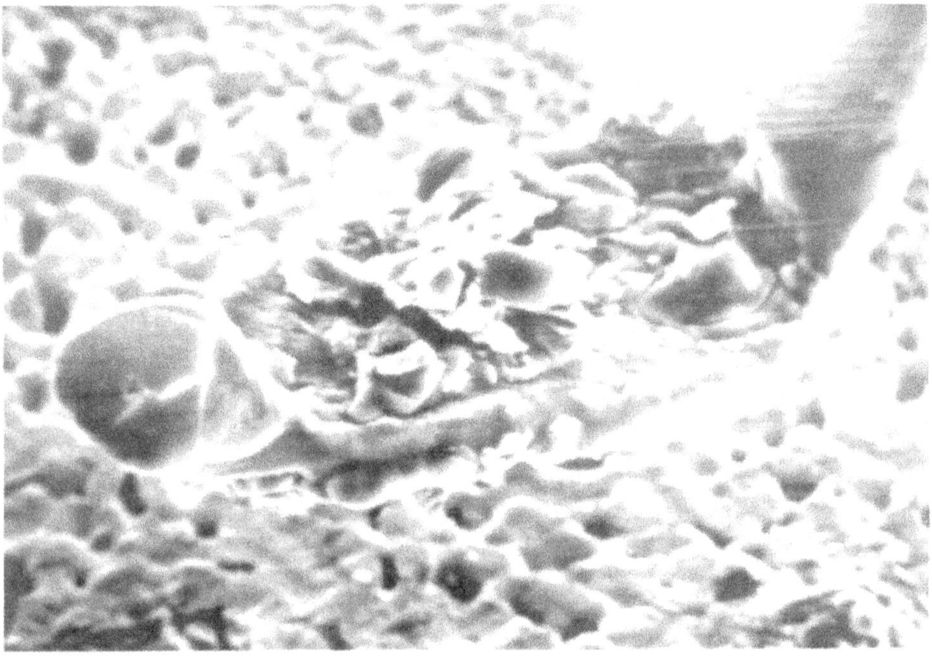

Figure 2.2 SEM micrograph - Aluminum wirebond, corrosion. (Courtesy of Motorola, Inc.)

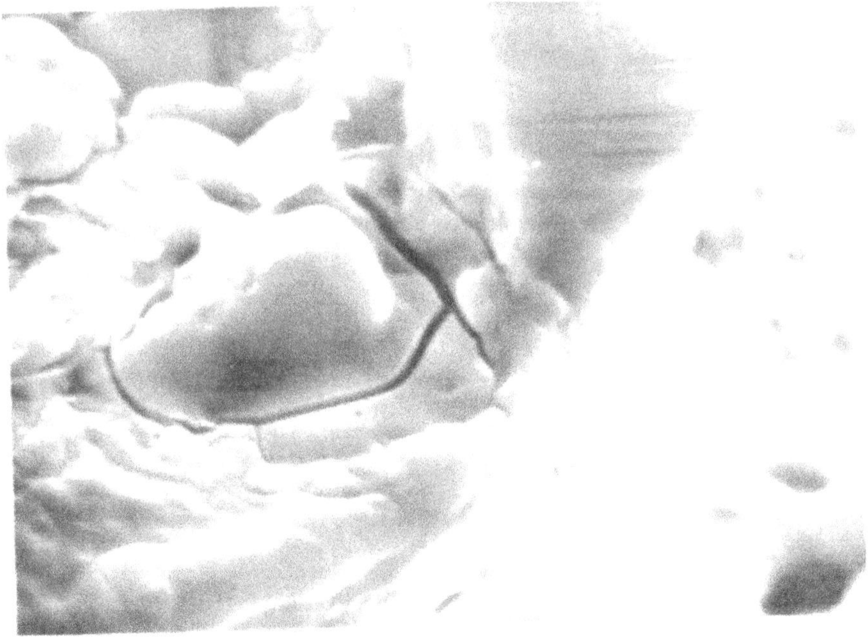

Figure 2.2 SEM micrograph - Alumium corrosion, magnified. (Courtesy of Motorola, Inc.)

Peck [1986] reviewed all the published literature on humidity conditions vs life at 85/85%RH for epoxy packages and found the time to failure due to corrosion to be exponentially dependent on steady-state temperature. The relationship between life and temperature and between life and humidity was calculated based on a regression analysis. The time to failure due to corrosion was modeled as

$$TF_{corr} \propto (RH)^{n_{corr}} e^{E_{a-corr}/K_B T} \tag{1}$$

where TF_{corr} is the time to failure due to corrosion, RH is the relative humidity in %, E_{a-corr} is the activation energy for corrosion in eV, K_B is Boltzmann's constant (8.617 x 10^{-5} eV/K), and T is steady-state temperature. The valid combinations of n_{corr} and E_{a-corr} are shown in Table 1.

Pecht modeled the time to failure due to corrosion as the sum of moisture ingress time and the time for corrosion attack and failure. The moisture ingress time for a hermetic package was determined based on its internal volume and the leak rate. The method for calculating the worst-case operation-independent moisture ingress time was based on standard testing procedures. After sufficient moisture ingress, a critical moisture content is attained inside the package, and corrosion can initiate when the non-operating sealed package is exposed to temperatures below the dew point. At this time, the moisture inside the package condenses and combines with any ionic contaminant present to provide a conductive path between adjacent metallic conductors. The conductive path serves as a medium for the transfer of ions in the corrosion process. When the package is operating, the heat dissipated inside the package will typically elevate the temperature above the dew point. Consequently, the electrolyte will evaporate and no longer provide an electrolytic path between conductors. Elevated temperature due to device power thus acts as a mechanism to slow the corrosion process [Pecht and Ko, 1990; RAC Report SOAR-3, 1985].

When a packaged component is in the off mode, it can equilibrate with the ambient RH. However, when the component reaches its operating temperature, the same moisture content results in a lower relative humidity. The low relative humidity at operating temperature may prevent the formation of an electrolyte and may negate exposure to corrosion entirely. The MTF due to corrosion is shorter for longer ON times, if device power dissipation is great. For small-power devices, the main time to failure (MTF) due to corrosion is not much affected by the ON:OFF ratio (Figure 2.3). The amount of moisture absorbed increases with an increase in OFF time (Figure 2.4), but the average amount of moisture remains the same as long as the ON:OFF ratio is the same (Figure 2.5). Higher-power dissipating devices evaporate moisture due to junction temperature rise regardless of environmental conditions (Figure 2.6) [Ajiki, 1979; Macheils, 1991; Shirley; 1991].

Table 1. Valid combinations of n_{corr} and E_{a-corr}

Activation energy (E_{a-corr})	Exponent (n_{corr})	Correlation coefficient
0.79	-2.66	0.986
0.81	-2.50	0.985
0.77	-3.00	0.987

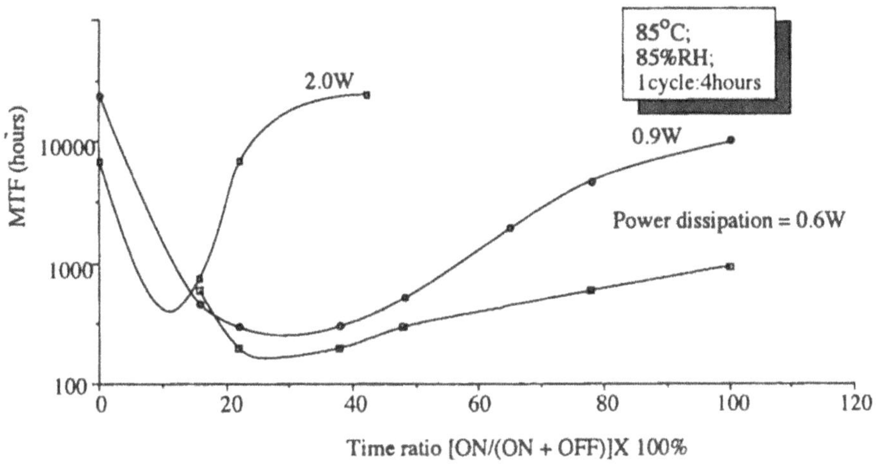

Figure 2.3 MTF is shorter for longer ON times, if device power dissipation is large [Ajiki, 1979]. MTF is not much affected by ON:OFF time ratio when device power dissipation is small.

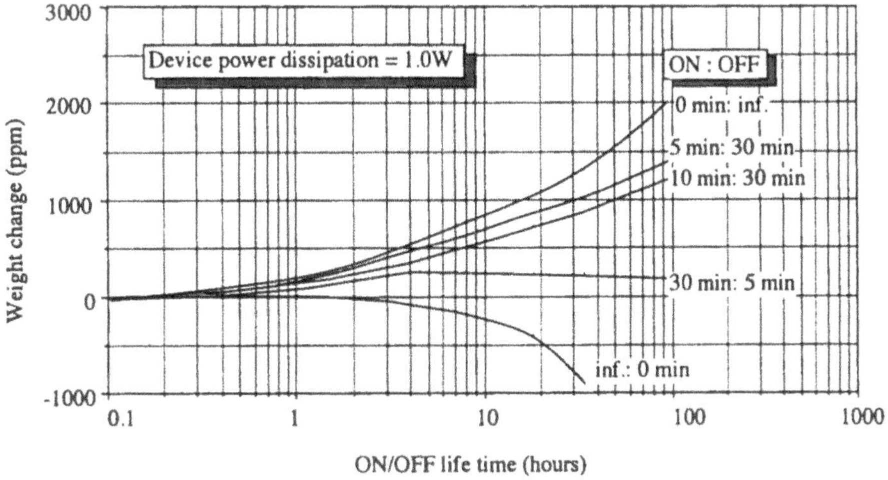

Figure 2.4 The amount of moisture absorbed increases with increase in the OFF time, i.e., the more time spent at lower temperature, the larger the amount of moisture absorbed [Ajiki, 1979].

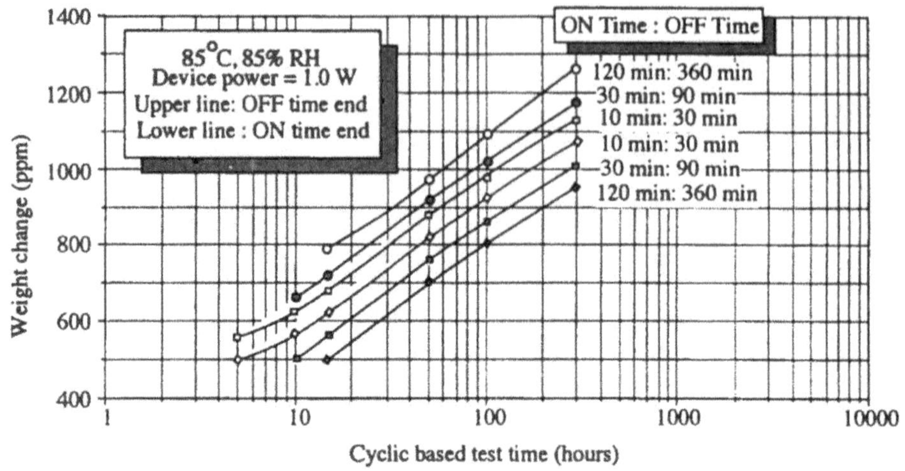

Figure 2.5 The average amount of moisture remains the same as ON:OFF ratio is the same [Ajiki, 1979].

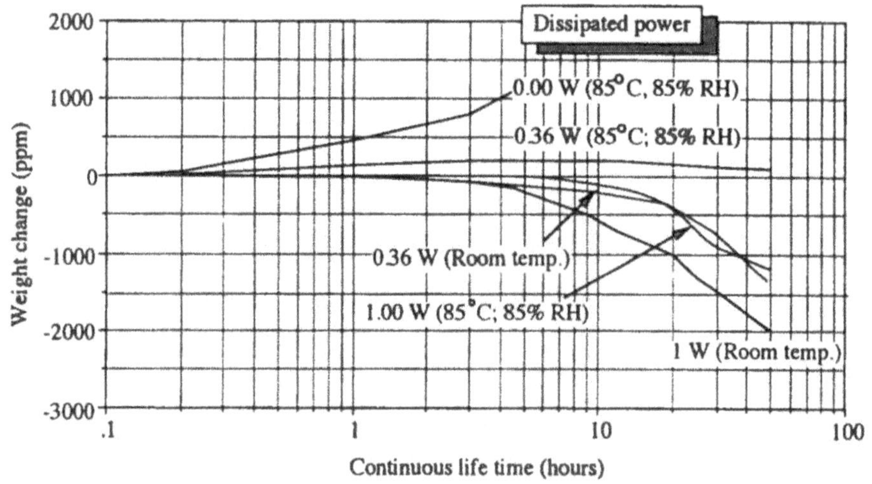

Figure 2.6 Higher power dissipation devices evaporate moisture due to junction temperature rise, irrespective of environmental conditions [Ajiki, 1979].

The time to failure as a result of corrosion is given by

$$TF_{corr} = \left(\frac{K_1 K_2 K_3}{K_4}\right)\left[\frac{w_{cond}^2 h_{cond} n_{chem} d_{cond} F \rho_{elec}}{4 M_{met} V_{met} Z_{elec}}\right] \tag{2}$$

where TF_{corr} is the corrosion time to failure, K_1 is the physical chemical properties index, K_2 is the coating integrity index factor, K_3 is the mission profile corrosion factor, K_4 is the environmental stress correction factor, w_{cond} is the conductor width, h_{cond} is the thickness of the conductor, M_{met} is the atomic weight of a metal conductor, n_{chem} is the chemical valency of the metallization, ρ_{elec}/Z_{elec} is the resistivity of the electrolyte, V_{met} is the voltage applied, Z_{elec} is the ionic charge of the electrolyte, and F is Faraday's constant. The effect of the environmental stress factor on the MTF is represented in Figure 2.7.

The only temperature-dependent K term is the environmental stress correction factor, K_4, used to determine the time to failure for various temperature and humidity conditions. The K_4 term is modeled as

$$K_4 = \frac{(RH_{ref})^{n_{corr}} \exp(E_{a-corr}/K_B T_{ref})}{(RH)^{n_{corr}} \exp(E_{a-corr}/K_B T)} \tag{3}$$

where RH_{ref} is a reference relative humidity (%), RH is the relative humidity, E_{a-corr} is an activation energy for corrosion (eV), K_B is Boltzmann's constant (eV/K), n_{corr} is a corrosion exponent, and T_{ref} is a reference temperature (°K).

Galvanic corrosion. Galvanic corrosion arises in chip and packages from an extensive use of precious metals (gold, palladium, silver) coupled with base metals (aluminum) [Howard, 1987]. Galvanic corrosion, a form of wet corrosion, occurs when two or more different metals, such as aluminum and gold (e.g., at a bonding pad), are in contact in the presence of an electrolyte. An electrolyte can be moisture or a combination of moisture and contaminants, such as salt spray from the ocean. Each metal is associated with a unique electrochemical potential. When two metals are in contact, the metal with the higher electrochemical potential becomes the cathode and the other becomes the anode. The anode is the active metal, where most corrosion occurs; the cathode is the noble metal, which is more corrosion-resistant. The electrical contact between dissimilar metals leads to the formation of a galvanic cell. Current flow, driven from the anode to the cathode by the potential difference between the two metals, liberates hydrogen and forms an alkali at the cathode.

The rate of galvanic corrosion is governed by the rate of ionization at the anode (the rate at which the anode material passes into solution), and this in turn depends on the difference in electrochemical potential between the two metals in contact. The larger this potential difference, the higher the rate of galvanic corrosion. For example, for a gold-zinc galvanic cell, the electrochemical potential with respect to a normal hydrogen electrode is $E_H = -0.76$, and $E_H = 0.34\text{-}0.52$ for a gold-copper cell. The rate of galvanic corrosion will be higher for the gold-zinc pair, all other factors remaining constant. Another example is the use of gold as a lead finish - the large electrochemical potential difference between gold and Kovar increases the rate of galvanic corrosion [Berry, 1987]. The conductivity of the corrosion medium will affect both the rate and the distribution of the galvanic attack. In solutions of high conductivity, the corrosion of the more active, anodic alloy will be dispersed over a relatively large area. In solutions with low conductivity, most of the galvanic attack will occur near the point of electrical contact between the dissimilar metals. Yet, even with the most incompatible metals, direct galvanic effects will not extend more than 5 mm from the contact area. When the surface area of the cathode is large in comparison with the area of the anode, the anodic current density is very large and galvanic corrosion is accelerated. However, when the area ratios are reversed, galvanic corrosion is not as pronounced. For example, when a very small anode of exposed base metal at the lead is in contact with a very large cathode of gold plating, galvanic corrosion occurs rapidly. Larger areas of exposed base metal at the leads corrode at a moderate rate. In addition, the closer the physical proximity of the two metals, the higher is the rate of galvanic corrosion [Davis, 1987]. Galvanic corrosion eventually ends in failure as an open.

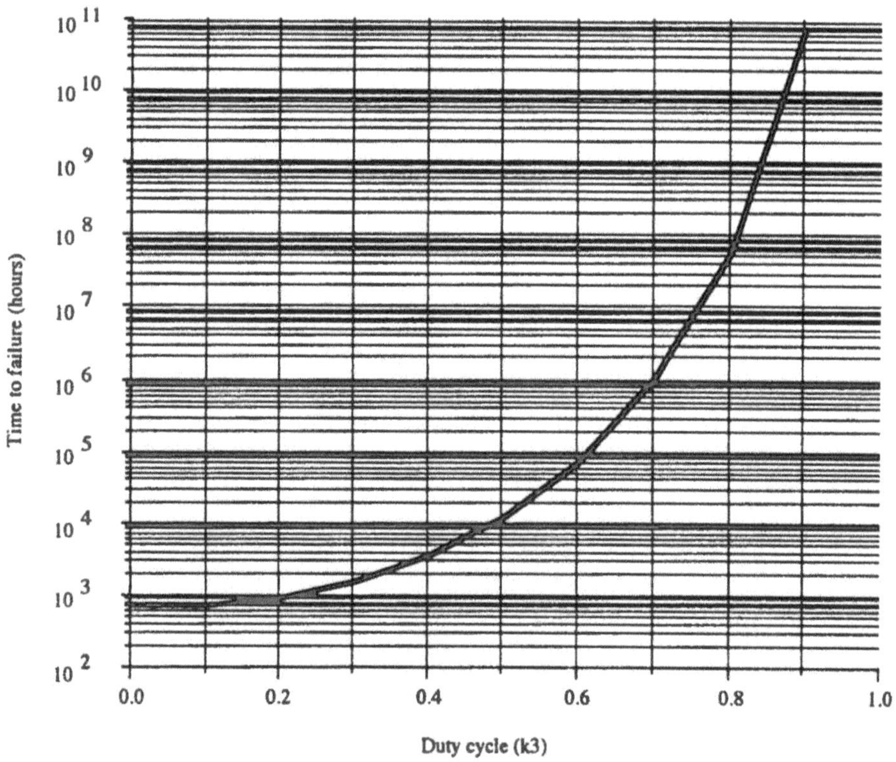

Figure 2.7 Time to failure due to corrosion vs. duty cycle [Pecht, 1990].

Various methods prevent and/or predict galvanic corrosion. First, galvanic corrosion can be minimized by metal-spraying the assembly or electroplating both metals before assembly. The cathodic metal can also be electroplated with an anodic finish to reduce the potential difference between the two metals [Dummer, 1962]. Second, in a galvanic series, metals are arranged in order of most anodic to most cathodic in a specific electrolyte. The galvanic series identifies which metal or alloy in a galvanic couple is more active. The separation between the two metals or alloys in the galvanic series gives an indication of the probable magnitude of corrosion [Davis, 1987]. Metals do not have a constant electrochemical potential, because the potential is affected by environmental factors; therefore, each galvanic series pertains only to a particular environment. A galvanic series can be formed by evaluating the corrosion of a metal when coupled to other metals, and arranging the metals according to increasing corrosion [Kucera, 1982]. Materials can also be selected in accordance with MIL-STD-1250, Table III, Galvanic Couples [Noon, 1987].

The rate of galvanic corrosion in the presence of a liquid electrolyte increases with an increase in temperature due to a more rapid rate of electron transfer at higher temperatures. Typically, corrosion products, such as aluminum hydroxide, are derived from reaction processes that are monotonically increasing functions of temperature. However, temperature is not the most significant factor. Although the corrosion rate depends in part on the steady-state temperature, it also depends on the magnitude and polarity of the galvanic potential, which are functions of electrolyte concentration, pH, local flow conditions, and aeration effects.

1.2 Electromigration

A major very large system integrated VLSI failure mode is mass transport, resulting from a momentum exchange between conducting electrons and the metal atoms in the conductor. The phenomenon of electrotransport, or "electromigration," is the result of high current density (typically on the order of 10^6 amperes/cm^2 in aluminum) in metallization tracks, which produces a continuous impact on the grains in the metallization, causing the metal to pile up in the direction of the electron flow and produce voids upstream with respect to the electron flow [Schnable, 1988]. The result is a net flux of metallization atoms that generally migrate in the same direction as the electron flux. Electromigration-induced damage in thin-film conductors usually appears in the form of voids and hillocks. Voids can grow and link together to cause electrical discontinuity in conductor lines, leading to open circuit failure. Hillocks can also grow and extrude materials, causing short-circuit failure between adjacent conductor lines on the same level, or between adjacent levels in multi-level interconnecting structures. Alternatively, hillocks can break through the passivation or the protective coating layers and lead to subsequent corrosion-induced failures. Void-induced open-circuit failures usually occur earlier than extrusion-induced short-circuit failures in thin-film conductors using aluminum-based metallurgies.

The role of temperature in electromigration is extremely complex, especially within normal operating temperatures 125°C. The lifetime limitation due to electromigration is a complex function of temperature and cannot be represented by a simple activation energy. The temperature acceleration can, however, be represented by an apparent activation energy that changes with operating temperature (Figure 2.8). Insufficient tests have been conducted on electromigration at temperatures less than 125°C, since such tests are difficult. Most tests have been performed at elevated temperatures in the neighborhood of 150°C or higher, with the results extrapolated to normal operating or room temperatures. Extrapolating failure rates from stress results at higher temperatures to provide reliability estimates at lower temperatures is not accurate, since the physics-of-failure phenomenon is not the same.

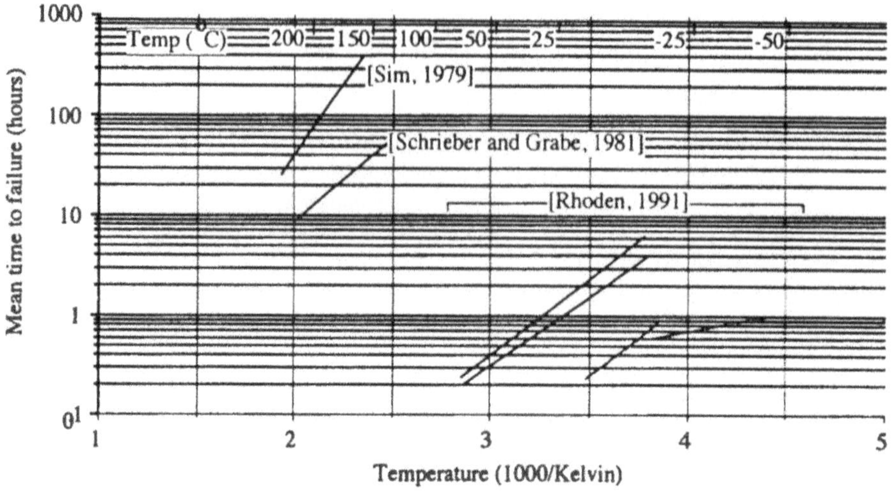

Figure 2.8 The lifetime due to electromigration is a complex function of temperature and can be represented by an apparent activation energy which changes with operating temperature.

Electromigration damage forms at sites of atomic flux divergence, of which the three main sources are structural defects, microstructural inhomogeneities, and local temperature gradients. Electromigration failures tend to be localized near sites of maximum temperature gradient [Lloyd, 1988; Schwarzenberger, 1988], even though typical field failures are characterized by structurally induced flux divergences, rather than temperature gradient-induced flux divergences [Shatzkes, 1986].

Modeling temperature effects on electromigration. The phenomenon of electromigration-induced mass transport is attributed to atomic flux resulting from electromigration in the lattice and at grain boundaries during the passage of current through a polycrystalline thin-film conductor.

Diffusivity *(D)*, exponentially dependent on temperature, is the only temperature-dependent term in the flux equation. The contribution due to electromigration in the lattice is [Huntigton, 1961]:

$$J_{l,met} = \frac{1}{K_B T} N_{l,met} D_{l,met} j_{met} \rho_{met} e Z^{*}_{l,met} \tag{4}$$

The contribution of electromigration in the grain structure is given by Huntington [1961]:

$$J_{b,met} = \frac{1}{K_B T} \frac{N_{b,met} w_{b,met}}{d_{ave,met}} D_{b,met}\, j_{met} \rho_{met} e Z^{*}_{b,met} \qquad (5)$$

where $J_{l,met}$ is the lattice electromigration atomic flux in the metallization, $N_{l,met}$ is the lattice atomic density of the metallization, $D_{l,met}$ is the lattice diffusivity of the metallization, j_{met} is the current density in the metallization, ρ_{met} is the resistivity of the metallization, $eZ^{*}_{l,met}$ is the effective lattice charge (e_0^{0} is the electronic charge of 1.6×10^{19}, $Z^{*}_{l,met}$ is the ionic charge in the lattice of the metallization), $w_{b,met}$ is the effective grain boundary width in the metallization, $d_{ave,met}$ is the average grain size of the metallization, ρ_{met} is the resistivity of metallization, $eZ^{*}_{b,met}$ is the effective grain boundary charge in metallization, $N_{b,met}$ is the grain boundary atomic density of the metallization, and $J_{b,met}$ is the grain boundary atomic flux in the metallization. The subscripts l and b represent lattice terms and grain boundary, respectively. The quantity w_b is the effective boundary width, is about 10 Å for mass transport. In aluminum, the transport is via the grain boundary, so grain boundary parameters such as diffusivity are important.

Electromigration damage can occur only where there is divergence in the electromigration flux, J, caused by variations in any of the parameters on the right-hand side of Equations 4 and 5. For instance, if the grain size of the metallization changes, perhaps due to a change in the substrate, the flux-carrying capacity of the track is altered and flux divergence results. Cross-section changes in themselves do not lead to flux divergences, because a reduction in the cross-sectional area leads to a higher current density in the remaining section. However, a section change can cause a change in local self-heating and, consequently, a temperature-induced divergence.

Temperature gradients along the stripe are important sources of flux divergence, because the flux depends exponentially on temperature [Schwarzenberger, 1988]. Temperature gradients may arise from changes, in the thermal properties of the substrate as the metallization passes over other features on the substrate, or by changes in self-heating caused by section changes, for instance, at steps over substrate features. Temperature changes can also lead to thermomigration (transport of material in a temperature gradient), though the magnitude of thermomigration is small compared to electromigration [Schwarzenberger, 1988].

At moderate temperatures (much lower than 0.5 $T_{m,cond}$, where $T_{m,cond}$ is the melting temperature of the materials), the atomic flux from electromigration in the lattice is minute compared with that from the grain boundaries, so grain boundary electromigration becomes the dominant mode of mass transport for temperatures much lower than 0.5 $T_{m,cond}$ [Ho, 1989[h]]. The mass transport in the grain boundaries in most metals occurs by vacancy diffusion [Kwok, 1981]. The relative contributions of the atomic flux due to lattice diffusion and grain boundary diffusion can be estimated from the ratio of $J_{l,met}$ and $J_{b,met}$, given by [Ho, 1989[h]]:

$$\frac{J_{l,met}}{J_{b,met}} = \frac{N_{l,met}}{N_{b,met}} \frac{d_{ave,met}}{w_{b,met}} \frac{D_{l,met}}{D_{b,met}} \frac{Z^{*}_{l,met}}{Z_{b,met}} \qquad (6)$$

The subscripts l and b denote the electromigration parameters of the lattice and grain boundary, respectively. The measured values of Z_l and Z_b usually do not differ by more than one order of magnitude. For thin films with 1 μm grain size at 0.5 $T_{m,cond}$, where $T_{m,cond}$ is the melting temperature of the materials [Ho, 1989[h]],

$$\frac{N_{l,met}}{N_{b,met}} \cong 1 \qquad \frac{d_{ave,met}}{w_{b,met}} \cong 10^{3} \qquad \frac{D_{l,met}}{D_{b,met}} \cong 10^{-7} \qquad \frac{J_{l,met}}{J_{b,met}} \cong 10^{-4} \qquad (7)$$

Attardo [1970], while conducting experiments on aluminum films 0.4 to 0.6 mil wide and 10 to 12 mil long on silicon wafers with 800 Angstroms of thermally grown silicon dioxide and 2000 Angstroms of sputtered quartz, found that mass transport during electromigration shifted from grain-boundary diffusion to lattice at threshold temperatures higher than 0.5 $T_{m,cond}$. The rate of electromigration was determined by the film's degree of preferred orientation.

Diffusion rates are highly anisotropic. The diffusion parallel to dislocations such as tilt boundaries proceeds at several orders of magnitude greater than the diffusion perpendicular to the dislocation. The vacancy flux in the grain boundary, with contributions from electromigration and diffusion via grain boundary, are given by

$$\tau_{v,met} = -D_{v,met} \nabla C_{v,met} + J_{b,met} \tag{8}$$

where $C_{v,met}$ is the vacancy concentration in the metallization, and $D_{v,met}$ is the vacancy diffusivity in the grain boundary [Ho, 1989[h]]. The local variation of the vacancy concentration is given by the equation

$$\frac{dC_{v,met}}{dt} = -\nabla . J_{v,met} + \frac{C_{v,met} - C''_{v,met}}{\tau_{v,met}} \tag{9}$$

where $C''_{v,met}$ is the vacancy concentration in the metallization at thermal equilibrium, and $\tau_{v,met}$ is the average lifetime of vacancy in metallization. Under a steady-state condition, $dC_{v,met}/dt=0$,

$$C_{v,met} - C''_{v,met} = \tau_{v,met} \nabla . J_{v,met} \tag{10}$$

The vacancy flux is represented as

$$\frac{dC_{v,met}}{dt} = D_{v,met}\left(\frac{d^2 C_{v,met}}{dx^2} - \frac{dC_{v,met}}{dx}\frac{Z\ eE}{K_B T}\right) - \left(\frac{D_{v,met}}{K_B T}\frac{dC_{v,met}}{dx} - \frac{D_{v,met} C_{v,met} Z^* eE}{(K_B T)^2}\right) \left(\frac{\Delta E_{a-gb}}{T}\left(\frac{dT}{dx}\right) + \frac{d(\Delta E_{a-gb})}{dx}\right) + \frac{C_{v,met} - C''_{v,met}}{\tau_{v,met}} \tag{11}$$

where ΔE_{a-gb} is the activation energy for the grain boundary diffusion [Attardo, 1970]. The solution of this equation provides the rate of vacancy buildup at the point of divergence and the maximum vacancy concentration that can be achieved during steady-state electromigration. It is evident from Equation 11 that the damage from vacancy supersaturation can arise from any number of discontinuities other than temperature, such as structural variations between boundaries or between regions of the stripe with different structures [Attardo, 1970]. The actual process of hole formation is a void growth process, and the time required for vacancy buildup to maximum supersaturation is orders of magnitude less than that needed for observation of holes. Rosenberg and Ohring [1971] calculated the vacancy supersaturations

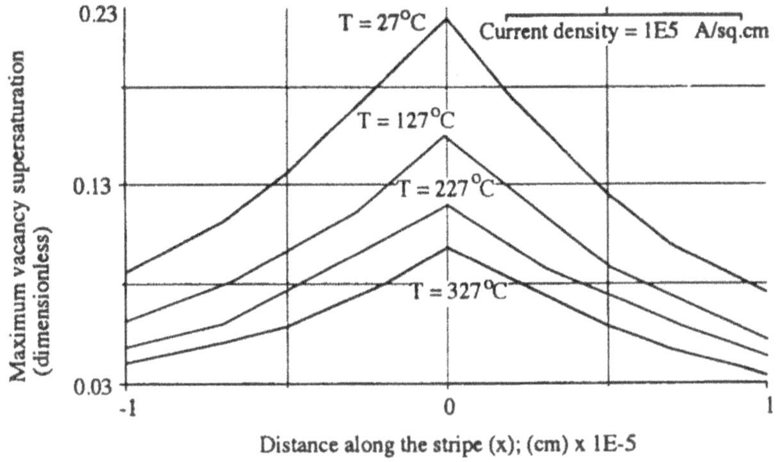

Figure 2.9 Temperature dependence of vacancy supersaturation distribution [Rosenberg, 1970].

by considering the case of two boundaries of differing diffusion characteristics located in the isothermal region $dT/dx = 0$ and joining at x=0, which is the accumulation site. The steady-state solution is represented in Figure 2.9, which demonstrates the effects of temperature on vacancy supersaturation. Assuming an aluminum product, with a DZ^* of $3 \pm 0.5 \times 10^{-2}$ cm^2/sec, and current density of 1×10^6 A/cm^2, Rosenberg and Ohring [1971] found the maximum vacancy supersaturation decreased from 27°C to 327°C. Once the maximum supersaturation is reached, the driving force for vacancy diffusion due to concentration gradient is lost. Rosenberg and Ohring observed that, under current densities of 1×10^6 A/cm^2 and steady-state temperature of 127°C, the maximum supersaturation was between 0.1 and 1.

Generally, electromigration has a linear dependence on temperature gradient, an exponential dependence on temperature for temperatures greater than 150°C, and a dependence on current density whose order varies from 2 to 14. The combinations of current density and temperature that will produce electromigration damage in titanium-platinum-gold metallizations have been characterized in Figure 2.10. The characteristics have been derived for not more than 0.14% cumulative failures (an average of eight failures/10^9 stripe hours) after twenty years, which is the three-sigma limit of the lifetime distribution. In Figure 2.11, the lifetimes are assumed to have a fourth dependence on the current density [English, 1974].

In an ideal case of a structurally uniform conductor with no temperature gradient, there is no flux divergence, so that electromigration damage will not occur [Ho, 1974, 1989]. The extent of deviation of vacancy concentration from equilibrium is proportional to the vacancy flux divergence. Whenever there is spatial variation in any of the parameters affecting grain boundary electromigration, which may include structural defects in the form of non-uniform conductor thickness or temperature gradient, a divergence in the atomic flux occurs, giving rise to a local vacancy supersaturation or depletion (void formation).

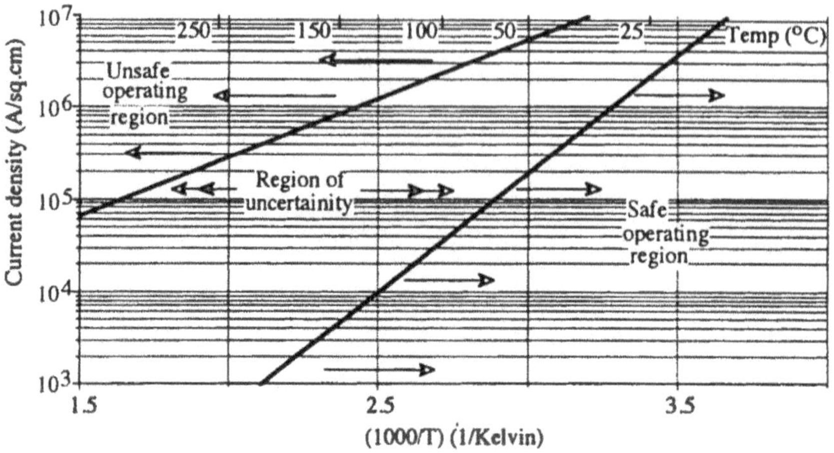

Figure 2.10 Combination of current density and temperature which will produce electromigration damage in Ti-Pt-Au metallization [English, 1974].

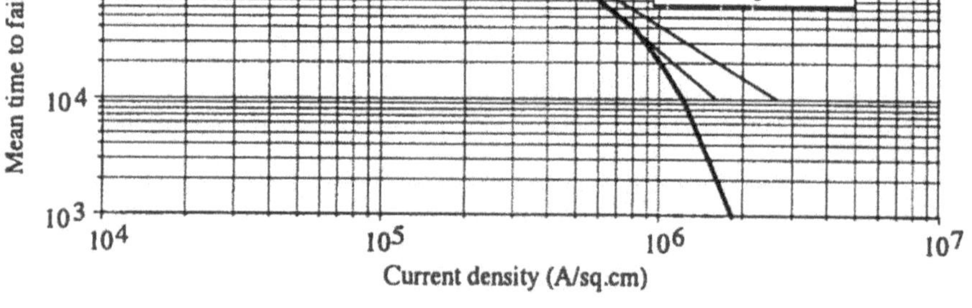

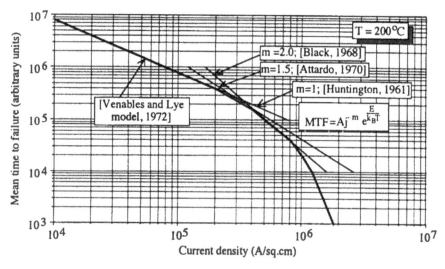

Figure 2.11 Mean time to failure varies with current density and temperature in a complex manner. Simple power laws can represent experimental observations over a small range of current densities [Venables, 1972].

Temperature dependence of electromigration in thin-film conductors. Steady-state temperature effects in the presence of cracks. The effect of cracks in the conductor metallization perpendicular to the current flow direction as correlated with temperature has been predicted by Sigsbee [1973]. Grain-boundary electromigration, internal heat generation, and current crowding at growing voids dominate the rate processes that lead to failure. Assuming that the bottom interface below the substrate is held at a constant temperature, T_e, and that the conductor experiences a temperature rise, ΔT_o, above T_e due to joule heating, for a crack perpendicular to the current direction, the current crowding is approximated by

$$J_{ce}(X) = \frac{J_{ceo}}{\left(1 - \frac{X}{w_{cond}}\right)} \tag{12}$$

where $J_{ce}(x)$ is the current density near the crack in the metallization, J_{ceo} is the current density in the bulk metallization; $J_{v,c}$ is the atomic flux of vacancies in cracks, $J_{v,co}$ is the atomic flux of vacancies in the bulk, evaluated in presence of a crack at initial temperature, and w_{cond} is the conductor width. Joule heating of the stripe causes an initial temperature rise, resulting in an instability in the stripe that causes the vacancies in the stripe to migrate along grain boundaries and precipitate on a suitable boundary, forming elongated voids. The crack grows long by the accumulation of vacancies flowing along nearby grain boundaries. The atomic flux of vacancies in the crack is represented as

$$J_{v,c} = J_{v,co}\left(\frac{1}{1 - L_{met}}\right)\left(1 - \frac{E_a \Delta T_o [\ln(1 - L_{met}) + L_{met}]}{K_B T^2 L_{met}}\right) \tag{13}$$

where $L_{met} = l_{crk,met}/w_{cond}$, $l_{crk,met}$ is the crack length in metallization, w_{cond} is the conductor width, K_B is Boltzmann's constant, T is the steady-state temperature, $w_{b,met}$ is the effective grain boundary width, $N_{a,met}$ is the atomic density of the conductor metallization, Z_{met}^* is the ionic charge, and ρ_{met} is the resistivity of the metallization. The second term in brackets is due to current crowding; the third term in brackets is due to self-heating. Assuming that only grain-boundary migration contributed to crack growth, the lifetime is calculated as

$$TF_{elec} = \frac{l_{crk,met} w_{cond} K_B T}{w_{b,met} N_{a,met} e Z_{met}^* \rho_{met} J_{ctip,met} D_{o,met} \exp\left(-\frac{E_a}{K_B T}\right) 2\left(1 + \left(\frac{0.2 E_a}{K_B T^2}\right) \Delta T_o\right)} \tag{14}$$

where $j_{ctip,met}$ is the current density at the crack tip, $D_{0,met}$ is the diffusion coefficient of the metallization, E_a is the activation energy, and TF_{elec} is the time to failure due to electromigration. This model neglects the effects due to the coefficient of resistance and the temperature sensitivity of e; Sigsbee showed that these factors had negligible effect on the life prediction estimate. The lifetimes were found to have a J_e^{-n} dependence, with n varying from unity at low ΔT_0 levels to 15 for high ΔT_o. The model has been shown to model grain-boundary electromigration for temperatures in the neighborhood of 260°C [Sigsbee, 1973].

The above grain-boundary grooving mechanism is typically noticed in silver metallizations [Venables and Lye, 1972].

Steady-state temperature effects in the presence of voids. The atomic flux is dependent on temperature; therefore, local temperature gradients will cause a divergence in atomic flux. The depletion of mass occurs wherever the electrons flow in the direction of increasing temperature. Conversely, the accumulation of mass occurs wherever the electron flow is in the direction of decreasing temperature [Venables and Lye, 1972]. Metallized stripes in good thermal contact with their substrates have negligible temperature gradients [Venables and Lye, 1972]. Electromigration in thin-metallized films is confined mainly to grain boundaries [Blech, 1967; Agarwala, 1970; Rosenberg, 1968]. Thus, the steady-state temperature dependence of electromigration is of the same magnitude as that of grain-boundary diffusion [Blech, 1967; Rosenberg, 1968].

Voids form as a consequence of flux divergences at non-symmetrical nodes, and grow with time, eventually coalescing to form a gap across the conductor. The effect is particularly severe in films with a small grain size, because a large number of nodes are available to act as nuclei for void formation. On the other extreme, if the grain size is comparable to stripe width, the probability that a single grain will cover the entire stripe increases. This introduces an additional source of flux divergence acting as a barrier to atoms migrating from the negative side and preventing the replacement of atoms transported away from connecting boundaries on the other side of the grain [Attardo, 1970; Blair, 1970].

The flow of current through stripes creates voids at grain-boundary nodes that are suitably oriented relative to the current flow direction and longitudinal temperature gradient. The resulting porosity increases as a function of the density of the grain-boundary nodes, current density, resistivity, and mobility of metal ions along grain boundaries, and is represented by

$$\frac{dp_{p,met}}{dt} = C_{p,vl} n_{gb,met} j_{met} \rho_{met} \mu_{gb,met} \tag{15}$$

where $p_{p,met}$ is the porosity of the metallization, $C_{p,vl}$ is Venables and Lye's constant of proportionality, $n_{gb,met}$ is the grain-boundary node density in the metallization, j_{met} is the current density in the metallization, ρ_{met} is the resistivity of the metallization, $\mu_{gb,met}$ is the mobility of metal ions along metallization grain boundaries, $j_{0,met}$ is the current density in a pore-free stripe, and $j_{p,met}$ is the current density in a metallization stripe with pores. Pore formation reduces the cross-sectional area of the metal stripe available for carrying the current, thereby increasing the local current density within the remaining section [Venables, 1972]:

$$j_{p,met} = \frac{j_{0,met}}{(1-p_{p,met})} \tag{16}$$

where $j_{0,met}$ is the current density in a pore-free stripe and $j_{p,met}$ is the current density in a metallization stripe with pores. The increased current density causes an increase in the current-enhanced motion of the metal atoms in the stripe, at the same time increasing the joule heating within the remaining conducting portions of the stripe. The local temperature in the stripe increases above the ambient temperature, T_o, by an amount that is proportional to joule heating, as given by [Venables, 1972]:

$$\Delta T = T - T_o = \frac{j_{met}^2 \rho_{met}}{h_{met}} \tag{17}$$

where ΔT_o is the temperature rise due to joule heating, h_{met} is the convective heat transfer coefficient of the metallization, j_{met} is the current density in the metallization, and ρ_{met} is the resistivity of the metallization. The temperature rise leads to a corresponding increase in the mobility, $\mu_{gb,met}$, of the metal atoms along grain boundaries [Venables, 1972]:

$$\mu_{gb,met} = \frac{D_{met}}{K_B T} = \left(\frac{D_{o,met}}{K_B T}\right) e^{-\frac{E_a}{K_B T}} \tag{18}$$

where $\mu_{gb,met}$ is the grain-boundary mobility of the metallization, D_{met} is the diffusivity of the metallization, $D_{0,met}$ is the diffusion coefficient of the metallization, and E_a is the activation energy for this process. The increase in temperature leads to a change in the resistivity, ρ_{met}, of the metal [Venables, 1972]:

$$\rho_{met} = \rho_{0,met}(1 + \alpha_{r,met}(T-T_o)) \tag{19}$$

where $\alpha_{r,met}$ is the temperature coefficient of resistivity for the metallization, ρ_{met} is the resistivity of the metallization, $\rho_{0,met}$ is the resistivity of the metallization at reference temperature, T is the steady-state temperature, and T_o is the reference temperature. At a constant total current, the increase in resistivity causes an additional increase in the local rate of joule heating and in the effective electric field (j_{met}, ρ_{met}) experienced by the atoms. Electromigration failure occurs where the grain-boundary migration and temperature gradient combine to create suitable conditions for porosity to develop until it exceeds the critical value, resulting in the melting of the stripe. Venables and Lye [1972] combined the above equations to give the time to failure as

$$TF_{elec} = \frac{1}{2C_{p,vl}n_{gb,met}} \left(\frac{\tau_o K_B T_o}{j_{o,met}\rho_{o,met}D_{o,met} e^{-\frac{E_a}{K_B T}}} \right) \int_{x_o}^{x_1} \frac{e^{\frac{-E_a x}{K_B T_o}}}{x^2(1-x+x\alpha T_o)} dx \tag{20}$$

where

$$\tau_o = \frac{\Delta T_o}{T_o} = \frac{j^2_{o,met}\rho_{o,met}}{h_{met}T_o} \tag{21}$$

and

$$x = \frac{\tau_o}{1 - p^2_{p,met} + \tau_o(1 - \alpha_{r,met} T_o)} = 1 - \left(\frac{T_o}{T}\right) \tag{22}$$

$$x_o = \frac{\tau_o}{1 - \tau_o(\alpha_{r,met} T_o - 1)} \quad @t=0$$
$$x_1 = 1 - \left(\frac{T_o}{T_{m,met}}\right) \quad @t=T_F \tag{23}$$

where $T_{m,met}$ is the melting temperature of the metallization, and TF_{elec} is the time to failure due to electromigration. They showed that time to failure vs current density varied complexly, and that a simple power-law dependence (as shown by Black) was inadequate to describe the experimental conditions over more than a small range of current densities. The results of Attardo [1970], Black [1968], and Blair [1970] were tangential to the results from the Venables and Lye model at a temperature of 210°C and current densities ranging from 1 x 10^4 A/cm^2 to 2 x 10^6 A/cm^2 (Figure 2.11). Venables and Lye [1972] showed that when the temperature dependence of times to failure due to electromigration was represented as an Arrhenius plot, although the curves appeared as accurate straight lines, the slopes yielded only apparent activation energy, which varied with test conditions, indicating that the time to failure was a complex function of baseline temperature and could not be represented by an Arrhenius plot to give an activation energy (Figures 2.12 and 2.13).

Steady-state temperature effects without assumption defect magnitudes. A general, but not universal, expression for the mean time to failure MTF (or t_{50}, which is the time to reach 50% failure of a group of identical conductor lines) is given in Equation 24. This is not a failure-rate expression, as the failure times typically follow a lognormal distribution:

$$TF_{elec} = A_{g,Black}\, j_{met}^{-n_{e,Black}}\, e^{\frac{E_a}{K_B T}} \tag{24}$$

where TF_{elec} is the time to failure due to electromigration; $A_{g,Black}$ is Black's electromigration coefficient, j_{met} is the current density in the metallization, and $n_{e,Black}$ is Black's electromigration exponent. This model is applicable only to conductor films that are wider than the average grain size of the aluminum film from which they are constructed. As the conductor width is reduced and approaches or becomes less than the average grain size, the structure begins to "bamboo", that is, most of the grain boundaries become normal to the electron flow. Black's relationship does not apply when bambooing starts [Black, 1982].

The electromigration lifetime test is carried out under a set of accelerated test conditions at

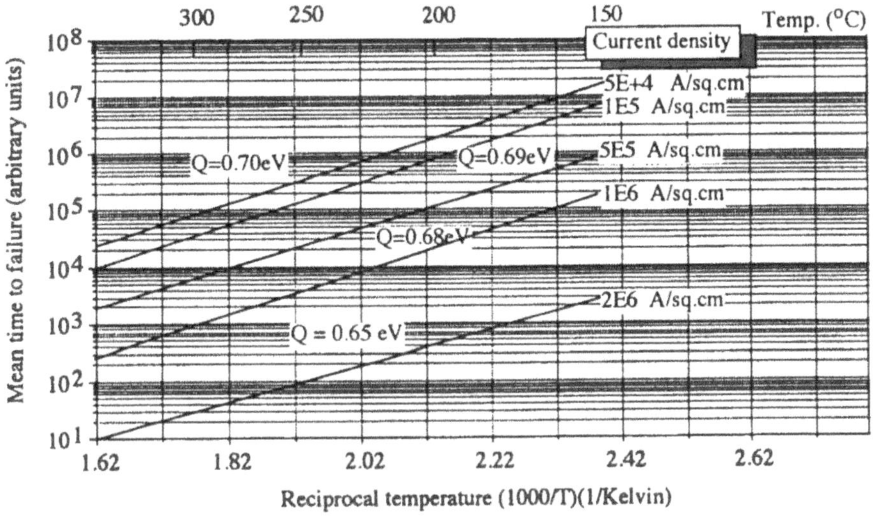

Figure 2.12 Temperature dependence has been represented by an apparent activation energy which changes with test conditions [Venables, 1972].

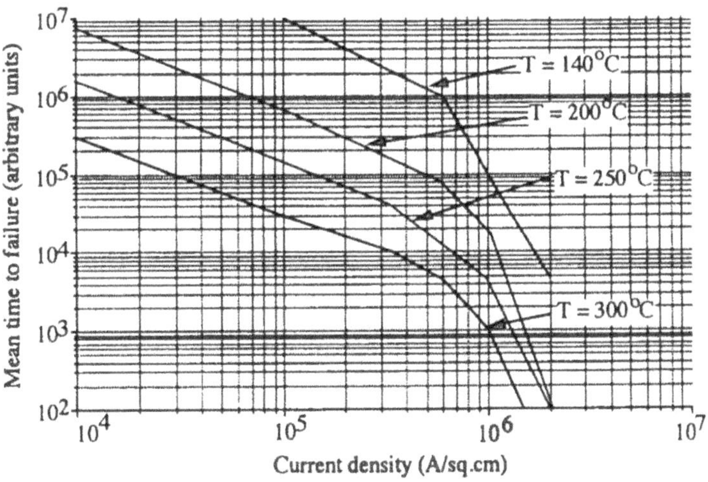

Figure 2.13 Influence of baseline temperature on mean time to failure [Venables, 1972].

elevated temperatures and with high current density stressing. The data are then extrapolated to device-operating conditions, with current-density stressing below 5×10^5 A/cm^2, using the Arrhenius-like empirical equation cited above (originally formulated by Black [1969a,b]). Values of $n_{e,Black}$ that have been reported are shown in Table 2.

Black characterized his data in the range of $0.5 \times 10^6 < j_{met} < 2.8 \times 10^6$, with an exponent of $n_{e,Black}=2$. Attardo reported $n_{e,Black}=1.5$ in the range $10^5 < j_{met} < 10^6$ A/cm^2. Blair reported a value of $n_{e,Black}=4$-5 in the range of $10^6 < j_{met} < 2 \times 10^6$ A/cm^2. Venables found that the simple power-law dependence of time to failure on current density could be used to describe experimental observations in a small range of current densities [Venables, 1972]. Lloyd [Shatzkes, 1986] treated electromigration failures by superimposing Fickian diffusion and mass transport due to electromigration force, and derived a modification of Black's equation:

$$TF_{elec} = \left(\frac{2C'_{vc,f}}{D_{o,met}}\right)\left(\frac{K_B}{Z^*_{met}e\rho_{met}}\right)^2 T^2 j_{met}^{-2} e^{\frac{E_a}{K_B T}} \quad (25)$$

where $C'_{vc,f}$ is the critical value of vacancy concentration at which failure occurs, $D_{o,met}$ is the diffusion coefficient for the metallization, ρ_{met} is the resistivity of the metallization, and j_{met} is the current density in the metallization. Equation 25 differs from Black's equation in that it has a T^2 pre-exponential term, but it fits Black's data equally well.

Table 3 shows the estimated MTF of titanium:tungsten/aluminum (Ti:W/Al) and titanium: wolfram/aluminum + copper (Ti: W/Al + Cu) film conductors at 85°C for 5×10^5 and 2×10^5 A/cm^2 current densities at 100% duty cycle. Even for the worst case $n_{e,Black} = 1$, it has been anticipated that the actual time to failure will be greater than those predicted in Table 3 [Ghate, 1981]. For current densities of 2×10^5 A/cm^2, temperatures of 125°C will not lead to failure in less than ten years with a typical exponent of $n_{e,Black} = 1.7$.

Few studies at lower temperatures have been performed on unpassivated stripes, but these have shown that the phenomenon of electromigration shifts from grain-boundary migration to surface migration, with detachment of the stripe from the chip, at temperatures in the neighborhood of 223 °K to 347 °K [Rhoden, 1991].

In order to improve the resistance of the conductor to surface electromigration, passivations consisting of glass overlays, metallic coatings, or natural oxides are used to cover the thin-film conductor. Typically, the electromigration lifetime of most conductors increases by one order of magnitude or more with complete surface coverage [Lloyd, 1983; Felton, 1985; Yeu, 1985]. The use of transition layers such as titanium nitride (TiN) [Grabe, 1983], chromium (Cr) [Levine, 1984], and titanium-tungsten (Ti-W) [Fried, 1982], has been reported to improve the lifetime of thin-film conductors by one order of magnitude. The transition layer provides a redundant structure, allowing void healing, and also acts as a metal diffusion layer.

Table 2. Reported values of n_e

Reference	value of $n_{e,Black}$
Chabra and Ainslie [1967]	1 - 3
Attardo [1972]	1.5
Danso and Tullos [1981]	1.7
Black [1983]	2
Blair et al. [1970]	6

Temperature-gradient dependence of electromigration. Lifetime temperature gradients exist both globally and locally in thin-film conductors, due to heat generation from joule heating and power dissipation from active devices on the chip. The global temperature gradient is small, except near electrodes or contact pads. Large local temperature gradients or hot spots can be caused by poor adhesion or contaminations at the interface between the metal film and the substrate, or by variations in the thickness of the metal film.

Temperature gradients are important sources of atomic flux divergence, since the flux depends exponentially on temperature. For example, at 200°C in aluminum, a 5°C change in temperature results in a change of more than 10% in the electromigration flux [Schwarzenberger, 1988]. Studies on electromigration damage due to temperature gradients have revealed that void formation occurs in regions where the electron flow is in the direction of increasing temperature, and hillocks form in locations where the electron flow is in the direction of decreasing temperature [Blech, 1967]. Temperature gradients can also lead to thermomigration, although compared to electromigration, thermomigration is small for aluminum tracks in ICs.

The temperature gradient dependence of electromigration failures has been confirmed by Lloyd [1988] and Schwarzenberger [1988]. Lloyd noticed that the electromigration failure location was typically near the location of the maximum temperature gradient, whereas the location of failures not caused by temperature gradients was randomly distributed for chromium/aluminum-copper (Cr/Al-Cu) conductors covered with polyimide passivation. He modeled the temperature of the stripe carrying current to produce significant joule heating as the balance between the heat generated in the stripe and the heat conducted away from the stripe to the surrounding thermal sinks, such as substrate, passivation, and the non-conducting portion of the metal stripe. The heat balance relation is given by

$$\rho_{met} j_{met}^2 = K_{met} \left(\frac{d^2 \Delta T}{dx^2} \right) - h \Delta T_0 \tag{26}$$

per unit volume, where K_{met} is the metallization's thermal conductivity, h is the heat transfer coefficient of the substrate, and ΔT_0 is the temperature rise due to joule heating [Lloyd, 1988].

Table 3. Estimated MTF values for electromigration at 85°C, as a function of the exponent used in Black's equation

Current density	Exponent	Ti: W/Al (years)	Ti: W/Al + Cu (years)
5×10^5 A/cm^2	$n_{r,Black} = 1.0$	4	12
	$n_{r,Black} = 1.5$	5	17
	$n_{r,Black} = 2.0$	8	24
2×10^5 A/cm^2	$n_{r,Black} = 1.0$	10	30
	$n_{r,Black} = 1.5$	23	68
	$n_{r,Black} = 2.0$	50	152

The term on the left of Equation 26 is the heat generated due to current passing through the metal element; the first term on the right is the heat conduction along the metal stripe away from the heating element; and the last term is the heat conduction to the environment. If the heat conduction through the oxide (of thickness l_{oxide}) is considered alone, the value of h is of the order

$$h \approx \frac{K_{oxide} K_{met}}{l_{oxide} l_{met}} \tag{27}$$

where K_{oxide} is the thermal conductivity of oxide, and l_{met} is the length of the metallization stripe. The heat sink is assumed to be at the ambient temperature. The solution to the heat conduction equation for a stripe with the origin in the center was given by Lloyd [1988] as

$$T(x) = A_{int} \left\{ 1 - \left[\frac{Cosh(B_{int} x)}{Cosh\left(\frac{B_{int} l}{2} \right)} \right] \right\} \tag{28}$$

where

$$B_{int} = \sqrt{h - \frac{\rho_{met} \alpha_{r,met} j_{met}^2}{K_{met}}} \tag{29}$$

$$\rho_{met} = \rho_{o,met} \left(1 + \alpha_{r,met} \Delta T \right) \tag{30}$$

$$A_{int} = \frac{\rho_{met} j_{met}^2}{B_{int}^2 K_{met}} \tag{31}$$

The location of failure was argued to be near the position of maximum flux divergence. The atomic flux is

$$J_{a,met} = \frac{D_{met} F_{diff}}{K_B T} \tag{32}$$

where A_{int} and B_{int} are intermediate factors, and F_{diff} is the diffusion force; the point of maximum flux divergence is

$$\frac{dJ_{a,met}}{dx} = \left(\frac{dJ_{a,met}}{dT} \right) \left(\frac{dT}{dx} \right) \tag{33}$$

and the condition of failure is when

$$\frac{d^2 J_{a,met}}{dx^2} = \frac{d}{dx}\left(\frac{dJ_{a,met}}{dT}\right)\left(\frac{dT}{dx}\right) = 0 \tag{34}$$

The location of electromigration failure calculated from Equation 34 is near the location of the maximum temperature gradient [Lloyd, 1988]. The points of maximum flux divergence move from the edges (for higher current densities) to the middle (for lower current densities). The change in the failure location was due to an exponential dependence of flux divergence on temperature, with only linear dependence on temperature gradient.

Figures 2.14 and 2.15 show the variation in the location of failure sites versus distance along the stripe, for low and high current densities. At high current densities (> 10^6 A/cm^2) accompanied by high joule heating, the highest flux divergence is very close to the location of highest temperature gradient, which is near the edge. The failure site is thus closer to the site of maximum temperature gradient. At lower current densities (< 10^6 A/cm^2), the failure location is more randomly distributed, since the effect of temperature gradient is small compared with other flux divergences induced by structural discontinuities.

The variation of flux between different regions of the conductor may lead to failure due to depletion in some regions. The time to failure is inversely proportional to the flux gradient within that region. The flux gradient, in terms of the temperature gradient, is given as

$$\frac{dJ_{a,met}}{dX} = \frac{N_{i,met} e j_{met} D_{0,met}}{K_B} e^{-\frac{E_a}{K_B T}} f(\rho_{met}, Z^*_{met}, T) \frac{dT}{dX} \tag{35}$$

where $N_{i,met}$ is the density of the ions in the metallization, e is the electronic charge, j_{met} is the current density in the metallization, $D_{0,met}$ is the diffusion coefficient of the metallization, T is the steady-state temperature, and x is the distance along the metallization stripe. Schwarzenberger et al. [1988] demonstrated the importance of temperature gradient as a source of electromigration flux divergence in metallization tracks and, therefore, the necessity of controlling the temperature profile in the integrated circuit to optimize its lifetime [Oliver and Bower, 1970].

Temperature dependence of electromigration in multilayered metallizations. The definition of electromigration failure has changed with the introduction of multilayered interconnection metallizations with sensitive electrical circuits [Onduresk, 1988]. Many studies on single-layer metallizations have used the opening of the conductor as a criterion for failure, ignoring functional failures, including the resistance change of the metallization. The open criterion for layered metal systems may not be achievable in a test environment if one of the layers is not susceptible to electromigration.

Ondrusek derived a void-formation model for multilayered metallizations that allows calculation of void length from measured resistance and temperature coefficients [Onduresk, 1988]. The temperature coefficient of resistance was investigated to verify that the refractory layer remains undamaged throughout the voiding process. The temperature coefficient, $\alpha_{r,mult}$, for metal is defined by the equation

$$R_{T,mult} = R_{i,mult}\left[1 + \alpha_{r,mult}(T - T_{ref})\right] \tag{36}$$

where $R_{T,mult}$ is the resistance of the multilayer metallization at temperature T and $R_{i,mult}$ is the

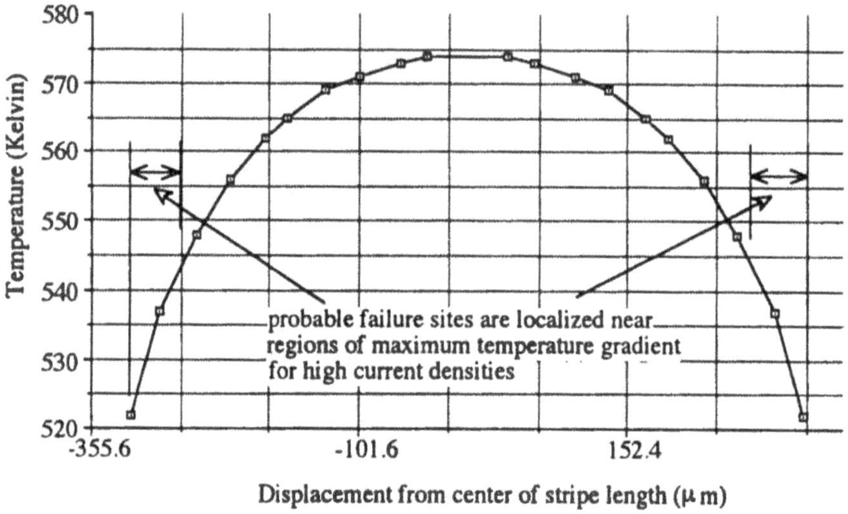

Figure 2.14 Location of failure sites versus temperature distribution along stripe for high current condition [Lloyd, 1988]

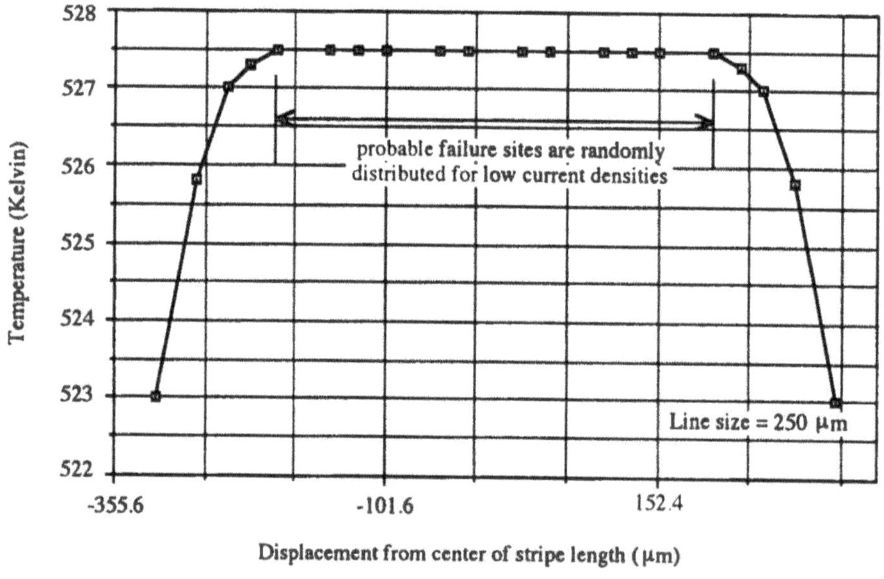

Figure 2.15 Location of failure sites versus temperature distribution along stripe length for low current condition [Lloyd, 1988].

resistance of the multilayer metallization at initial temperature. $\alpha_{r,mult}$ is the temperature coefficient of resistance of the multilayer metallization, T is any steady-state temperature value, and T_{ref} is the reference temperature. In a simplified case of a void extending through the aluminum layer, all of the current must pass through the refractory metal layer. The total resistance of the metal stripe is the sum of the initial component due to the refractory material-metal/aluminum sandwich plus the additional series refractory metal resistance that results from the void. This is represented by the following expressions:

$$R_{T,mult} = R_{1,mult} + R_{2,mult} \tag{37}$$

where

$$R_{1,mult} = R_{im,met}\frac{L_{met}-L_{void,met}(t)}{L_{met}}[1+\alpha_{r,met}(T-T_{ref})] \tag{38}$$

and

$$R_{2,mult} = R_{ir,met}\frac{L_{void,met}(t)}{L_{met}}[1+\alpha_{r,refct}(T-T_{ref})] \tag{39}$$

Taking the derivative of Equation 25 with respect to temperature gives

$$\frac{\partial R_{T,met}}{\partial T} = R_{im,met}\alpha_{r,met}\frac{L_{met}-L_{void,met}(t)}{L_{met}}+R_{ir,met}\alpha_{r,refct}\frac{L_{void,met}(t)}{L_{met}} \tag{40}$$

where $R_{im,met}$ is the resistance of the refractory in a multilayered metallization, $r_{r,refct}$ is the temperature coefficient of resistance of the refractory in a multilayered metallization. Deviation from this model can occur due to intermetallic compound formation.

Reducing electromigration damage in metallization stripes. Solutions to reducing electromigration do not lie in reducing steady-state temperature. The basic requirement for reducing electromigration damage is to reduce the local divergence of atomic flux. This can be accomplished, in principle, by reducing the magnitude of the atomic flux and/or the inhomogeneity of the parameters controlling the mass transport. The magnitude of atomic flux is determined by the electromigration driving force and the grain-boundary diffusivity. Thus, to reduce the atomic flux, the option is to reduce either the driving force and/or the diffusivity.

Reducing the driving force has some inherent difficulties, since it requires either a change in the scattering process responsible for the effective charge or a reduction in the current density. Because the current density is dictated by device functional requirements, the only choice is to reduce the grain-boundary diffusivity. The most common approach is solute addition, which also produces improvements in conductor properties through grain structure modification. Common examples are the addition of copper or other solute elements such as manganese, magnesium, and titanium to aluminum stripes [Ho, 1989[h]]. There is a strong correlation between microstructure and electromigration lifetime in thin metal lines, which is particularly noticeable in VLSI technology when the line width and thickness of the metal lines are reduced to submicron range. Longer lifetimes have been reported in large-grained aluminum [Attardo and Rosenberg, 1970] and aluminum-copper films with bamboo structure [Vaidya, 1980, Pierce, 1981] (Figure 2.16). Annealing aluminum-copper lines at elevated temperature induces grain growth, increasing the electromigration lifetime. Structural modification of aluminum-copper films by adding titanium and chromium at about 400°C results in the formation of intermetallic compounds such as Al_3Ti and Al_7Cr; this improves electromigration lifetimes due to microstructure changes that reduce damage formation and block void growth using a redundant barrier that maintains current continuity [Kwok, 1987].

Line width has a strong influence on electromigration lifetimes. Agarwala [1970] reported a linear relationship between line width and lifetime in aluminum lines with widths down to 5μm. Subsequent studies have indicated that as line width sinks below a critical value, lifetime levels off or reaches a minimum and then increases, reversing the trend in wider lines (Figure 2.17). The critical width decreases with decreasing film thickness. Kwok found the critical line width to be about 0.75 μm for a 0.5-μm line thickness in aluminum-copper lines (Figure 2.18) [Kwok, 1989]. The critical line width is sensitive to the thickness of the metal lines. Lifetime increases by a factor of 5 as line width increases from 1μm to 2 μm. The lifetime of aluminum-copper-silicon, chromium-silver-chromium, and aluminum/ titanium lines levels off beneath critical line widths of around 2.0 μm, 1.5μm, and 1.2μm, respectively. The line width dependence is much stronger for aluminum-copper and aluminum-copper-silicon lines than for chromium-silver-chromium lines. The probability of alignment failure will cause defects across a wide line is lower than for a narrow line; thus, it is more difficult for a crack to propagate across a wide line, for which the expected lifetime increases. The dependence of lifetime on line width is also a function of pattern technique, metallurgy, and metal deposition conditions. It was found that aluminum-copper lines patterned by chemical etch reached a minimum lifetime for line widths in the neighborhood of 5.5μm. Aluminum-copper lines of the same size, patterned by metal lift-off, have a minimum lifetime for widths in the neighborhood of 3.5μm. Electromigration lifetime increases with decreasing thickness. The critical width decreases from 2.5 to 1.5μm when the film thickness decreases from 1.1 to 0.8μm in aluminum-copper-silicon lines.

Predicted interconnection failure rates at current density and temperature use conditions typically vary by several orders of magnitude, as they strongly depend on accelerated test data and model parameter selection. Generally, the failure time depends inversely on the temperature gradient and current density. Temperature acts as an strong accelerator of electromigration above temperatures of 150°C. Electromigration failures in structurally uniform conductors cannot be accelerated in reasonable time frames at temperatures lower than 150°C.

The simple Black-type equation can be used for extrapolation of failure to use conditions only over a small range of current densities. Moreover, the current exponent changes over various densities, while the actual dependence of lifetime is a complex function of current density and temperature that cannot be represented by an activation energy. The lifetime can be represented by an apparent activation energy that changes with operating conditions. Electromigration damage forms at sites of maximum atomic flux divergence, of which the three main sources are structural defects, microstructural inhomogeneities, and local temperature gradients. Because typical field failures are characterized by structurally induced flux divergences introducing geometric configurations, and metallization grain structures, reducing these structural variations (but not steady-state temperature) can reduce electromigration damage.

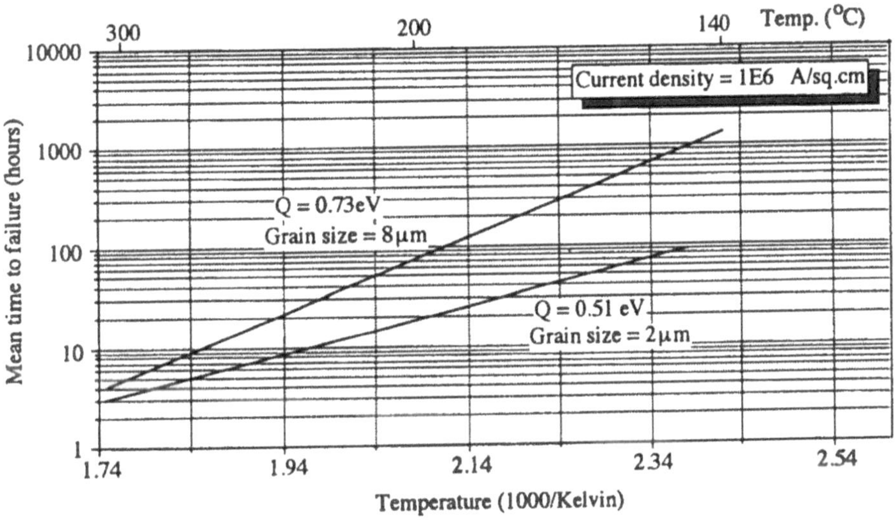

Figure 2.16 Effect of temperature on MTF vs grain size. [Attardo, 1970].

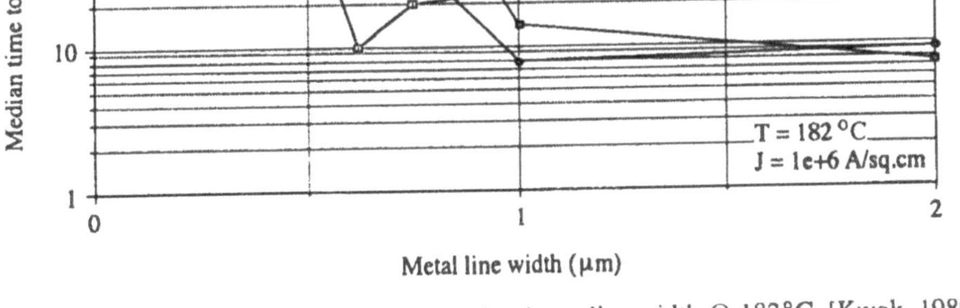

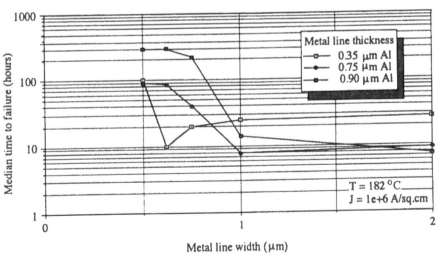

Figure 2.17 MTF due to electromigration vs. aluminum line width @ 182°C [Kwok, 1989].

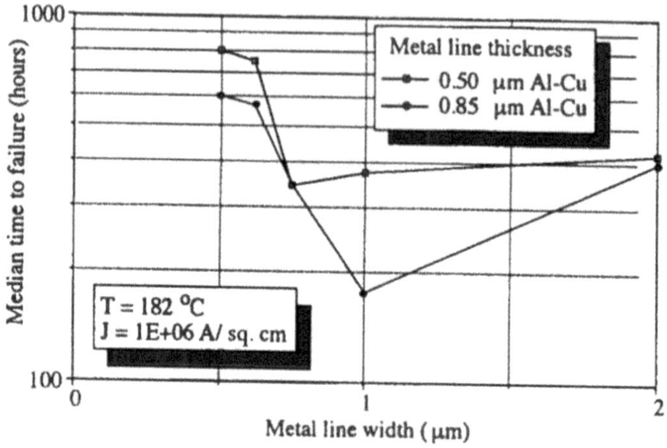

Figure 2.18 MTF due to electromigration vs. Al-Cu line width @ 182°C [Kwok, 1987].

1.3 Hillock Formation

Hillocks in die metallization can form as a result of electromigration or extended periods under temperature cycling conditions (thermal aging) [LaCombe Christou, 1982; Thomas, 1983]. Hillock formation as a result of electromigration often occurs in the electron flow upstream from the area of voiding, but hillocks grow fastest at the downstream edge closest to the source of the migrating metallization. Voiding and hillock formation sometimes occur on top of each other at temperatures in the neighborhood of 140°C to 200°C [Thomas, 1983].

Hillock formation due to extended periods under temperature cycling conditions (thermal aging) is believed to be due to a self-diffusion process that occurs in the presence of strains within the metallization [LaCombe Christou, 1982]. These strains may be due to a mismatch in the thermal expansion coefficients of the metallizations gold and titanium-tungsten, and the underlying refractory layer of silicon and silicon dioxide. Hillock growth is more extensive in films deposited on room temperature substrates than on heated substrates, indicating that the effect may vary with grain size. Coating the metal with silicon nitride prevents failures, voids, and hillocks for at least 500 hours at 360°C. Hillocks form at random in aluminum films heated to temperatures around 400°C during fabrication, and can cause electrical shorts between adjacent lines and fracture of the overlying dielectric film. In double-level metallized devices, hillocks can result in shorts between the underlying and overlying metal layers. Hillocks, which can be caused by electromigration, can result in thin dielectric sites that are susceptible to subsequent breakdown of the intermetal dielectric. Hillock formation is largely remedied by using alloyed metal, such as aluminum-copper, and by improvements in the technologies of passivation and packaging. Hillock formation is more a function of temperature change and temperature, and is slightly dependent on steady state temperature (in the neighborhood of 400°C).

1.4 Metallization Migration

Metal migration occurs between biased lands under conditions conducive to electrocrystallization. Dendritic growth is a common cause of failure. Conditions for metal migration include a level of around 10 A/cm^2 current density at the tip of the dendrite through spheroidal and parabolic diffusion, sufficient liquid medium such as condensed water, applied voltage that exceeds the sum of anodic and cathodic potentials in equilibrium with the electrolyte, and materials with defects that allow water condensation to satisfy the current density requirement.

For a given material and pore frequency distribution, the ionic current density or rate of mass transport per unit area is proportional to the fraction of the pore area containing condensed water. DiGiacomo [1982] proposed a model to predict failure as a function of environmental conditions and the physical properties of the package, and verified the predictions on the basis of migration failure data from accelerated tests. The fractional area in which condensation occurs can be represented by

$$A_{fr,cnd} = \frac{1}{2} erfc\left(\frac{1}{2\sigma_{sd}}\right) \ln\left(\frac{r_{av,pore}}{r_{pore}}\right)^2 \tag{41}$$

where

$$r_{av,pore} = \frac{2\gamma_{st} V_{mol}}{K_B T \ln\left(\frac{p_{sat,men}}{p_{sat,flat}}\right)} \tag{42}$$

where r_{st} is the surface tension, V_{mol} is the molar volume, $p_{sat,men}$ is the saturated vapor pressure above the meniscus, and $p_{sat,flat}$ is the saturated vapor pressure above a flat surface.

Given the definition of the error function,

$$erfc\left(\frac{1}{2\sigma_{sd}}\right) = 1 - \frac{2}{\sqrt{\pi}} \int_{0}^{\left(\frac{1}{2\sigma_{sd}}\right)} e^{-u^2}\, du \tag{43}$$

The average current density is expressed as a function of ionic concentration, electric field, diffusivity, temperature, fractional condensed area, and overpotential, using the Butler-Volmer equation of electrode kinetics and assuming diffusion control [DiGiacomo, 1982; Barton and Bockris, 1962; Price, 1958; Adamson, 1990]:

$$j_{av,mmig} = \frac{(Z_{met}F)^2 C_{i,bulk} D_{met} V_{ov,pt}}{2K_B T} erfc\left(\frac{1}{2\sigma_{sd}}\right) . \ln\left(\frac{K_B T\,(\ln(RH)) r_{av,pore}}{2\gamma_{st} V_{mol,vl}}\right)^2 \tag{44}$$

where $j_{av,mmig}$ is the average current density during metal migration, F is Faraday's constant, $C_{i,bulk}$ is the ionic contamination in bulk, D_{met} is the diffusivity of the metallization, $V_{ov,pt}$ is the overpotential proportional to V_{app}/r_{melp}, $r_{av,pore}$ is the average pore radius, V_{app} is the applied voltage, RH is the relative humidity, Z_{met} is the valence of metal ions, and $V_{mol,vl}$ is the molar volume.

To produce a metal dendrite, the current density at the whisker's tip should be orders of magnitude higher than the average current density [DiGiacomo, 1982]. The growth is possible only through parabolic and spheroidal diffusion focusing the ionic current on the dendrite tip, which can be expressed in terms of the radius of curvature:

$$j_{tip,mmig} = \frac{(Z_{met}F)^2 C_{i,bulk} D_{met}}{2K_B T} \frac{V_{app}}{r_{d_{dndt}}} erfc\left(\frac{1}{2\sigma_{sd}}\right) . \ln\left(\frac{K_B T(\ln RH) r_{av,pore}}{2\gamma_{st} V_{mol,vl}}\right)^2 \tag{45}$$

where r_{dndt} is the radius of curvature of the dendrite, $j_{tip,mmig}$ is the ionic current on the dendrite tip, and $V_{mol,vl}$ is the molar volume. For dendritic growth, $j_{tip} > j_{critical}$, the critical value of the current density, or a mass transport rate of $\partial Q_{tip}/\partial t > \partial Q_{critical}/\partial t$. The mass transport rate is represented as $Q_{tip} = j_{tip}/ZF$. Integrating the equation with respect to time and substituting for $D = D_o \exp(-\Delta RH/K_B T)$ gives the critical value of the mass transport rate that is, the number of ions that must be transported to achieve dendritic growth across the gap as $\partial Q_c/\partial t = l\rho_d/Mt$, where l = distance between electrodes, ρ_d = density of dendrite, and M = atomic weight. The time to failure can be given as [DiGiacomo, 1982]:

$$TF_{mmig} = \frac{\left(\frac{d_{dndt}}{M_{at} C_{i,bulk}}\right)\left(\frac{2r_{dndt}}{Z_{met} F D_{o,met}}\right) l_{d,c}}{\beta_{an,ox}(V_{app} - V_{th}) \frac{1}{K_B T} e^{-\frac{E_a}{K_B T}} erfc \frac{1}{2\sigma_{sd}} \ln\left(\frac{K_B T \ln(RH) r_{av,pore}}{2\gamma_{st} V_{mol,vl}}\right)^2} \tag{46}$$

where d_{dndt} is the density of the dendrite, M_{at} is the atomic weight, $l_{d,c}$ is the distance between electrodes, $\beta_{an,ox}$ is the fraction of anode surface that is susceptible to oxidation, and $D_{o,met}$ is the diffusion coefficient of the metallization.

The average current, j, is expressed as a function of ionic concentration, electric field diffusivity, temperature, and condensed area. $\beta_{an,ox}$ is the fraction of metal surface at the anode that is susceptible to metal oxidation and, therefore, promotes metal migration in regions satisfying current density requirements. $\beta_{an,ox}$ varies from metal to metal with differences in device passivation and solubility, and also with testing time and temperature. DiGiacomo [1982] performed tests on silver migration between tinned silver-palladium leads in encapsulated packages with a polyimide surface coating, aluminum cap, and silicone rubber or epoxy backseal. The test was carried out under temperature/relative humidity conditions under bias. A failure was defined as a resistance of less than 5×10^6 Ω.

Metallization migration involves the formation of aluminum growths beneath the silicon dioxide layer. Metallization migration occurs during deposition of conductors on silicon dies due to the combined effects of elevated temperature and electrical stress. It has been reported that triangular aluminum growths causing local short circuits form beneath the silicon dioxide layer when conductor metallization is deposited at high temperature on a silicon substrate [Bart, 1969; Lane, 1970]. Lane [1970] found that triangles form within minutes after deposition of an 8,000-Angstrom-thick aluminum layer on silicon at temperatures from 500°C to 577°C. The time for growth formation decreases with increasing temperature. The high temperatures (500°C to 577°C) at which this failure phenomenon occurs makes it a recessive mechanism in the normal operation of microelectronic devices.

1.5 Contact Spiking

Contact spiking is the penetration of metal into the semiconductor in contact regions at temperatures typically above 400°C, causing increased leakage current or shorting. This failure mechanism is accompanied by solid-state dissolution of the semiconductor material into the metal or alloy of the metallization, with the metal semiconductor interface moving vertically or laterally into the semiconductor. Contact spiking in chips is observed at high chip temperatures or localized high contact temperatures. Localized high temperatures may cause failure of the chip-to-substrate bond, thus increasing the thermal resistance and producing a thermal runaway of the device or a large magnitude of electrical overstress.

Contact spiking can occur when the device is exposed to high temperatures during fabrication. This failure mechanism can be minimized by using silicon containing aluminum alloys, such as Al-1%Si, or using barrier metals such as titanium-tungsten (Ti-W) [Chang, 1988; Farahani, 1987; T.I. 1987]. It is not possible to characterize such interdiffusion failure mechanisms by an activation energy because of their irregular behavior.

Migration of aluminum along silicon defects has been observed in NMOS LSI devices (logic gates) due to contact migration, also known as electrothermomigration. Failures are accelerated by elevated ambient temperatures in the neighborhood of 400°C. Contact migration is a major cause of failure in GaAs devices [DeChairo, 1981; Christou 1982; Ballamy, 1978; Christou, 1980]. This failure mechanism may be dominant during VLSI manufacture and packaging when temperatures exceed 400°C.

1.6 Constraint Cavitation of Conductor Metallization

This failure mechanism is marked by the opening of the conductor metallization through the formation of slit-like voids or edge voids. Hinode and Owada, while examining stress-driven diffusive voiding (SDDV) failures in 0.9 μm wide Al-2% Si metallization stripes deposited on thermally oxidized silicon substrate, aged for various times at temperatures in the range of 200-295°C, found that open circuit failures occurred typically in passivated stripes. SDDV failure rate increases with an increase in aging time and temperature until a critical storage temperature, above which the failure rate decreases with a further increase in temperature. The time to failure rises with an increase in line width. Lifetime dependence on line width is characterized as $t \alpha w^{2.7}$ (for line widths in the range of 1-2.5 μm). Stress-driven diffusive voiding failures also manifest themselves in the form of resistance increase. Metallization stripes failing in high resistance mode are not broken; however, the resistance of the line increases stepwise during aging. Typical resistance increase steps are in the range of 1 kΩ to 50 kΩ. The resistance increase steps increase in magnitude as aging temperature rises. [Hinode and Owada, 1987; McPherson and Dunn, 1987].

There seems to be substantial disagreement in the microelectronic community as to the source of the stresses causing these voids. Various theories have been proposed to explain this failure mechanism, including:

- large silicon nodules in aluminum lines, causing metal voiding during room temperature storage;
- Coble and Nabaro-Herring creep of the conductor metallization;
- compressive stress of the plasma-enhanced silicon nitride passivation, combined with intrinsic stresses in nitrogen-contaminated aluminum metal, causing voids at elevated temperatures [Klema, 1984]; and
- compressive stresses of the overlying passivation, induced by cooling.

Nitrogen contamination induced voiding. Klema et al. [1984], while evaluating MOS integrated circuits, found that Al/Si metallization films deposited under conditions of nitrogen contamination coupled with subsequent silicon nitride passivation resulted in open metal stripes. The opens occurred primarily at steps in narrow metal lines.

Failures were explained by the tendency of the metallization film to lower its energy. Energy stored in metallization includes free surface energy, equivalent to the surface tension; interface energy; axial or compressive strain; and grain boundaries or surface energy. Grain boundaries were treated as cuts in a larger crystal. The atoms on the cut surface are chemically bonded to the solid on one side and not on the other, and thus sit in equilibrium potential wells that are at a higher potential than the atoms in the bulk lattice. This extra potential energy constitutes the grain boundaries' potential or surface free energy. Smaller grain size thus indicates a greater potential for chemical reaction to minimize the overall film surface energy. Reactive species such as nitrogen accumulate at grain boundaries by grain- boundary diffusion both during and after deposition, and reduce metallization film free energy principally by compound formation and strain relief. Compound formation involves the spurious reactive species chemically bonding to the available atomic orbitals of higher energy grain-boundary edge atoms to produce compounds of host metal with typically large negative free energies. For aluminum metallization, the compounds include aluminum nitride (AlN) and aluminum oxide. Compound formation during sputter deposition thus reduces the driving force for grain growth. However, compounds such as aluminum oxide and aluminum nitride stop grain boundary motion and increase metallization resistivity and hardness.

Metal voiding was determined to be the result of a combination of intrinsic metal stress produced by the increased brittleness of the metallization due to a nitrogen-contaminated sputtering process and the thermal mismatch between the metallization and the overlying passivation. The failure rate due to stress-driven diffusive voiding was found to be dependent on steady state temperature for temperatures below 180°C, and inversely dependent on steady state temperature for temperatures above 180°C [Klema, 1984].

Silicon nodule formation theory. Curry et. al [1984], while evaluating 64k dynamic rams, observed a mechanism of metallization failures in both the flat and steep areas of the metallization. The failures had two distinct morphologies: clean, sharp breaks; and large, irregularly spaced voids. The observed failure mechanism was not affected by voltage and current density, and could be generated by high-temperature storage, though no failures were noticed in thermal cycling. The silicon content of the metal films was much greater than the solid solubility of silicon and aluminum, with the precipitates of around a quarter of a micrometer in diameter. The failures were attributed to the presence of impurities and silicon defects in aluminum films. The mechanisms responsible for the failures were assumed to be grain-boundary diffusion, temperature-assisted creep, and hydrogen-embrittlement fracture enhancement.

The failure of the metallization due to silicon nodule formation was also recognized by O'Donnell et al. [1984], who stated that the presence of large silicon nodules in aluminum in comparison to the cross-sectional area of the aluminum metallization, were large enough to restrict the current flow. Based on experimental evidence, they proposed the silicon nodule formation theory to explain the ph... on of constraint cavitation. The theory is based on the fact that silicon alloys with aluminum at relatively low temperatures, and the solubility increases with temperature. The higher the temperature, the more silicon can be dissolved in the aluminum metallization before precipi... n. At normally high wafer-processing temperatures (aluminum is annealed at temperatures in the neighborhood of 475°C), more silicon can be dissolved in aluminum than at room temperature. Thus, as the wafer cools, the excess silicon precipitates and forms nodules, epitaxial mounds, and epitaxial layers. The nodules form preferentially at dislocations in grain boundaries and at stress points. The integrated circuits metallization is considered a one-dimensional diffusion path, since in most applications the metallizations were long stripes. The diffusion of silicon in aluminum for thin films was found to be forty times that of bulk aluminum at elevated temperatures. This behavior was attributed to the polycrystalline nature of the aluminum metallization, which provided an enhanced diffusion path due to the larger number of grain boundaries. Silicon dissolved into the aluminum at the contact points, diffused throughout the strip, and was transported outward in all directions. The diffusion length of silicon into aluminum is expressed by the following equation:

$$L_{si,tr} = \left[D_{si,met} t_{ann}\right]^{1/2} \quad (47)$$

where $L_{si,tr}$ is the length silicon can be transported through the metallization; $D_{si,met}$ is the diffusivity of silicon in metallization, t_{ann} is the annealing time. In supersaturated aluminum films, silicon modulus form in cooling and further exposure to high temperature of glass deposition causes precipitates to act as nucleation sites. Large nodules form by adsorbing smaller ones, and deplete the adjacent aluminum of its silicon. The size and shape of these precipitates depend on the crystalline structure of the aluminum and the rate of cooling. Slow cool-down results in large precipitates, while quick cooling results in the formation of numerous small precipitates. The temperature required to form appreciable precipitates is 200°C or higher, but the process is time and temperature dependent and can continue at room temperature, although at a much reduced rate.

Coble and Nabaro-Herring creep theory. The failures in the form of voids and cracks in long metallization lines less than 4 microns wide have also been attributed to Coble and Nabarro-Herring creep of the aluminum metallization [Turner, 1985]. The source of this creep is a thermal expansion mismatch between the aluminum and the underlying silicon and silicon dioxide. The mechanism is complicated by the presence of silicon precipitates in the metallization. Turner et al. [1985] noted that the current density in lines failing from this mode was much less in those experiencing electromigration failures; moreover, the failure rate characteristic decreased with time, unlike with electromigration failures. Turner tried to refute the earlier theory of Curry et al., which attributed voids in the metallization to silicon nodule formation. They argued that the silicon nodule precipitate, being resistive, gives the metallization the electrical appearance of an open and that the precipitate would be removed in the top removal of the glass, leaving the appearance of a void or an open.

Turner et al. explained the failure of the metallization by reference to the thermal expansion mismatch between the aluminum and the silicon chip. The coefficient of thermal expansion of aluminum is about 26 ppm/°C, while that of silicon is about 3 ppm/°C, and that of silicon dioxide (SiO_2) is about 0.5 ppm/°C. Aluminum alloys are typically deposited onto a substrate heated to about 300°C, then alloyed at 425°C, packaged at temperatures between room temperature and 350°C, burned in at 125°C, and operated at less than 70°C. Every time the temperature of the metal-substrate system is changed, the significant thermal expansion mismatch induces a stress in the system; if the temperature is raised, the metal is held in compression. If the system is cooled, the metal is pulled in tension. This stress is equal in both dimensions of the interface plane. The force generated by the thermal expansion mismatch is proportional to the length of the line, while the strength of the line is proportional to its width. At high temperatures, the stress in the aluminum is relieved by grain-boundary migration, in which the atom in one lattice site jumps to an adjacent vacancy. In order for an atom to jump, it must have enough energy to overcome the energy separating the two sites. In equilibrium, this occurs frequently, but there is no migration because the electron jumps are at random; thus, there is no net mass flow. However, if the metal line is subjected to a stress field, some mass flow will occur to relieve the stress. Since grain boundaries have the highest vacancy density, migration along grain boundaries is most rapid. According to Gibbs, the stress relaxation rate by grain-matrix diffusion of thin polycrystalline films in tension is given by

$$\frac{d\sigma_{met}}{dt} = \frac{-B_{gm,fct} V_{at,vl} E D_{l,met}}{d_{ave,met} h_{met} K_B T f_{diff,fct}} \sigma_{met} \tag{48}$$

where σ_{met} is the stress in the metallization, E is the elastic modulus, $V_{at,vl}$ is the atomic volume, $B_{gm,fct}$ is the geometric factor (approximately equal to 10) for grain matrix diffusion, $d_{ave,met}$ is the average grain size in the metallization, h_{met} is the metallization thickness, $f_{diff,fct}$ is the correlation factor for diffusion, and $D_{l,met}$ is the lattice diffusivity of the metallization. The stress relaxation rate by grain-boundary migration (Coble creep) is

$$\frac{d\sigma_{met}}{dt} = \frac{-B_{gm,fct} V_{at,vl} E D_{b,met} \delta}{d_{ave,met} h^2_{met} K_B T} \sigma_{met} \tag{49}$$

where $D_{b,met}\delta$ is the combined grain boundary diffusion parameter, and $B_{gm,fct}$ is the geometric factor for grain boundary migration [Turner, 1985]. These two mechanisms work together in a thin film on substrate.

The stress may be relieved by two mechanisms - grain-boundary migration activated by grain-matrix diffusion, and plastic deformation. Because there are few vacancies within the grain, grain-matrix diffusion is relatively slow. Plastic deformation is fast, but operates only when the stress exceeds the yield stress of metal. Since grain-boundary migration can rapidly reduce the grain-boundary normal stress at higher temperatures, the metal rarely reaches the yield stress. Thus, at higher temperatures, grain-boundary migration rapidly relieves the grain-boundary normal stress, while grain-matrix diffusion slowly relieves the intra-grain stress. In unpassivated films, the edges of the lines act as vacancy sinks. In passivated films, the vacancies are trapped beneath the glass; if these vacancies collect in one location, they will form a void. Such voids are the first step in the formation of open metal lines, weakening the metal line and concentrating line stress in the metal around the void. At lower temperatures, neither grain-boundary diffusion nor grain-matrix diffusion can progress at appreciable rates [Turner, 1985].

McPherson and Dunn Model. The McPherson and Dunn model [McPherson and Dunn, 1987] assumes vacancy to be an elemental unit of failure. The movement of vacancies results in clustering, which results in void formation, void growth, and conductor failure. The three primary sources of vacancies in VLSI Al-1%Si metallization are:

- Vacancies become super-saturated during metallization annealing at around 450°C.
- The solid solubility of silicon in aluminum at room temperature is 0.5%, therefore, excess silicon during cool-down from the annealing temperature precipitates, forming new nodules or increasing the size of existing nodules.
- Grain boundaries serve as potential sources and sinks for vacancies. Under passivation stress, gradients develop in the metallization transverse and parallel to the grain boundaries. Stress gradients serve as the driving force for vacancy migration.

The McPherson and Dunn model [1987] does not account for the effect of local yielding and work hardening during stress relaxation, and thus does not predict whether voiding will continue until the metallization stripe is totally severed. Stress concentration developed locally due to a void can result in local yielding. Further power cycling below recrystallization temperature can result in work hardening of metal and produce breakage. The later stage of stress-driven diffusive voiding can be dominated by rapid electromigration or fusing due to local rises in current. The time to failure is represented by McPherson and Dunn [1987] as

$$TF_{sddv} = B_o \sigma_{met}^{-1} e^{\frac{E_a}{K_B T}} \tag{50}$$

where TF_{sddv} is the time to failure due to stress-driven diffusion voiding. Then,

$$B_o = \frac{\ln\left(\frac{1}{f_c}\right)}{\left(\frac{(\phi(b,t) - \phi(a,t)A_o}{N_{v,met}(t)}\right)} \tag{51}$$

where f_c is the $N_{v,met}(t{=}TF_{sddv})/N_{vo,met}$, $N_{v,met}$ is the vacancy density in the metallization, at time t, A_o is the area of surface bounding the volume of interest, and Φ is the back diffusion due to concentration gradient. The transport of vacancies in metallization is assumed to be Fickian and is represented by:

$$J_{v,met}(x,t) = \mu_{v,met} d_{v,met}(x,t)F - D_{met}\left(\frac{\partial d_{v,met}(x,t)}{\partial x}\right) \tag{52}$$

$$J_{v,met}(x,t) = D_{met}\left(1 - \frac{\frac{\partial d_{v,met}(x,t)}{\partial x}}{\beta_p d_{v,met}(x,t)\sigma_{met}}\right) \beta_p d_{v,met}(x,t)\sigma_{met} \tag{53}$$

where $J_{v,met}$ is the vacancy flux in the metallization, $\mu_{v,met}$ is the vacancy mobility in the metallization, $d_{v,met}$ is the density of vacancies in the metallization, F is the force acting on the vacancies due to stress gradients ($F=\beta_p\ \sigma_{met}$ where β_p is the proportionality factor), and D_{met} is the diffusivity factor. The first term in Equation 52 represents drift transport due to stress gradients, second term represents back diffusion due to concentration gradient. The model assumes that the force for vacancy transport is derived from and proportional to the tensile stress, σ, in the metallization, and that the drift component is much larger than the back diffusion component (Φ). Further, the temperature dependence of Φ is assumed to be negligible compared to the exponential temperature dependence of the activation energy term.

Passivation constraint theory. Hinode [1989], while examining stress-driven diffusive voiding failures in aluminum-copper-silicon lines, found that stress-causing aluminum transport has two origins: thermal expansion mismatch between the aluminum and passivation films and compressive stress of the passivation. In fine lines, voids form due to thermal expansion mismatches, and in wide lines, voids form mainly due to compressive stress of the passivation. In wide lines (in the neighborhood of 30 μm), stress relaxation at higher temperatures is in the form of bulge formation in the passivation containing the metallization, which causes a large deformation at the metallization edges. The temperature dependence of the void growth rate is attributed to the temperature dependence of the deformation rate of passivation and the mobility of the aluminum. In fine lines (in the neighborhood of 2 μm), the passivation has a lower compressive stress, and bulge formation in the passivation becomes less dominant. A thermal expansion mismatch between the metallization and the passivation is the rate-controlling process in fine lines. Voids in fine lines disappear when the lines are heated to a higher temperature and reappear during the cooling period [Hinode, Asano, and Homma, 1989]. Several models have been proposed based on the passivation constraint theory of stress-driven diffusive voiding.

Yue model. Yue et al. [1985] noticed that metallization voids were a result of certain device fabrication conditions, and that the density of these voids had a strong functional dependence on the compressive stress of the passivation film. They examined the effect of the overlying passivation and the rate of cooling, and found that the rate of cooling from the passivation temperature had a significant effect on the void formation mechanism. The wafers, when cooled from 350°C to room temperature, developed a stress in the passivation, due to the difference in the coefficients of thermal expansion between the passivation and silicon. The maximum compressive stress was reached at room temperature and was independent of the cooling rate. The cooling rate of 22°C/min. allowed void formation, but a rapid temperature quench (approximately 65°C/min) did not allow sufficient time for the diffusion to form voids in aluminum. Yue hypothesized that since the missing aluminum from the voids was not found at any other location, as in hillock formation, the void space was a result of the coalescing of vacancies under a driving force. This driving force is comprised of large internal stresses and stress gradients developed in aluminum under cooling and enhanced by large applied passivation stress. In this process, aluminum covered by passivation is in a state of tension at room temperature. Annealing at 350°C causes the stress to become compressive, inducing multiaxial and localized stress gradients in the aluminum, especially near the line edges. A biaxial state of stress, consisting of both tension and compression, then coexists in the metal. Intergranular vacancy diffusion can occur from the boundaries in tension to the boundaries in compression via diffusion creep. The coalescing of voids and vacancies can be expected to be greatest at the corners. The metal void density can be analyzed by line-fitting $d_{vd,met}$ (the metal void density) with the following expression:

$$d_{vd,met} \; \alpha \; \left| \frac{\sigma_{psv,met}}{G_{met}} \right|^{n_{sddv}} \tag{54}$$

where G_{met} is the shear modulus of the metallization, and $\sigma_{psv,met}$ is the passivation-induced stress in the metallization, n_{sddv} is the stress-driven diffusive voiding exponent, and $d_{vd,met}$ is the metal void density. The steady-state creep, the strain rate, $d\varepsilon/dt$, of bulk metal is related to the stress in the metal, $\sigma_{psv,met}$, by the following [Yue, 1985]:

$$\frac{d\varepsilon}{dt} \alpha \left|\frac{\sigma_{psv,met}}{G_{met}}\right|^{n_{sddv}} \tag{55}$$

where $\sigma_{psv,met}$ is the passivation-induced stress in the metallization and G_{met} is the shear modulus of metallization. For n_{sddv}=1, diffusion creep processes, such as Nabarro-Herring or Coble creep, take place; for the values of n_{sddv}=4 to 7, dislocation creep dominates. The volume of voids is proportional to the product of $d\varepsilon/dt$ (strain rate) and the fixed time and volume of the line. The stress in the metallization is not a result of silicon formation, since even pure aluminum films yield metal void formation [Yue, 1985]. The reduction in the density and size of the voids in hot substrate-sputtered aluminum/silicon films is explained through the grain-size dependence of diffusion creep. In diffusion creep, the steady-state rate is related to the grain size by

$$\frac{d\varepsilon}{dt} \alpha \left|\frac{b_{vt}}{d_{ave,met}}\right|^{m_{sddv}} \cdot \left|\frac{\sigma_{psv,met}}{G_{met}}\right|^{n_{sddv}} \tag{56}$$

where b_{vt} is Burger's vector, $d_{ave,met}$ is the average grain size, m_{sddv} is the grain size exponent for stress-driven diffusive voiding, and n_{sddv}=2,3 for Nabarro-Herring and Coble creep, respectively.

Yost model. Yost [1988, 1989], in line with the passivation constraint theory of Yue [1985], proposed that aluminum-silicon conductor lines in integrated circuits have a tendency to fail through nucleation and growth of vo:˙s, assuming two dominant morphologies: a wedge-shaped form and a narrow, crack-shaped form. The conductor lines are typically passivated with a glass layer at approximately 427°C and cooled to room temperature. The coefficient of thermal expansion of glass is an order of magnitude smaller than that of the conductor, so the cooling process imparts tensile stress to the conductor line. This tensile field is not uniform, and the stress gradients cause diffusive mass transports. The passivated conductor lines are in a state of mechanical constraint that prevents large-scale grain-boundary sliding. With extensive mass transport and insufficient grain-boundary sliding, void growth is possible. Typically, stress-driven diffusive voiding of the conductor metallization has been characterized as intergranular cracking without the extensive deformation usually associated with tensile creep [Yost, 1988]. Relaxation due to dislocation glide and climb will occur throughout the conductor line network at a rate determined by the local stress, temperature, and rate of resulting strain accommodation in the passivation layer and interface.

Yost [1989] conceptually cooled 3-μm wide conductor lines rapidly from 400°C to 306°, 213°, 119°, and 25°C. The effect of temperature o.. .he rate of growth of voids is shown in Figure 2.19, where Y_0 is the stress relaxation distance on both sides of the grain boundary, and 2b is the conductor width. As the growth temperature (i.e., the final temperature to which the metallization system, along with the passivation, is cooled from 400°C) increases, voids reach their final size - the size when far-field stress is balanced by the stress relaxation due to voiding - more rapidly. The cracks, predominantly oriented perpendicularly to the line length, usually nucleate in the center of the conductor and progress to open circuit failure. This tyı.. of cracking was present without nitrogen embrittlement of the metallization and progressed in the absence of current and corrosive agents [Yost, 1988]. Yost et al. [1988] demonstrated analytically and experimentally that stress-driven mass transport along the grain boundaries leads to the formation of either wedge-shaped voids or slit (crack)-shaped voids at the conductor cc ·ter. The void shape was dependent upon the granular orientation of the metallization with respect to the applied stress.

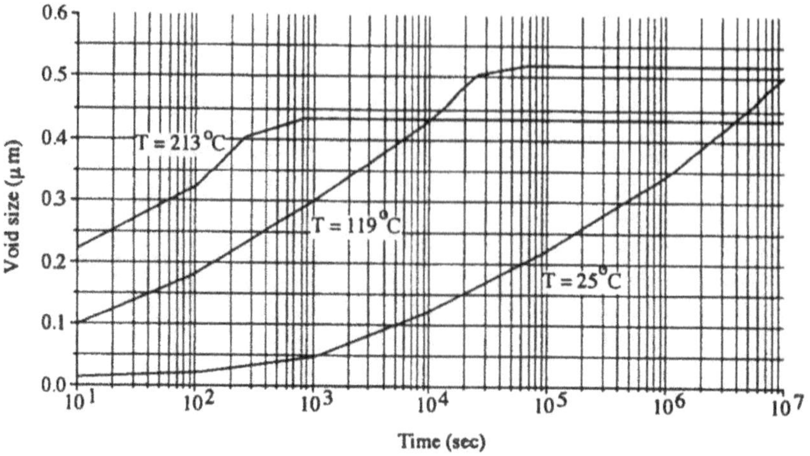

Figure 2.19 Growth kinetics of wedge voids at various aging temperatures [Yost, 1989].

The temperature dependence of the surface diffusion coefficient has been modeled by

$$\log_{10} D_{s,met} = 5.07 - \frac{1.09x10^4}{(273+T)} + \frac{2.89x10^6}{(273+T)^2} - \frac{3.24x10^8}{(273+T)^3} \tag{57}$$

where $D_{s,met}$ is the surface-diffusion coefficient for metallization. The material stack-up used for the analysis was silicon, boron phospho-silicate glass (BPSG), aluminum-silicon, phospho-silicate glass (PSG). The time to failure decreases with a decrease in conductor line width. This behavior is a direct consequence of the higher stress states that develop in narrower lines. The calculations also show an increasing growth rate as the temperature increases. The calculated increase is due to the diffusion coefficient, which increases exponentially with temperature, and the average far-field stress, $\sigma_{\infty,met}$, which decreases quadratically with temperature (Figure 2.20). In its present form, the model does not include the effect of temperature-dependent stress; it includes the temperature-dependent stress term derived from the differential thermal expansion between the aluminum and the glass passivation layer, but does not include any stress relaxation that would occur as a highly stressed conductor is aged.

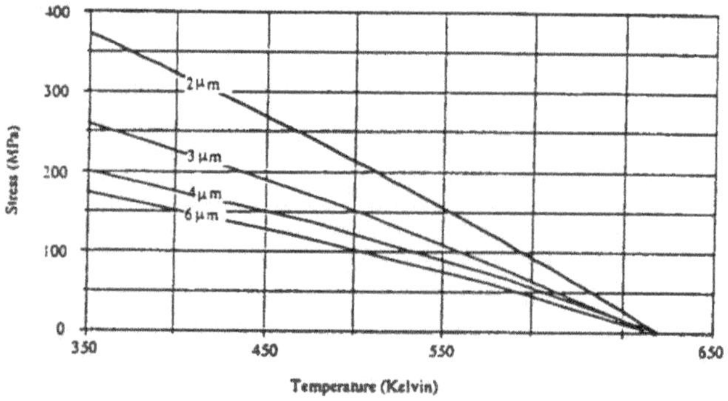

Figure 2.20 Far field stress in aluminum as a function of temperature [Yost, 1988].

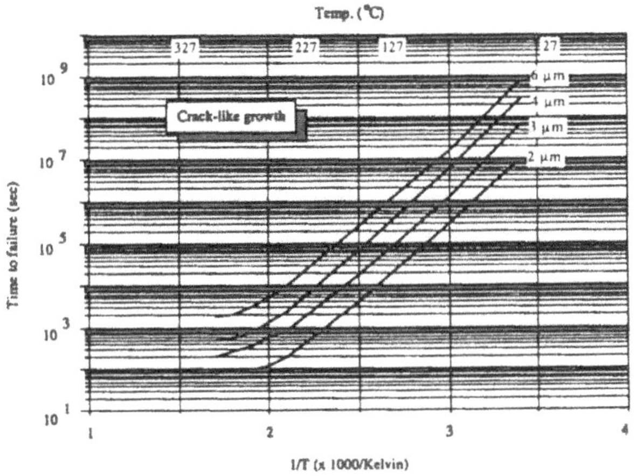

Figure 2.21a Time to failure for crack-like growth [Yost, 1989]. The results presented here are very conservative, because the models do not take into account stress relaxation effects at high temperature.

At least two types of stress relaxation can occur. A homogenous, or mean field, relaxation can take place throughout the entire conductor line and passivation layer; a local stress relaxation process can occur as a result of the void process itself. The extent of the relaxation process depends on the metal grain size and the geometry of the debond region. The inclusion of stress relaxation effects reduces the stress and significantly increases the time to failure at high temperatures. The rate of growth for crack-like voids is slightly faster than for wedge-shaped voids. The predicted time to failure for crack growth in 3-μm conductor lines is 1.5 years at 27°C, while the time to failure for wedge-like growth in 3-μm conductor lines is 2.3 years at 27°C. These results are consistent with experience [Yost, 1988]. The time to failure decreases with an increase in temperature. The temperature dependence of time to failure for crack- and wedge-like growth is given by Figures 2.21a and 2.21b, respectively.

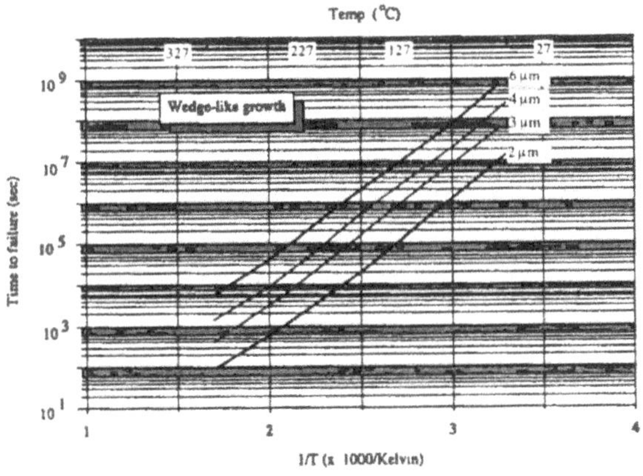

Figure 2.21b Time to failure for wedge-like growth [Yost, 1989]. The results presented here are very conservative, because the models do not account for stress relaxation effects at high temperatures.

Okabayashi model. More recent work on stress-driven diffusive voiding [Okabayashi, 1991] has revealed a more complex dependence of lifetime on temperature than the simple Arrhenius model characterized by Yost [1988, 1989]. Diffusion is exponentially dependent on temperature and has been considered the rate-determining step in the Yost formulation. Okabayashi [1991] gave an analytical model for open-circuit failure for both wedge- and slit-shaped voids, and accounted for the stress relaxation occurring during void growth. Assumptions in the model are that

- the void shape does not change during growth;
- the metallization stress within a distance of ±L/2 from a void is uniformally relaxed as the void grows;
- atoms diffuse from the void to the grain boundary intersecting the void, and are accommodated in the grain boundary to relax the metallization stress;
- atomic flux is proportional to the nth power of the metallization stress;
- metallization stress is isotropic; and that
- Young's modulus and Poisson's ratio are independent of steady-state temperature.

Okabayashi [1991] modeled the void volume increase rate by

$$\frac{dV_{vd,vl}}{dt} = 2J_{a,met}V_{at,vl} \tag{58}$$

where $V_{vd,vl}$ is the void volume, $J_{a,met}$ is the atomic flux in metallization, and $V_{at,vl}$ is the atomic volume. The flux is represented as

$$J_{a,met} = \frac{A_{sddv,ok}h_{met}D_{sd}}{K_BT}\left(\frac{\sigma_{met}}{G_{met}}\right)^{n_{ok}} \tag{59}$$

where $A_{sddv,ok}$ is Okabayashi's coefficient for stress-driven diffusive voiding, h_{met} is the metallization thickness, D_{sd} is the self-diffusion coefficient, σ_{met} is the metallization shear modulus, and n_{ok} is Okabayshi's exponent. Approximating the void volume to be equal to a^2H $tan\,\psi$, the rate of void growth was calculated from Equations 58 and 59 as

$$a_{crk}(t)\frac{da_{crk}(t)}{dt} = \frac{A_{sddv,ok}V_{at,vl}D_{sd}}{K_BT\,\tan\psi_{vd,hf}}\left(\frac{\sigma_{met}(t)}{G_{met}}\right)^{n_{ok}} \tag{60}$$

where a_{crk} is the crack size along line-width direction, $V_{at,vl}$ is the atomic volume, D_{sd} is the self-diffusion diffusivity, $\Psi_{vd,hf}$ is the half-angle of void, and n_{ok} is Okabayashi's exponent [1991]. Okabayashi showed n_{ok}=5 dependence on stress. The line stress has been approximated as

$$\sigma_{met}(t) = \sigma_{met}(0) - \frac{a_{crk}(t)^2E\,\tan\psi_{vd,hf}}{W_{met}L_{rlx,vd}} \tag{61}$$

where $L_{rlx,vd}/2$ is the length over which stress due to void formation is relaxed, and W_{met} is the metallization width. The time to failure for the metallization can be calculated by substituting Equation 61 for D_{sd}, in Equation 60:

$$TF_{sddv} = \frac{K_B G_{met} L_{rlx,vd} w_{met} T\, e^{\frac{E_a}{K_B T}}}{2A_{sddv,ok} E D_{0,sd}(n_{ok}-1)} \left[\left(\frac{\sigma_{met}(0)}{G_{met}} - \frac{E w_{met} \tan\psi_{vd,hf}}{G_{met} L_{rlx,vd}}\right)^{1-n_{ok}} - \left(\frac{\sigma_{met}(0)}{G_{met}}\right)^{1-n_{ok}}\right] \quad for\ n_{ok} \neq 1 \tag{62}$$

$$TF_{sddv} = \frac{K_B G_{met} L_{rlx,vd} w_{met} T\, e^{\frac{E_a}{K_B T}}}{2A_{sddv,ok} E D_{0,sd}} \left[\ln\frac{\sigma_{met}(0)}{G} - \ln\left(\frac{\sigma_{met}(0)}{G} - \frac{E w_{met} \tan\psi_{rd,hf}}{G_{met} L_{rlx,vd}}\right)\right] \quad for\ n_{ok} = 1 \tag{63}$$

where TF_{sddf} is the time to failure due to stress-driven diffusive voiding, $L_{rlx,vd}/2$ is the length over which stress due to void formation is relaxed, and $D_{o,sd}$ is the diffusion coefficient for self-diffusivity.

$$\sigma_{met}(t) = \sigma_{o,met}(T_{psv,dp} - T)\left(\frac{w_{o,met}}{w_{met}}\right)^{m_{ok}}\left(\frac{h_{o,met}}{h_{met}}\right)^{p_{ok}} \tag{64}$$

where $\sigma_{0,met}$ is the metallization stress due to a temperature change of 1°C; $T_{psv,dp}$, is the passivation deposition temperature; T is the steady-state temperature; $w_{0,met}$ is the normalization width of the metallization; w_{met} is the metallization width; $h_{0,met}$ is the normalization thickness of the metallization; h_{met} is the metallization thickness; p_{ok} is Okabayashi's metallization thickness exponent, equal to 0.5 and m_{ok} is Okabayashi's metallization width exponent, equal to 0.5.

The lifetime dependence on temperature is represented in Figures 2.22a and 2.22b. The lifetime reaches a minimum at a certain temperature, T_m. This can be explained by the lifetime dependence on $D_{sd}\sigma_{met}^{n}$. The increase in lifetime when T increases or decreases from T_m (the temperature at which the minima occurs) results from a decrease in σ_{met} for $T \geq T_m$ or a decrease in D_{sd} for $T \leq T_m$, respectively. The lifetime increases with an increase in the metallization width.

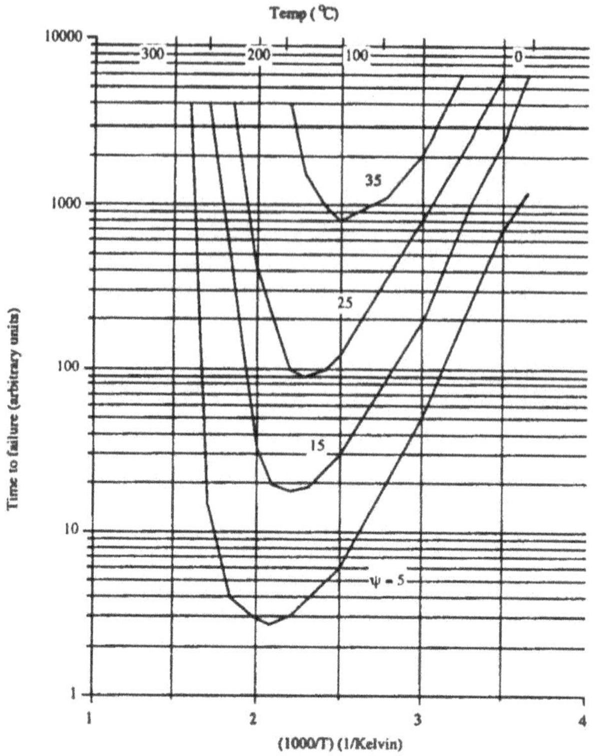

Figure 2.22a Temperature dependence of lifetime under stress-driven diffusive voiding for various notch angles [Okabayashi, 1991].

Kato and Niwa model. Kato and Niwa theoretically estimated the stresses in an aluminum metallization under passivation arising from the different coefficients of thermal expansion of the aluminum track and passivation film, using Eshelby's method-of-inclusion problem [Kato et al., 1990; Niwa et al., 1990]. The phenomenon of stress relaxation was found to be different at low and high temperatures. Diffusion was practically inoperative at low temperatures (defined as the temperature at which the time for diffusional relaxation is greater than 10^3 seconds; see Equations 82, 83, and 84) and only instantaneous plastic deformation by dislocation glide was considered to be the relaxation mechanism. [Kato et al., 1990; Niwa et al., 1990]. Kato et al. found that the stresses did not become hydrostatic after relaxation due to plastic deformation. The stresses and strains after plastic deformation as a function of aspect ratio are as follows.

For $r_{met} > 0$ and $r_{met} < 1$:

For aspect ratios r_{met} = metallization thickness / metallization width, such that $r_{met} > 0$ or $r_{met} < 1$. The strains after relaxation by plastic deformation are:

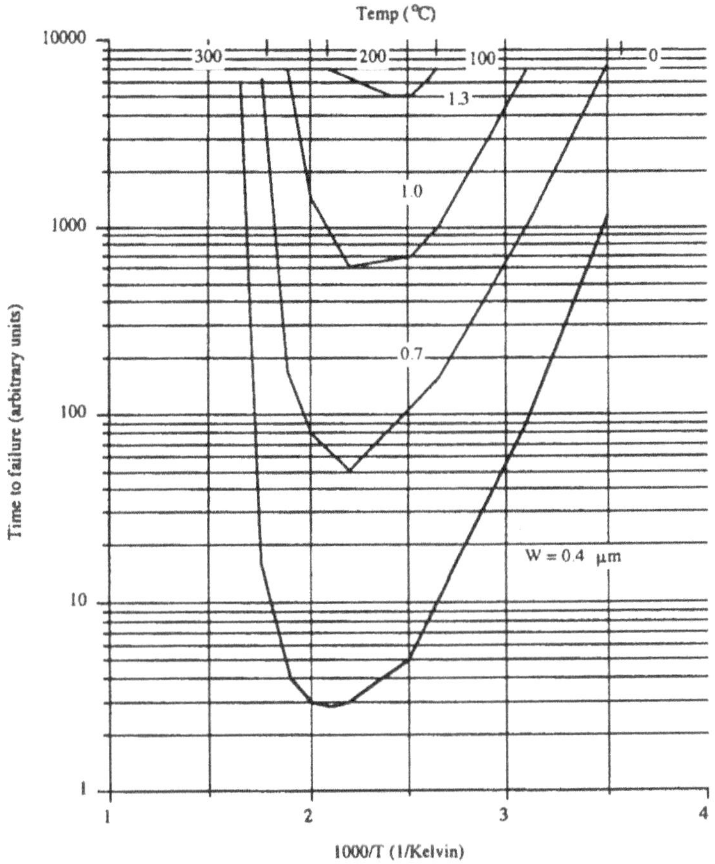

Figure 2.22b Lifetime due to stress-driven diffusive voiding versus temperature for different line widths [Okabayashi, 1991].

$$\epsilon_{11}^{P} = \frac{-((C_6+C_2C_7)\epsilon^T + C_1(C_3+C_2C_5))}{(C_2(2C_3+C_2C_5)+C_4)} \tag{65}$$

and

$$\epsilon_{22}^{P} = C_1 + C_2\epsilon_{11}^{P} \tag{66}$$

where

$$C_1 = \frac{(1+r_{met})^2}{4r_{met}^2(1-\nu_{met})+r_{met}(5-4\nu_{met})+2}\left(\frac{2(1+\nu_{met})\epsilon^T}{1+r_{met}} + \frac{(1-\nu_{met})\sigma_y}{G_{met}}\right) \tag{67}$$

$$C_2 = \frac{2(1+r_{met})^2(1-\nu_{met})+r_{met}}{4r_{met}^2(1-\nu_{met})+r_{met}(5-4\nu_{met})+2} \tag{68}$$

$$C_3 = \frac{2r_{met}^2(1-\nu_{met})+r_{met}(5-4\nu_{met})+2(1-\nu_{met})}{(1+r_{met})^2} \tag{69}$$

$$C_3 = \frac{2r_{met}^2+r_{met}(5-4\nu_{met})+4(1-\nu_{met})}{(1+r_{met})^2} \tag{70}$$

$$C_3 = \frac{4r_{met}^2(1-\nu_{met})+r_{met}(5-4\nu_{met})+2}{(1+r_{met})^2} \tag{71}$$

$$C_6 = -\frac{2r_{met}(1+\nu_{met})}{1+r_{met}} \tag{72}$$

$$C_6 = -\frac{2(1+\nu_{met})}{1+r_{met}} \tag{73}$$

were, G_{met} is the shear modulus of the metallization, ν_{met} is Poisson's ratio for the metallization, ϵ^T is the thermal strain given by Equation 86, σ_y is the yield strength of the metallization, and r_{met} is the aspect ratio of the metallization line ($r_{met} = a_2/a_1$; $2a_2$ is the metallization thickness and $2a_1$ is the metallization width) [Kato et al., 1990; Niwa et al. 1990]. Stresses resulting from plastic deformation are:

$$\sigma_{11}^R = -\frac{G_{met}}{(1-\nu_{met})}\left(\frac{2(1+\nu_{met})}{(1+r_{met})}\epsilon^T + \frac{r_{met}(1-2\nu_{met})+2(1-\nu_{met})}{(1+r_{met})^2}\epsilon_{11}^R + \frac{r_{met}(1-2\nu_{met})-2\nu_{met}}{(1+r_{met})^2}\epsilon_{22}^R\right) \tag{74}$$

$$\sigma_{22}^R = -\frac{G_{met}}{(1-\nu_{met})}\left(\frac{2r_{met}(1+\nu_{met})}{(1+r_{met})}\epsilon^T + \frac{r_{met}(1-2\nu_{met})-2\nu_{met}r_{met}^2}{(1+r_{met})^2}\epsilon_{11}^R + \frac{r_{met}(1-2\nu_{met})-2r_{met}^2(1-\nu_{met})}{(1+r_{met})^2}\epsilon_{22}^R\right) \tag{75}$$

$$\sigma_{33}^R = -\frac{G_{met}}{(1-\nu_{met})}\left(2(1+\nu_{met})\epsilon^T + \frac{-2r_{met}-2(1-\nu_{met})}{(1+r_{met})}\epsilon_{11}^R + \frac{-2-2r_{met}(1-\nu_{met})}{(1+r_{met})}\epsilon_{22}^R\right) \tag{76}$$

where σ_{ii} for i = 1, 2, 3 are the stresses in the metallization line along its width, thickness, and length, respectively.

For $r_{met} \approx 1$:

For aspect ratios r_{met}, such that $r_{met} \approx 1$, the strains after relaxation by plastic deformation are:

$$\epsilon_{11}^P = \epsilon_{22}^P = -\frac{1}{2}\epsilon_{33}^P = -\frac{(1+\nu_{met})\epsilon^T + \frac{(1-\nu_{met})\sigma_y}{G_{met}}}{(4\nu_{met}-5)} \tag{77}$$

where G_{met} is the shear modulus of the metallization, ν_{met} is Poisson's ratio for the metallization, ϵ^T is the thermal strain given by Equation 86, σ_y is the yield strength of the metallization, and r_{met} is the aspect ratio of the metallization line ($r_{met} = a_2/a_1$); $2a_2$ is the metallization thickness and $2a_1$ is the metallization width) [Kato et al., 1990; Niwa et al. 1990]. Stresses resulting from plastic deformation are:

$$\sigma_{11}^P = \sigma_{22}^P = -\frac{G_{met}}{(1-\nu_{met})}((1+\nu_{met})\epsilon^T(1-2\nu_{met})\epsilon_{11}^P) \tag{78}$$

$$\sigma_{33}^P = \frac{-2G_{met}}{(1-\nu_{met})}((1+\nu_{met})\epsilon^T+(-2+\nu_{met})\epsilon_{11}^P) \tag{79}$$

where σ_{ii} for i = 1, 2, 3 are the stresses in the metallization line along its width, thickness, and length, respectively.

For $r_{met} \approx 0$:

For aspect ratios, r_{met}, such that $r_{met} \approx 0$, the strains after relaxation by plastic deformation are:

$$\epsilon_{11}^P = \epsilon_{33}^P = -\frac{1}{2}\epsilon_{22}^P = -\epsilon^T - \frac{(1-\nu_{met})\sigma_y}{2G_{met}(1+\nu_{met})} \tag{80}$$

where G_{met} is the shear modulus of the metallization, ν_{met} is Poisson's ratio for the metallization, ϵ^T is the thermal strain given by Equation 86, σ_y is the yield strength of the metallization, and r_{met} is the aspect ratio of the metallization line ($r_{met} = a_2/a_1$); $2a_2$ is the metallization thickness and $2a_1$ is the metallization width) [Kato et al., 1990; Niwa et al. 1990]. Stresses resulting from plastic deformation are:

$$\sigma_{11}^P = \sigma_{33}^P = \sigma_y \qquad \sigma_{22}^P = 0 \tag{81}$$

where σ_{ii} for $i = 1, 2, 3$ are the stresses in the metallization line along its width, thickness, and length, respectively.

At higher temperatures (defined as the temperature at which the time for diffusional relaxation is less than 10^3 seconds; see Equations 82, 83, and 84), further relaxation of stresses after plastic deformation was possible due to diffusional relaxation [Kato et al., 1990; Niwa et al., 1990]. The stress state after diffusional relaxation (a function of metallization grain structure) was completely hydrostatic. Bamboo-structured metallization lines showed negligible diffusional relaxation at lower temperatures, while equiaxed grain structures demonstrated diffusional relaxation at all temperatures. The relaxation times for lattice, interfacial, and grain-boundary diffusion are given by

$$t_{rlx,lat} = \frac{L_{g,met}^2 K_B T}{8ED_{lo,met} V_{at,vl}} \tag{82}$$

$$t_{rlx,int} = \frac{a_2 L_{g,met}^2 K_B T}{8ED_{lo,met} w_{I,met} V_{at,vl} (1 + r_{met})} \tag{83}$$

where $t_{rlx,int}$ is the relaxation time for interfacial diffusion, $L_{g,met}$ is the grain length along the metallization axis, $t_{rlx,met}$ is the relaxation time for lattice diffusion, $D_{lo,met}$ is the diffusion coefficient for lattice diffusion, $V_{at,vl}$ is the atomic volume of metallization, $w_{I,met}$ is the effective width of interfacial diffusion, $V_{at,vl}$ is the atomic volume, $L_{g,met}$ is the grain length along the metallization axis, and $a_2 = H_{met}/2 = ½$ of the metallization thickness.

$$t_{rlx,gb} = \frac{d_{ave,met}^3 K_B T}{32ED_{b0,met} w_{b,met} V_{at,vl}} \tag{84}$$

where $d_{ave,met}$ is the average grain size of the metallization, $D_{b0,met}$ is the diffusion coefficient for grain-boundary diffusion, $w_{b,met}$ is the thickness of the grain boundaries, $V_{at,vl}$ is the atomic volume, and $t_{rlx,gb}$ is the relaxation time for grain-boundary diffusion. Further relaxation of hydrostatic stresses was conceived to be due to several mechanisms, including decohesion or sliding at the aluminum line-passivation interface, and nucleation and growth of voids [Kato et al., 1990; Niwa et al., 1990].

The completely relaxed stress state was represented by

$$\sigma_{11}^R = \sigma_{22}^R = \sigma_{33}^R = \sigma^R = \frac{-6G_{met} r (1 + \nu_{met}) \epsilon^T}{(2r_{met}^2 + r_{met} + 2) - 4r_{met}\nu_{met} - 2(1 - r_{met}^2)\nu_{met}^2} \tag{85}$$

and

$$\epsilon_{11} = \epsilon_{22} = \epsilon_{33} = \epsilon^T = \Delta\alpha(T - T_{psv,dp}) \tag{86}$$

where G_{met} is the shear modulus of the metallization line (Pascal or N/m^2); r_{met} is the aspect ratio of the metallization line [r_{met} = (metallization thickness)/(metallization width)]; v_{met} is Poisson's ratio for the metallization line; and $T_{psv,dp}$ is the thermal strain resulting from the mismatch between the passivation and the metallization.

The time to failure due to stress-driven diffusive voiding was considered to be the time during which the area fraction, A, increases from A_{min} to A_{max} (area fraction = r_o^2/l^2; r_o is the void radius, and 2l is the void separation).

$$For \quad \sigma_{met} \geq \frac{\Gamma_{sf,ar}}{r_{o,vd}}$$

$$TF_{sddv} = \frac{K_B T L_{int,vd}^3}{20 D_{b0,met} w_{b,met} \sigma_{met} V_{at,vl}} \tag{87}$$

$$L_{int,vd} = \frac{a_2}{r_{met}} \qquad for\ r_{met} \geq 1$$

$$L_{int,vd} = a_2 \qquad for\ 0 \leq r_{met} \leq 1$$

$$For \quad \sigma_{met} < \frac{2\Gamma_{sf,ar}}{r_{0,vd}} \tag{88}$$

no void growth

where $\Gamma_{sf,ar}$ is the surface energy per unit area, $w_{b,met}$ is the grain-boundary thickness, $L_{int,vd}$ is the intermediate distance between voids, and $r_{0,vd}$ is the void of the radius.

The stress in metallization is given by the following:

for equiaxed grain structure:

$$\sigma_{met} = \sigma^R$$

for bamboo grain structure: (89)

$$\sigma_{met} = \sigma_{33} \qquad t_{relax} \leq 1000\ seconds$$

$$\sigma_{met} = \sigma^R \qquad t_{relax} \geq 1000\ seconds$$

where σ_R is the hydrostatic stress after diffusional relaxation given by Equation 85, and σ_{33} is the stress along the metallization length given by Equations 76, 79, and 81.

2. EFFECT OF HYDROGEN (H_2) AND HELIUM (He) AMBIENTS ON METALLIZATION VS TEMPERATURE

Reductions in the solid-state transport have been found possible through the interaction of active gases with thin-film metal conductors and bi-metallic diffusion couples (see Table 4). Studies indicate that a hydrogen ambient results in an improvement in electromigration rates, and stabilizes the intermetallic formation due to interdiffusion.

Pasco and Schwarz [1983] considered several pathways for gaseous ambients to affect mass transport mechanisms in metallization, including external surface diffusion, grain-boundary diffusion of metal atoms, and diffusion of vacancies into the bulk from dislocations and from defects present on external and internal surfaces. They studied the effects of heating rates in pure hydrogen (H_2) and helium (He) 8.5%H_2, relative to an inert helium environment, on the electromigration of Al-2%Cu conductors. For a heating rate of 1°C/min, the electromigration rate decreased by a factor of 10 for the hydrogen-vs helium-ambient. A decrease in the electromigration rate by a factor of five is achieved with only an 8 1/2% H_2 in the ambient. At higher heating rates (5°C/min), an increase in the degree of damage before the onset of failure increases with an increasing hydrogen constant. Table 5 shows the effect of the heating rate on hydrogen and helium ambients [Pasco and Schwarz, 1983]. The activation energy is constant, while the pre-exponential area fraction varies, indicating that the ambient effects are independent of heating rate, since the activation energy is constant.

The process of electromigration is a combination of two processes: nucleation of voids, and growth of voids. Davis [1967] has shown that the nucleation of dislocation loops, with quenching in the vacancy supersaturation, is a low-temperature process. At low temperatures in the neighborhood of 25°C, the process of nucleation of voids is transport-limited, and at moderately high temperatures in the neighborhood of 100°C, the process is limited by a low driving force, due to the effectively lower supersaturation of vacancies. Dislocation loops can be precursors to the formation of microscopic voids in electromigration. Under current stressing, ions are transported along the grain-boundary generally, in the direction of the electron flow for aluminum alloys. In addition, vacancies flow in the opposite direction, producing local vacancy supersaturation. These vacancies can cluster to form dislocation loops and microscopic voids, given sufficient time at lower temperatures. The effect of the active ambient is to promote uniform nucleation and an increase in the nucleation rates. The ambient increases the rate of electromigration damage, up to the point where the growth of voids becomes the rate-limiting factor. This increase in the nucleation rate is actually beneficial, since many small voids are less likely to be harmful than a few large voids.

Void growth is decreased by the hydrogen ambient through a number of mechanisms [Pasco and Schwarz, 1983]. Adsorbed hydrogen can become bound to vacancies, decreasing their migration rates, pinning them, or effectively reducing their concentration. In addition, the segregation of hydrogen to grain boundaries decreases the grain-boundary energy and diffusion rates. These effect a decrease in the overall transport and electromigration rates, due to the hydrogen ambient [Pasco, 1983]. Higher heating rates decrease the time for nucleation and shift the electromigration damage to higher temperatures, which is reflected in the lowered pre-exponential area fraction for higher heating rates (Table 5) [Pasco and Schwarz, 1983]. In addition, hydrogen promotes uniform nucleation and, therefore, increases the amount of the damage that can be sustained before the rapid onset of failure in all cases.

Black [1978] observed that the cracking of the passivating glass overlayer as a result of hillock formation during electromigration damage could lead to a sudden increase in electromigration rates, due to the release of the compressive stresses in the conductor. These sudden increases in the electromigration rate can be avoided if the device is packaged in hydrogen ambient.

Table 4. Effect of H_2, He, Ar, and N_2, ambients vs. temperature

Atmosphere	Stripe composition	Failure mechanism/ Failure definition	Observation	Reference
argon, hydrogen	aluminum	electromigration /resistivity	Electromigration rate ($dR/R_0\,dt$) decreased by a factor of 10 for hydrogen; atmosphere compared to argon at j= 0.7 x 10^6 A/cm^2 and 200°C; j current density in metallization	Shih and Ficalora [1981]
hydrogen(H_2), helium(He)	aluminum aluminum-copper	electromigration /resistivity	Al stripes: H_2 improves (decreases) electromigration rate by a factor of 13 to 16, as compared with helium. Al-Cu stripes: H_2 improves electromigration rate by a factor of 4 to 7, compared with helium environment.	Sardo [1981]
nitrogen, hydrogen, and water	aluminum	hillock formation /reflectivity	Hillock formation at elevated temperatures during film deposition was greatly decreased for high partial pressures of H_2 and H_2O, relative to that of N_2.	McLeod and Hartsough [1977]
hydrogen incorporated during E-beam deposition	aluminum	hillock formation	TTF of 40 hours at 7 x 10^6 A/cm^2 for hydrogen incorporated samples; an anneal at 550°C produced fewer hillocks in the E-beam film.	Meyer [1983]
hydrogen, argon, oxygen, nitrogen, helium, and air	gold aluminum thin-film diffusion couple	intermetallic compound formation	Au-Al intermetallic compound is formed at 350 to 400°C for all environments except H_2.	Shih and Ficalora [1978]
hydrogen, air	Cu-Sn, Ag-Sn, Ni-Sn thin-film diffusion couple	intermetallic compound formation	Rates of intermetallic formation reduced dramatically in H_2 environment compared to air. Reduction of rate by a factor of 3 to 10 observed.	Shih and Ficalora [1979]

oxygen, argon	Au-Al thin-film diffusion couple	intermetallic compound formation	After 250°C anneal in 0.5 atm of either O_2 or Ar, SEM photography was used to evaluate the character of the intermetallic formation. Grain-boundary diffusion appears to be the dominant failure mechanism in O_2 environment; surface diffusion dominated in Ar environment.	Shih and Ficalora [1978]

3. TEMPERATURE DEPENDENCIES OF FAILURE MECHANISMS IN THE DEVICE OXIDE

In this section, the temperature dependence of mechanisms in the oxide of MOSFETs (metal oxide semiconductor field effects transistors), are discussed. In MOS devices the oxide forms the dielectric of the capacitor that controls charge transfer between the source and the drain of the transistor. The presence of oxide charges thus crucially affects the functionality of the MOSFET. To get a feel for the magnitude of oxide charge that will result in a charge anomaly, the surface density of electrons ($\approx (10^{15})^{2/3}$ or 10^{10}, where typical dopant density is 10^{15} per cm^{-3}) is the same order of magnitude as the surface density of silicon atoms ($\approx (5 \times 10^{22})^{2/3}$, where silicon atom density is 5×10^{22} per cm^3). Thus a surface charge density of 10^{-5} times the atomic density can cause the MOSFET to deviate from ideal behavior [Muller, 1986].

Table 5. Summary of kinetic parameters for electromigration and extent of damage sustained by Al-2%Cu thin stripes (current stressed at 3×10^6 A/cm^2). An almost constant activation energy at different heating rates suggests that the phenomenon is temperature-independent in hydrogen and helium ambients.

Heating rate	He (Helium)	He-8½%H_2	H_2
1 K/min	$E_a = 0.67$ eV	$E_a = 0.67$ eV	$E_a = 0.67$ eV
	$A = 7.0 \times 10^3\ s^{-1}$	$A = 1.4 \times 10^3\ s^{-1}$	$A = 6.7 \times 10^2\ s^{-1}$
	$\Delta R/R_{o\ failure} = 0.67$	$\Delta R/R_{o\ failure} = 0.96$	$\Delta R/R_{o\ failure} = 1.2$
5 K/min	$E_a = 0.67$ eV	$E_a = 0.67$ eV	$E_a = 0.67$ eV
	$A = 1.4 \times 10^3\ s^{-1}$	$A = 4.2 \times 10^3\ s^{-1}$	$A = 1.1 \times 10^4\ s^{-1}$
	$\Delta R/R_{o\ failure} = 0.28$	$\Delta R/R_{o\ failure} = 0.56$	$\Delta R/R_{o\ failure} = 0.62$

Four distinct types of charge exist in oxides, including the fixed oxide charge, oxide trapped charge, interface-trapped charge, and mobile charge. A fixed interface charge is positive and is trapped in a thin layer of non-stoichiometric silicon dioxide (SiO_x). The oxide trapped charge is both positive and negative, and is located in traps distributed throughout the oxide layer. A mobile charge results from alkali-metal ions such as sodium and potassium, widely distributed in various metals and chemicals. Interface-trapped charges are located in the forbidden gap energy levels. The temperature dependence of various mechanisms for charge buildup in the oxide are discussed in this section, including slow trapping, time-dependent dielectric breakdown, electrostatic discharge, and electrical overstress. These mechanisms are a mix of wearout and overstress mechanisms. While slow trapping and time dependent dielectric breakdown are wearout mechanisms that will cause device failure from cumulative charge accumulation, which results in variation of the threshold voltage, electrostatic discharge will cause overstress failure due to sudden charge buildup which results in device burnout.

3.1 Slow Trapping (Oxide Charge Trapping and Detrapping)

Slow trapping is a failure mechanism, observed only in standard MOS transistors and certain types of memory devices, programmed by the transport of charge from the source or the drain through the gate oxide to the gate interface. High temperatures, combined with a high electric field, provide electrons with enough energy to cross the silicon-silicon dioxide interface [Nicollian, 1974; Woods, 1980]. Interstitial states at the silicon-silicon dioxide (Si-SiO_2) interface trap electrons and hold them in the oxide, permanently shifting the threshold voltage of the device. The presence of electrons permanently trapped at the oxide interface decreases the speed at which the device can be programmed by creating a field that opposes further electron flow through the oxide interface. The failure mechanism of slow trapping decreases the circuit speed and causes functional failures.

The trapped charge within the MOS oxide results in a C-V curve identical to that of the ideal structure, but shifted along the voltage axis by an amount equal to the flatband voltage shift. The flatband voltage shift is a function of the location of the oxide-trapped charge with respect to the silicon-silicon dioxide interface. The charges in the oxide induce equal and opposite charges divided between the silicon substrate and the metal gate. The closer the charge is to the silicon-oxide interface, the greater the charge induced in the silicon. The charge in the silicon alters the charge stored at thermal equilibrium, and thus alters the flatband voltage. The maximum value for the flatband voltage shift occurs when the charge is located at the silicon-oxide interface ($x_m = t_{ox}$), because the charge induced is contained entirely in silicon. In contrast, if the charge is located adjacent to the metal-oxide interface ($x_m = 0$), there is no effect on the flatband voltage (Equation 90). The flatband voltage shift of C-V curves is expressed by Balland and Barbottin [1989] and Muller and Kamins [1986] as

$$\begin{aligned}\Delta V_g = \Delta V_{fb} &= -\frac{1}{C_{ox}}\int_0^{t_{ox}}\frac{x'}{t_{ox}}\rho(x')dx' \\ &= -\frac{1}{\epsilon_o\epsilon_{ox}}\int_0^{t_{ox}}x'\rho(x')dx' \\ &= -\frac{1}{\epsilon_o\epsilon_{ox}}x_m Q_{ot}\end{aligned} \tag{90}$$

where ΔV_g is the gate voltage, ΔV_{fb} is the flat band voltage shift, C_{ox} is the oxide capacitance, t_{ox} is the oxide thickness, $\rho(x')$ is the charge distributions in the oxide thickness, ϵ_0 is the free spare permittivity, ϵ_{0x} is the relative permittivity of oxide, x' is the distance along the oxide thickness (from metal to the SiO_2 interface), Q_{ot} is the density of oxide trapped charge, x_m is the centroid of trapped charge in oxide.

The gate bias voltage shift of the I-V curve is also a function of the distribution and location of the oxide-trapped charge. When a positive voltage is applied to the metal gate, the gate bias voltage shift is given by:

$$\Delta V_g^+ = -\frac{x_m Q_{ot}}{\epsilon_o \epsilon_{ox}} \tag{91}$$

When a negative voltage is applied to the metal gate, the gate bias voltage shift is given by

$$\Delta V_g^- = -\frac{t_{ox}}{\epsilon_o \epsilon_{ox}}\left(1 - \frac{x_m}{t_{ox}}\right) Q_{ot} \tag{92}$$

If the trap distribution in the oxide is uniform parallel to the interfaces, the C-V curve shifts without distortion. For a non-uniform trap distribution parallel to the interfaces, the C-V curve distorts.

First-order trapping model. Balland and Barbottin [1989] related the drift and deformation in C-V and I-V characteristics to trap parameters in the oxide (N_T is the spatial density of traps; E_T, is the fundamental energy level of traps; and σ_c is, the capture cross-section of traps). The first-order model assumed that the capture cross-coefficient and trap density remain constant during trapping. Trap-filling kinetics when re-emission is negligible can be predicted based on the following assumptions:

- The oxide has only one form of electron traps.
- The electron traps possess a single discrete level, E_T.
- The electron traps do not act as generation-recombination centers; they can exchange electrons only with the conduction band.
- During the trap-filling phase, no re-emission takes place; i.e., $e^{th}_n=0$, $e^{opt}_n=0$.

The oxide-trapped charge resulting from trap filling is given by

$$\begin{aligned} Q_{ot} &= e n_{TT}(t) \\ &= q_e \left(1 - e^{-t_{inj}/\tau_c}\right) \int_0^{t_{ox}} N_{trp}(x) dx \end{aligned} \tag{93}$$

where q_e is the absolute value of the electronic charge, N_{trp} is the spatial density of the interfacial traps, t_{inj} is the length of the injection phase (seconds), τ_c is the capture time constant (seconds), and t_{ox} is the oxide thickness (cm). The capture time constant is given by the equation [Balland and Barbottin, 1989]

$$\tau_c = \frac{e v_{dft}}{A_{c,e} J_{inj} v_{th}} \tag{94}$$

where v_{th} is the drift velocity (cm s^{-1}), v_{th} is the thermal velocity (cm s^{1}), $A_{c,e}$ is the capture cross-section for electrons (cm^2), and J_{inj} is the density of injected current (A cm^{-2}).

Trap emptying. Trap emptying is represented as a separate process once the traps have been filled partially or totally [Balland and Barbottin, 1989]. The emptying kinetics of traps when re-trapping is negligible can be predicted based on the following assumptions:

- Only one type of carrier takes part in the exchange.
- Only free electrons are released thermally or optically by traps.
- Traps can exchange electrons only with the conduction band.
- The density of electrons is zero at the onset of depopulation.
- At all times, the density of free electrons is small compared to the density of carriers still trapped.

The charge trapped at time t decreases exponentially with time:

$$Q_{ot}(t) = Q_{ot}(0) \exp\left(-t\, c_{th,\, em}\right) \tag{95}$$

where $c_{th,em}$ is the thermal emission coefficient (sec^{-1}), t is the time in seconds, and $Q_{ot}(0)$ is the density of the oxide-trapped charge integrated over the thickness of the oxide per unit area of silicon-silicon dioxide interface at time t=0, i.e., at the onset of trap emptying given by Equations 96 and 97.

Two cases can exist at time t=0. First, the traps may be filled to saturation; and second, the traps may not be filled to saturation. If the traps are filled to saturation at time t=0 [Balland and Barbottin, 1989],

$$Q_{ot}(0) = e \int_0^{t_{ox}} N_{trp}(x)dx \tag{96}$$

If the traps are not filled to saturation at time t=0,

$$Q_{ot}(0) = e \int_0^{t_{ox}} n_{trp,fl}(x)dx \tag{97}$$

where e is the absolute value of the electronic charge (1.6×10^{-19} Coulomb), N_{trp} is the spatial density of trap possessing level E_T, $N_{trp,fl}$ is the spatial density of filled traps possessing level E_T (cm^{-3}), n_T is the spatial density of filled traps of energy E_T, and t_{ox} is the oxide thickness [Balland and Barbottin, 1989]. The thermal emission coefficient is a function of temperature given by

$$C_{th,em} = \left(\frac{2\sqrt{3}(2\pi)^{\frac{3}{2}} m_n^* K_B^2}{h^3} \right) A_{c,e}\, T^2 e^{-\frac{E_a}{K_B T}} \tag{98}$$

where m_n* is the effective mass of the electrons in silicon dioxide, h is Planck's constant (6.62 x 10^{-34} J.s), and E_a is the activation energy.

Hickmott model. Hickmott [1975] presented a model for thermally stimulated ionic conductivity due to trapped charges. The model is based on the assumption that at the start of the measurement, all the ions are present in the ion traps near one of the interfaces. When the temperature of the MOS capacitor is increased, the emission of ions from traps takes place at an increasing rate, which results in an increase in ionic current. After most of the ions have been emitted, the ionic current decreases. The flux of the ions emitted at time t, based on the assumption that ion traps corresponding to one type of ions have a single energy, E_a, is represented by

$$-\frac{dN_{ion,trp}(t)}{dt} = C_{pr,hm}\, N_{ion,trp}(t)\, e^{-\frac{E_a}{K_B T(t)}} \tag{99}$$

where $N_{ion,trp}$ is the density of the ion traps, $C_{pr,hm}$ is Hickmott's constant of proportionality, and E_a is the activation energy.

$$J_{ion}(t) = e\, \frac{dN_{ion,trp}(t)}{dt} \tag{100}$$

where J_{ion} is the current density due to the emission of ions from charge traps. The density of the ions still trapped follows from the integration of Equation 99:

$$N_{ion,trp}(t) = N_{ion,trp}(o)\, e^{C_{pr,hm} \int_0^t -\frac{E_a}{K_{B}T(t')} dt'} \tag{101}$$

and the current density is represented by

$$J_{ion}(t) = e\, C_{pr,hm} N_{ion,trp}(0)\, e^{-\frac{E_a}{K_B T(t)} - C_{pr,hm} \int_0^t e^{\frac{E_a}{K_B T(t')}} dt'} \tag{102}$$

The solution to Equation 102 depends on how the temperature varies as a function of time, and exists for hyperbolic variation of temperature ($1/T = 1/T_0$ - at) [Hickmott, 1975, Stagg 1977] and linear variation of temperature with time ($T = T_0$ + bt) [Hillen, 1981]. Typically, temperature transients and temperatures in the neighborhood of 175°C or greater result in failures due to slow trapping [Gottesfeld, 1984].

3.2 Gate Oxide Breakdown

Two forms of oxide breakdown can result from electrostatic discharge and electrical overstress or from time-dependent breakdown. These occur during operation within rated conditions of voltage, temperature, and power dissipation.

Electrostatic discharge. An integrated device can be considered equivalent to a circuit with multiple paths to the ground. When one pin is grounded, potentials sufficient to cause dielectric breakdown, junction shorts, or cracks between isolated regions may be discharged through the device. Electrostatic discharges (ESD) typically last less than 50 μs.

Electrostatic pulses can arise from contact with air, skin, glass, or charge-carrying particles. They can damage the gate oxides, causing cracking of the device oxide or melting of small amounts of the device, creating minute explosions on the device surface and resulting in voids, cratering, and subsequent short circuits or open circuits. Increased temperature causes a significant reduction in the electrostatic discharge resistance of the component [Kuo, 1983; Hart, 1980]. In extreme cases, the die may vaporize due to electrostatic pulses.

Typically, ESD failures result in fracture of the gate oxide in MOS devices, since voltage is in excess of the breakdown voltage (in bipolar devices, breakdown bulk occurs predominantly in the device). High currents through the breakdown site cause localized heating and usually produce a metal silicon alloy through the gate fracture site, forming a resistive short across the gate. The short can be a gate to a drain short, source short, or substrate short, depending on the structure of and imperfections in the oxide. The most likely sites for ESD damage in defect-free oxides are the source or drain sites, depending on the polarity of the transient and biasing of the device. Gate-to-substrate shorts are more prevalent in devices with pre-existing oxide defects in the form of geometrical or dopant irregularities [McAteer, 1989]. Typical PN-junction ESD damage occurs in reverse-biased conditions in the form of degraded PN-junction characteristics. The failed PN junction is characterized as cracked glass across the junction on the surface of the chip. Figure 2.23 to Figure 2.25 show the various manifestations of ESD damage in integrated circuits.

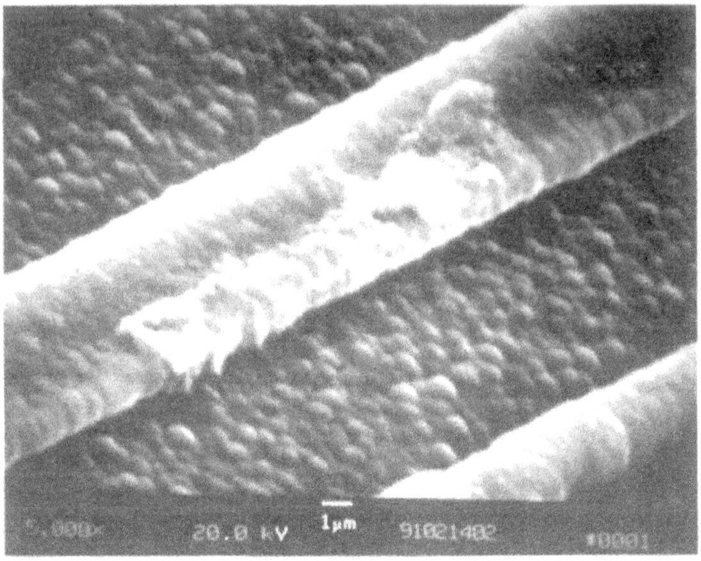

Figure 2.23 SEM micrograph shows electrostatic discharge damage to a metallized runner which caused failure indicated by a latched signal. The ESD event originated in the silicon substrate of an integrated circuit and erupted outward along one side of the runner causing two distinct craters along with the disturbed metal. (Courtesy of Motorola, Inc.)

The electrostatic discharge failure voltage decreases with an increase in temperature in the range of 25°C to 125°C [Moss, 1982; Hart, 1980]. Recent work has shown that simple low-voltage electrostatic pulses can damage gate oxides to varying degrees in NMOS without actually causing complete gate oxide breakdown [Amerasekera, 1986; 1987]. Damage appears in the form of reduced saturation drain current and gate voltage. Scherier [1978] has shown that MOSFETs are most susceptible to damage due to electrostatic discharge. Total breakdown of n-channel MOSFETs occurs around 100 V, while TTL NAND gate devices show degradation around 1.0 kV. Latent damage can also occur, with microcracks that remain dormant until aggravated.

Wunsch-Bell model. Wunsch and Bell [1968] characterized the failure due to pulse voltages as a localized temperature rise during avalanche breakdown. The conductivity of the least doped polar side of the junction becomes intrinsic and a negative resistance region appears. The resistance continues to decrease with increasing temperature, so the current is crowded into the localized hot-spot area, resulting in thermal runaway. The tiny cross-section involved will finally reach the melting point of silicon if sufficient energy is available. The mechanism is characterized by

$$\frac{P_{gt}}{A_{gt}} = \sqrt{\pi k_{tc,chp} d_{chp} C_{p,chp}} \frac{(T_{m,chp} - T_{ref})}{t^{1/2}} = k_{wb,pr} t^{-1/2} \tag{103}$$

where P_{gt}/A_{gt} is the power density or the power per unit junction area, d_{chp} is the density of the chip, $C_{p,chp}$ is a specific chip, $K_{tc,chp}$ is the thermal conductivity of the chip, $T_{m,chp}$ is the melting point of the chip, T_{ref} is the reference temperature, t is the time in seconds, and K_{wb} is a constant of proportionality. The equation holds in the quasi-adiabatic region (=pulse duration of 100 ns for silicon). P_{gt} is the power in the gate area, and A_{gt} is the gate area. Generally, reverse-bias damage occurs more easily than forward bias damage because of the order-of-magnitude difference in voltage or the correspondingly higher ration of current required to reach damaging power densities. A generalized equation of ESD damage is represented as

$$\frac{P_{gt}}{A_{gt}} = k_{1,wb} t^{-1} + k_{2,wb} t^{-1/2} + k_{3,wb} \tag{104}$$

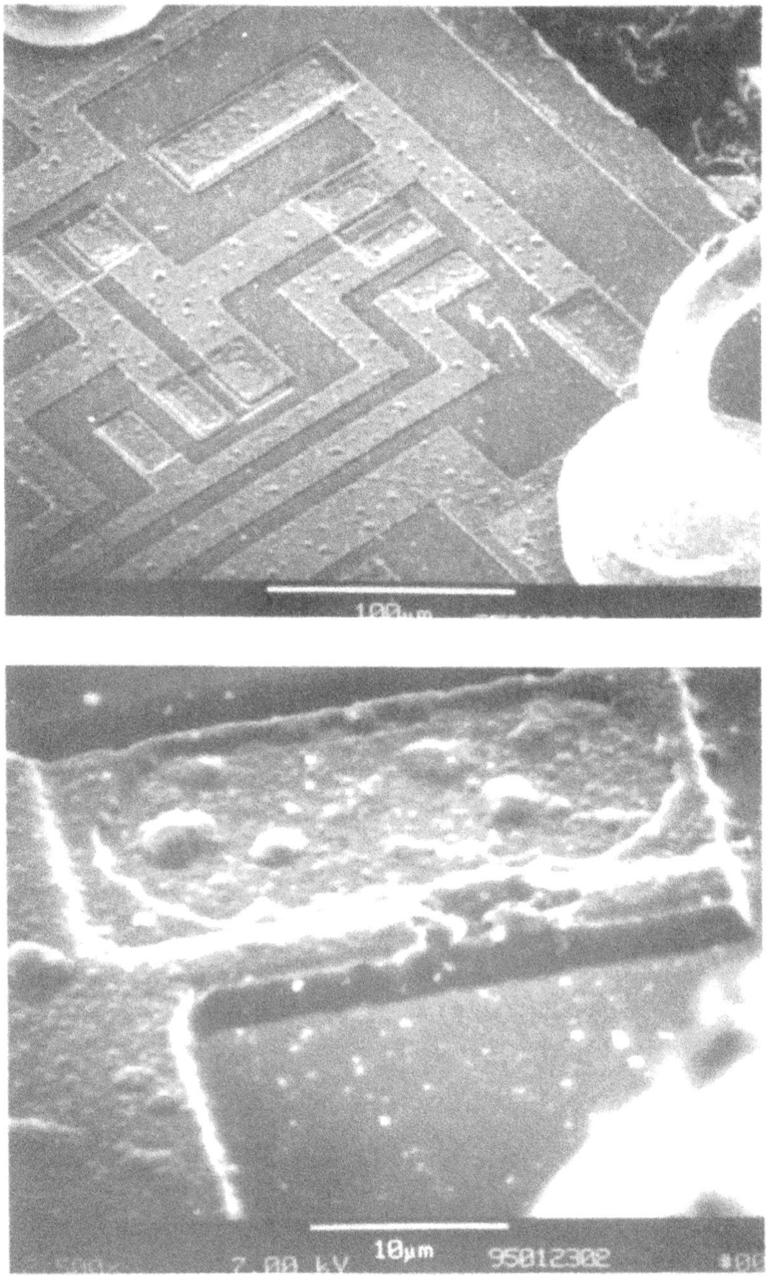

Figure 2.24 SEM micrograph (upper) shows the site of an ESD artifact on an IC indicated by an arrow. SEM micrograph (lower) details the ESD damage evidenced by a large crater in one edge of the metallized runner. (Courtesy of Motorola, Inc.)

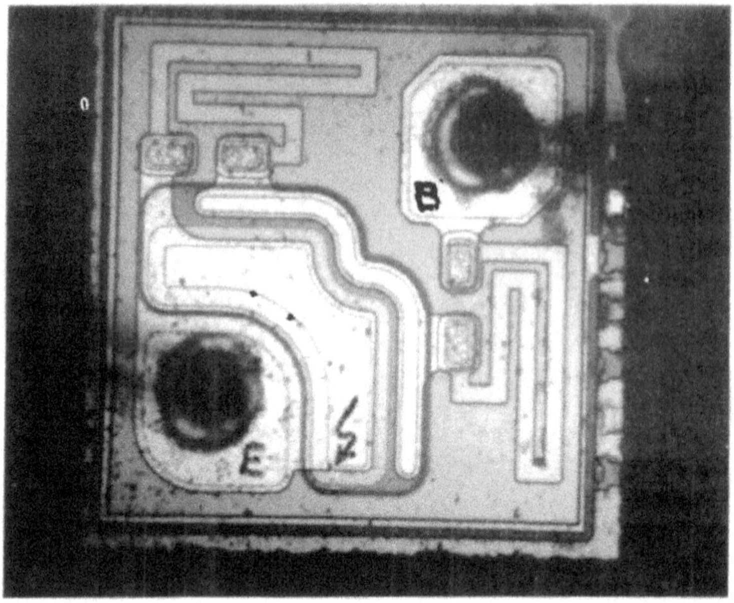

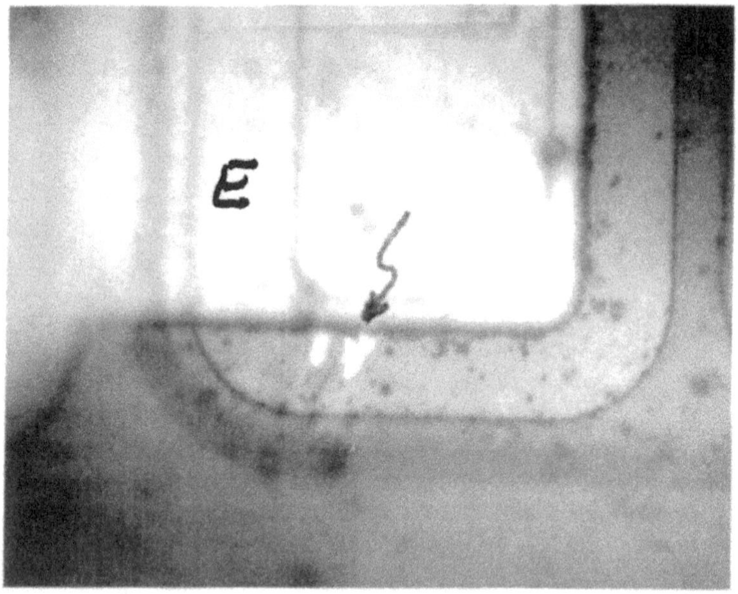

Figure 2.25 Photomicrograph (upper) shows the site of an ESD subsurface artifact on a digital switch transistor marked with an arrow. Photomicrograph (lower) provides a magnified view of the polyp-like pattern of discoloration caused by ESD-induced breakdown between the emitter and collector. (Courtesy of Motorola, Inc.)

where t is the time, and $K_{1,wb}$, $K_{2,wb}$, $K_{3,wb}$ are Wunsch-Bell constants of proportionality.

The first term in the equation is for short pulse widths in the adiabatic range. The second term is the intermediate duration transients, in which some heat dissipation occurs. The last term is for steady-state conditions under which the failure becomes independent of pulse width. Equation 103 shows that device operating temperature affects the damage threshold - the higher the device operating temperature, the lower the damage threshold. The Wunsch and Bell model is an approximation that assumes that all the power dissipated in the device occurs at the junction, with a resulting one-dimensional heat flow. In actuality, some power is dissipated in bulk silicon and the resulting heat flow is not one-dimensional.

Speakman's model. Speakman [1974] considered both the power dissipated at the junction and in bulk silicon, assuming that all the power is concentrated in a plane perpendicular to the current path. Speakman addressed static discharge through a reverse-biased PN junction, represented as a discharge circuit consisting of a charged capacitor discharging through a resistance. For a human-body discharge, the resistances comprise those from the body, the contact, and the device. The device parameters, including device voltage (V_d) and device resistance (R_d), are complex and dynamic during transient condition. Speakman assumed the device parameters to be constant during the transient. Since approximately 99% of the power is dissipated in five time constants, the average power dissipated is obtained by integrating the instantaneous power dissipated over the time constants:

$$P_{av,rpn} = \frac{1}{5\tau_{rc}} \int_0^{5\tau} V_{dv}\, i(t)\, dt + \frac{1}{5\tau_{rc}} \int_0^{5\tau} R_{by}\, i(t)^2\, dt \tag{105}$$

where V_{dv} is the device voltage, $i(t)$ is the discharge current, $P_{av,rpn}$ is the average power in a biased pn junction, R_{by} is the body resistance (device resistance and contact resistance are negligible compared to body resistance), and τ_{rc} is the time constant. The discharge current waveform for the RC circuit is a decaying exponential represented by

$$i(t) = I_p\, e^{-\frac{t}{\tau_{RC}}} \tag{106}$$

$$I_p = \frac{V_{by} - V_{dv}}{R_{by} + R_{dv} + R_{ct}} \tag{107}$$

$$\tau_{rc} = \left(R_{by} + R_{dv} + R_{ct}\right) C_{by} \tag{108}$$

where V_{by} is the discharge voltage on the human body; R_{by} is the body resistance; R_{dv} is the device resistance; R_{ct} is the constant resistance; I_p is the peak current of discharge waveform voltage (volts); and C_{by} is the body capacitance. The average power represented by Equation 105 can be modified by substituting for I (t) from Equation 106:

$$P_{av,rpn} = \frac{1}{5\tau_{rc}}\int_{o}^{5\tau} V_{dv} I_p e^{-\frac{t}{\tau_{Rc}}} dt + \frac{1}{5\tau_{rc}}\int_{o}^{5\tau} R_{by} I_p^2 e^{-\frac{t}{\tau_{rc}}} dt$$
$$= \frac{V_{dv} I_p}{5}(1-e^{-2}) + \frac{R_{by} I_b^2}{10}(1-e^{-10}) \tag{109}$$
$$\approx \frac{V_{dv} I_p}{5} + \frac{R_{by} I_p^2}{10}$$

$$R_{dv} = R_{sh,bs}\, X\, \frac{t_{be}}{l_{emtr}} \tag{110}$$

where l_{emtr} is the emitter length (transistor emitter), t_{be} is the base-emitter separation (transistor base), and $R_{sh,bh}$ is the base-sheet resistance (transistor base). $A_{x,be}$ is x-sectional area of base-emitter junction. For a bipolar transistor, the resistance component is given by

$$A_{x,be} = D_{bs}\, l_{emtr} \tag{111}$$

where D_{bs} is the depth of the transistor's base, and l_{emtr} is the length of the emitter region.

Metallization-open voltage threshold for ESD failures. Typically, vaporized metal lines are indicators of electrical overstress (EOS) failures, but sufficiently large electrostatic discharge (ESD) pulses can also result in such failures. Open-circuit sites typically are at points of constriction, such as oxide steps. ESD failures causing metallization vaporization are uncommon, however, due to the presence in the discharge current path of other energy-absorbing elements that reduce the transient current through the metal below damaging current densities [McAteer, 1989]. The condition for melting non-stepped metallization stripes under adiabatic conditions is

$$J_{met}^2\, t_{m,met} = k_{pr,int} \tag{112}$$

where J_{met} is the current density in metallization, (A/cm^2), $t_{m,met}$ is the time to melt (in seconds), and $k_{pr,int}$ is a constant (Equation 112). Thus,

$$K_{pr,int} = \frac{(\Delta T)H_{c,chp} + H_{f,chp}}{0.239\rho_{met}} \tag{113}$$

where ΔT is the temperature rise (= 660°C - 25°C = 635°C), $H_{c,chp}$ is the heat capacity per chip per unit volume (≈ 0.637 cal/cm^3 °C @ 240°C), $H_{f,chp}$ is the heat of fusion of chip per unit volume (248.5 cal/cm^3), and ρ_{met} is the volume resistivity of the metallization (≈ 5.1 x 10^6 Ω cm). The time constant of the RC circuit can be calculated, based on resistance and capacitance (R = 1500 Ω; C = 100 pF) values for a standard human body model. The time to stripe melting can be approximated as five time constants:

$$t_{m,met} = 5\,\tau_{rc} \tag{114}$$

where $t_{m,met}$ is the time to melt for the metallization, and τ_{rc} is the time constant. The average current in the circuit can thus be calculated from

$$J_{met}^{2} = \frac{I^{2}_{met,un}}{A^{2}_{met}} = \frac{k_{pr,int}}{T_{m,met}} = \frac{k_{pr,int}}{5\tau_{rc}}$$
$$I^{2}_{met,av} = \frac{k_{pr,int}\,A^{2}_{met}}{5\tau_{rc}} \qquad (115)$$

The peak discharge current can be estimated from the modified Speakman model (Equation 105, since the first term containing V_{dv} does not apply):

$$P_{av,rpn} = I^{2}_{met,av}R_{av,t} = \frac{R_{by}I_{p}^{2}}{10}$$
$$I_{p}^{2} = \frac{10(I_{met,av})^{2}R_{av,t}}{R_{by}} \qquad (116)$$

where $R_{av,t} = (R_{by} + R_{met} + R_{series})$, $R_{av,t}$ is total average resistance, R_{met} is metallization, and R_{series} is any series load resistance in the circuit. The ESD voltage threshold then required for metallization melting is

Devices are typically protected against ESD by protection structures that short the ESD pulse

$$V_{esd,th} = I_{p}\,R_{av,t} \qquad (117)$$

voltage to ground or to the supply voltage to limit the current entering the critical junction. The damage voltages for protected devices are in the range of 3,000 to 9,000 V, depending on protection, as compared to being as low as 100V for unprotected devices. However, on-chip protection can reduce device performance.

Time-dependent dielectric breakdown. Time-dependent dielectric breakdown (TDDB) is the formation of low-resistance dielectric paths through localized defects in dielectrics, such as in MOS devices only thermally grown or other oxides . Failures typically occur at weaknesses in the oxide layer, due to poor processing or uneven oxide growth. Various studies have demonstrated a correlation between the low breakdown strength and the presence of stacking faults in the oxides [Lin, 1983; Liehr, 1988]. Other studies have attributed early oxide breakdown to charge accumulation in the oxide [Lee, 1988; DiStefano, 1975; Harari, 1978; Ricco, 1983; Holland, 1984] and to local thinning and discontinuities in the oxide caused by metal precipitates [Honda, 1984, 1985, Wendt, 1989]. The mechanism is characterized by sudden, usually permanent, d.c. conduction in the dielectric of MOS capacitors. Typically, thin FET dielectric materials exhibit this breakdown failure mechanism, depending on latent defect density, temperature, electric field intensity, the ratio of the device operating potentials to the intrinsic dielectric strength, and the distances between the conductors in electronic packages defined by technology limits or electrical requirements.

Fowler-Nordheim Tunneling-based Models for TDDB. Fowler-Nordheim tunneling models are based on the failure of oxides due to TDDB, as a consequence of charge accumulation in the oxide. Breakdown in the oxide occurs when a critical charge density is triggering the breakdown process [Lee, 1988; DiStefano, 1975; Harari, 1978; Ricco, 1983, Holland, 1984]. The critical charge density is proportional to the total number of electrons injected in the oxide and the probability of holes being generated by these electrons being trapped in the oxide

[Chen, 1985] where

$$Q_{cr,ox} = I_{fn,ox}\, t_{bd}\, \alpha_h\, \eta_h \tag{118}$$

$$I_{fn,ox} \propto e^{-\frac{B}{E_{ox}}} \qquad \alpha_h \propto e^{-\frac{H}{E_{ox}}} \tag{119}$$

where $Q_{cr,ox}$ is the critical charge density in the oxide, $I_{fn,ox}$ is the Fowler-Nordheim current in oxide, t_{bd} is the time to oxide breakdown, α_h is the hole-generation coefficient, and η_h is the hole-trapping efficiency (constant). Based on experimental data, the time to breakdown has been shown to be an exponential function of the reciprocal of the electrical field (E_{ox}) [Moazzami, 1988; Lee, 1988]:

$$t_{bd} = \tau_{ox}\, e^{\frac{G_{ox,bd}}{E_{ox}}} = \tau_{ox}\, e^{\frac{G_{ox,bd}}{V_{ox}}} \tag{120}$$

where τ_{ox} is the intercept of the $\ln(t_{bd})$ versus $1/E_{ox}$ plot, $G_{ox,bd}$ is the slope of the $\ln(t_{bd})$ versus $1/E_{ox}$ plot ($G = B+H$, in Equation 119,) X_{ox} is the oxide thickness, and V_{ox} is the voltage across the oxide. In the case of defective oxides, the defects are modeled as localized oxide thinning. The electric-field dependence of time to breakdown for defective oxides is modeled as [Lee, 1988]:

$$t_{bd} = \tau_{ox}\, e^{\frac{G_{ox,bd}\, X_{eff,ox}}{V_{ox}}} \tag{121}$$

where $X_{eff,ox}$ is the effective thickness oxide. This concept is also used to model both asperities at the interface and localized areas having modified chemical composition, which may increase the charge-trapping rate or reduce the barrier height at the silicon/silicon dioxide (Si/SiO_2) interface. The natural logarithm of time to breakdown, $\ln(t_{bd})$, is a linear function of $1/E_{ox}$, and a non-linear function of E_{ox} [Moazzami, 1988; Lee, 1988]. The electric-field acceleration factor is defined as the tangential slope of the $\log_{10} t_{BD}$ vs E_{ox} plot [Lee, 1988]:

$$\gamma_{ef,acc} = -\frac{d(\log_{10}(t_{bd}))}{dE_{ox}} = \frac{G_{ox,bd}}{(\ln\, 10)\, (E_{ox})^2} \left(\frac{decades}{MV/cm}\right) \tag{122}$$

where $\gamma_{ef,acc}$ is the electric field acceleration parameter, t_{bd} is the time to oxide breakdown, E_{ox} is the electric field across the oxide, and G is the slope of the $\ln(t_{bd})$ vs $1/E_{ox}$ plot. Recent studies have modeled the pre-exponential term, τ_{ox}, in Equations 120 and 121 as an exponential function of temperature with an activation energy E_b [Moazzami, 1989; Moazzami, 1990]:

$$\tau_{ox}(T) = \tau_{ox,o}\, e^{-\frac{E_b}{K_B}\left(\frac{1}{T}-\frac{1}{300}\right)} \tag{123}$$

where $\tau_{ox,o}$ is the room temperature value of the pre-exponential; $\tau_{ox}(T)$ is the pre-exponential as a function of temperature; E_b is the activation energy of the pre-exponential, $\tau(T)$; and K_B is Boltzmann's constant. The slope of the plot of $\ln(T_{BD})$ vs $1/E_{ox}$ $G_{ox,bd}$ is also modeled as a Taylor expansion of a temperature exponential [Moazzami, 1989; 1990]:

$$G_{ox,bd}(T) = G^0_{ox,bd}\left(1 + \frac{\delta_{ox,bd}}{K_B}\left(\frac{1}{T} - \frac{1}{300}\right)\right) \tag{124}$$

$$\delta_{ox,bd} = \frac{K_B}{G^0_{ox,bd}}\,\frac{d\,G_{ox,bd}(T)}{d\,(1/T)} \tag{125}$$

where $G^0_{ox,bd}$ represents the room temperature value of the slope of the ln (t_{bd}) vs $1/E_{ox}$ plot. Moazzami et al. [1989] represented $G_{ox,bd}$ by Equation 123, to allow t_{bd} to follow the Arrhenius relationship. The time to breakdown is thus represented by an apparent activation energy,

$$t_{bd}(T) = \tau_{ox,o}\, e^{\frac{G^0_{ox,bd}X_{eff,ox}}{V_{ox}}\left(1 + \frac{\delta_{ox,bd}}{K_B}\left(\frac{1}{T}-\frac{1}{300}\right)\right) - \frac{E_b}{K_B}\left(\frac{1}{T}-\frac{1}{300}\right)}$$

$$t_{bd}(T) \propto e^{\frac{E_{tb,apt}}{K_bT}} \tag{126}$$

represented as follows:
where $E_{tb,apt}$ is the apparent activation energy for time-dependent dielectric breakdown.

$$E_{tb,apt} = \frac{G_{ox,bd}}{E_{ox}}\,\frac{X_{eff,ox}}{X_{ox}}\,\delta_{ox,bd} - E_b \tag{127}$$

The following values for $E_{tb,apt}$ and $\delta_{ox,bd}$ were determined from experimental data:

$$\begin{aligned} &\delta_{ox,bd} = 0.0167\ eV && \text{for} \quad 25^o\ C < T < 125^o\ C \\ &E_b = 0.28\ eV && \\ &\delta_{ox,bd} = 0.024\ eV && \text{for} \quad T > 150^o\ C \\ &E_b = 0.28\ eV && \end{aligned} \tag{128}$$

Thermodynamic Models for TDDB. McPherson [1985] proposed a thermodynamic model based on the assumption that when the dielectric breaks down, it undergoes an irreversible phase transition that transforms the material from an insulating phase to a conducting phase. The driving force for this transformation is the difference between the free energies of the conducting phase and of the insulating phase. The rate at which the reaction occurs is controlled by the free energy of activation associated with the growth of the conductive poly filament. McPherson represented a dielectric stored at a fixed field by a reaction rate constant, $C_{rr,ox}$,

$$C_{rr,ox} \propto \exp\left(\frac{-\Delta G_{gb,ox}}{K_BT}\right) \tag{129}$$

where $\Delta G_{gh,ox}$ is the Gibbs free energy for oxide breakdown processes and $C_{rr,ox}$ is the reaction rate constant for oxide breakdown, activation is associated with the breakdown processes. By treating the components as reactants and the broken-down components as reaction products, the time to failure of the dielectric is represented by McPherson as

$$t_{bd} \propto \frac{1}{C_{rr,ox}} \propto \exp\left(\frac{\Delta G_{gh,ox}}{K_B T}\right) \tag{130}$$

where t_{bd} is the time to failure, K_B is Boltzmann's constant, and T is the steady-state temperature. The internal energy, $E_{int,ox}$, of the dielectric under applied field stress is represented as

$$E_{int,ox} = E\,(\,S_{ox}, V_{ox,vl}, N_{dp,ox}, P_{ox}) \tag{131}$$

where S_{ox} is the entropy of dielectric system, $V'_{ox,vl}$ is the dielectric volume, $N_{dp,ox}$ is the number of dipoles induced and/or oriented, and P_{ox} is the dielectric polarization. The Gibbs free energy is obtained from internal energy, using the Legendre transformation:

$$\begin{aligned} G_{gh,ox}(T,E_{ox},\mu_{ox},P_{pr,ox}) &= E_{int,ox} - \left(\frac{\partial E_{int,ox}}{\partial S_{ox}}\right)_{V,N,P} - \left(\frac{\partial E_{int,ox}}{\partial V_{ox,vl}}\right)_{S,N,P} - \left(\frac{\partial E_{int,ox}}{\partial N_{dp,ox}}\right)_{S,V,P} - \left(\frac{\partial E_{int,ox}}{\partial P_{ox}}\right)_{S,V,N} \\ &= E_{int,ox} - TS_{ox} + P_{pr,ox}V_{ox,vl} - \mu_{ox}N_{dp,ox} - E_{int,ox}P_{ox} \\ &= H_{ox} - TS_{ox} + \mu_{ox}N_{dp,ox} - E_{int,ox}P_{ox} \end{aligned} \tag{132}$$

where H_{ox} is the enthalpy of the dielectric. The Gibbs free energy is expressed in terms of intensive parameters including temperature, T is electric field, E_{ox} is the electric field across oxide (or dielectric), μ_{ox} is the oxide chemical potential (or dielectric), $P_{pr,ox}$ is the pressure constant during dielectric stressing). T is the temperature. On rearranging,

$$-\,TS_{ox} - E_{int,ox}P_{ox} - \mu_{ox}N_{dp,ox} = K_B T f(T) g(E_{int,ox}) h(\mu_{ox}) \tag{133}$$

where *f, g, h,* are functions of *T, E,* and μ, respectively. Using thermodynamic relationships, the entropy, dielectric volume, and number of dipoles can be characterized by

$$S_{ox} = -\left(\frac{\partial G_{gh,ox}}{\partial T}\right)_{E,\mu} = K_B g h\left(f + T\frac{df}{dT}\right) \tag{134}$$

$$P_{ox} = -\left(\frac{\partial G_{gh,ox}}{\partial E_{int,ox}}\right)_{T,\mu} = -K_B T f h\left(\frac{dg}{dE_{int,ox}}\right) \tag{135}$$

$$N_{df,ox} = -\left(\frac{\partial G_{gh,ox}}{\partial \mu_{ox}}\right)_{T,E} = -K_B T f g\frac{dh}{d\mu} \tag{136}$$

Substituting the thermodynamic relationships into the above equation for Gibb's free energy,

$$\frac{T}{dT}\frac{df}{f} + \frac{E_{int,ox}}{dE_{int,ox}}\frac{dg}{g} + \frac{\mu_{ox}}{d\mu_{ox}}\frac{dh}{h} = 0 \tag{137}$$

Solving Equation 136,

$$G_{gb,ox}(T, E_{int,ox}, \mu_{ox}) = H_{ox} + K_B T \sum_m \sum_n \frac{C_{mn} E^{n}_{int,ox} \mu^{m-n}_{ox}}{T^m} \tag{138}$$

Considering the linear terms, McPherson represented the Gibbs free energy as

$$\begin{aligned} \Delta G_{gb,ox} &= \Delta H_{ox} + K_B T \left(\frac{C_{01}}{\mu_{ox}} + \frac{C_{11}}{T} \right) S_{ox} \\ &= \Delta H_{ox} + K_B T \left(B + \frac{C}{T} \right) S_{ox} \end{aligned} \tag{139}$$

where ΔH_{ox} is the change in enthalpy required to activate the polyfilament growth at breakdown, B and C are constants, $S_{ox} = (E_{bd,ox} - E_{app,ox})$. Thus,

$$t_{bd,f\%} = A_{ox,ef}\, e^{\left(\frac{\Delta H_{ox}}{K_B T}\right)} e^{(\gamma_{ef,acc}(T) S_{ox})} \tag{140}$$

where $\gamma_{ef,acc}$ is the electric field acceleration parameter, $t_{bd,f}\%$ is the time to dielectric breakdown of f% of population, $A_{ox,ef}$ is the oxide dielectric breakdown coefficient given by

$$\gamma_{ef,acc} = B + \frac{C}{T} \tag{141}$$

The effective activation energy of the field is given by

$$\Delta H_{eff,ox} = \Delta H_{ox} + K_B(\gamma_{ef,acc} T\, S_{ox}) \tag{142}$$

Empirical Models TDDB. Anolick [1981], Crook [1979], and Berman [1981] have proposed models to predict the time to failure based on these parameters. Secondary effects, especially dielectric thickness, electrode shape, materials, and other processing parameters may also affect the dielectric breakdown mechanism. Anolick and Nelson [1979] modeled the time to failure as

$$t_{bd,f\%} = A_{ox,ef}\, e^{\frac{\Delta H_{ox}}{K_B T}} e^{\gamma_{ef,acc}(V_{bd,ox}(f) - V_{app,ox})} \tag{143}$$

where $v_{bd,ox}$ is the breakdown voltage of the oxide or dielectrics, and $V_{app,ox}$ is the applied voltage across the oxide. Table 6 compares various models of time-dependent dielectric breakdown.

Time-dependent dielectric breakdown has been a significant failure mechanism in metal-gate MOS integrated circuits, silicon-gate transistors, and integrated circuits with two or more levels of polysilicon, and circuits with trench capacitors. The dielectric breakdown phenomenon has also been shown to be an important factor affecting the early life of dynamic memory devices [Barrett, 1978]. TDDB is a function of voltage stress in thin-gate oxides, and also a function of temperature (Figure 2.26). TDDB has both a dominant voltage dependence and a weak temperature dependence. Typically, TDDB has a very low thermal activation energy (say, 0.3 eV), but a very large voltage acceleration (10^7 /megavolt/cm) [Crook, 1978; 1979, Schnable, 1988]. The activation of TDDB is a function of stressing the electric field. The field acceleration is itself a function of temperature [McPherson, 1985]. The apparent activation energy of TDDB decreases from 1eV at low field stressing (E_b-E_s > 5 MV/cm) to 0.3 eV at higher fields (E_b-E_s < 3 MV/cm) (where E_b is the dielectric breakdown strength and E_s is the stressing electric field). The field acceleration also reduces from 6 decades/MV/cm at 25°C to 2 decades/MV/cm at 150°C [McPherson, 1985] (see Figure 2.27). The projected failure rate due to time-dependent dielectric breakdown under temperature stress was found to be too low to be of any importance for 10V and 55°C. At 10V and 200°C operation, the projected lifetime was 30 x 10^6 years, which represents a failure rate of 10^{-2} failures in 10^9 years [McPherson, 1985; Boyko, 1989; Swartz, 1986]. For this reason, device screening procedures involve stressing the devices at higher voltages and room temperature, rather than at normal biases and higher temperatures [Crook, 1978,1979; Schnable,1988].

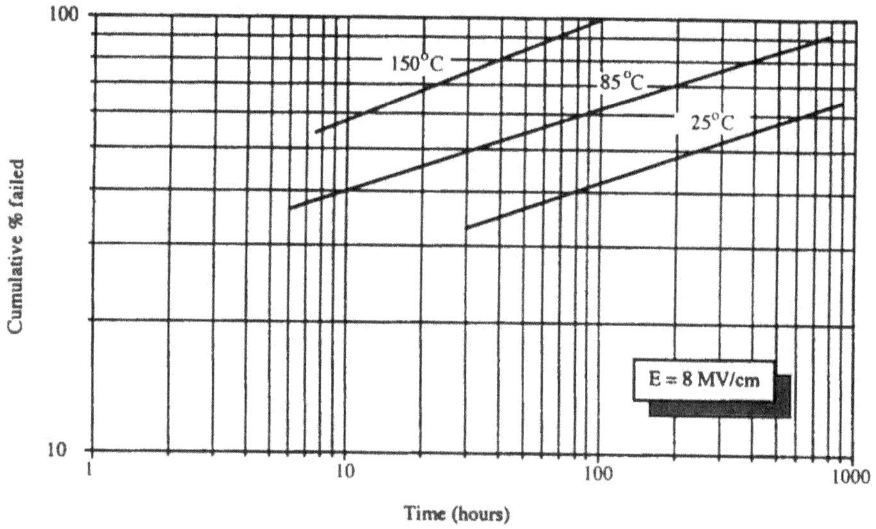

Figure 2.26 Temperature dependence of TDDB for electric field of 8 Mv/cm [McPherson, 1985].

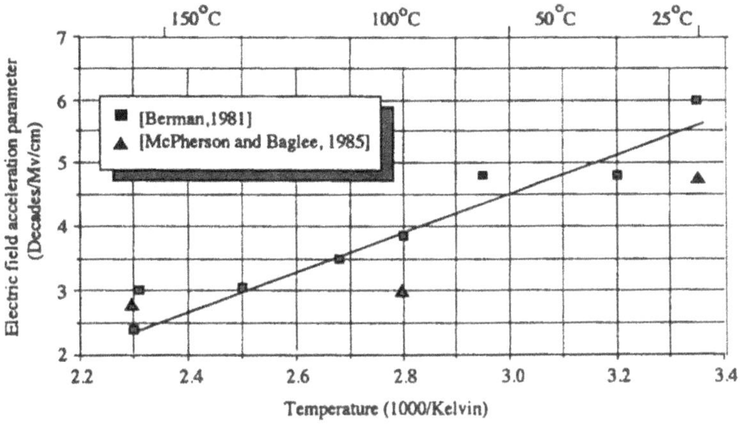

Figure 2.27 The electric field acceleration parameter is inversely dependent on temperature [McPherson and Baglee, 1985].

Table 6. Comparison of various time dependent dielectric breakdown models

Reference	Oxide thickness	Experimental conditions	Observations	Model predictions
Anolick Nelson [1979]	700 Å	E_S = 1.3 MV/cm $E_B - E_S$ = 7 MV/cm	$(\Delta H_{ox})_{50\%}$ = 2.1 eV $\gamma_{ef,ox} = B + \frac{C}{T}$	$(\Delta H_{ox})_{50\%}$ = 1.8 eV $\gamma_{ef,ox} = B + \frac{C}{T}$
Crook [1979]	1100 Å	E_S = 3.5 MV/cm $E_B - E_S$ = 3 MV/cm	$(\Delta H_{ox})_{50\%}$= 0.3 eV $\gamma_{ef,ox}$ = 7 @ 25°C	$(\Delta H_{ox})_{50\%}$ = 0.34 eV $\gamma_{ef,ox}$ = 6 @ 25°C
Berman [1981]	≥400 Å	linear ramp	$(\Delta H_{ox})_{50\%}$= 0.29 (E_B - E_S) $\gamma_{ef,ox} = -5.4 + \frac{0.29}{T}$	$(\Delta H_{ox})_{50\%}$ = 0.29 (E_B - E_S) $\gamma_{ef,ox} = -5.4 + \frac{0.29}{T}$
Hokari [1982]	100 Å	E_S = 5 -7 MV/cm $E_B - E_S$ = 5 MV/cm	$(\Delta H_{ox})_{50\%}$ = 1.0 eV @ 6 MV/cm $\gamma_{ef,ox}$ = 1.7 @ 250°C	$(\Delta H_{ox})_{50\%}$= 1.0 eV @ 6 MV/cm $\gamma_{ef,ox}$ = 1.5 @ 250°C
McPherson [1985]	100 Å	E_S = 6 - 8 MV/cm $E_B - E_S$ = 3 - 5 MV/cm	$(\Delta H_{ox})_{50\%}$ = 0.3 - 1.0 eV $\gamma_{ef,ox} = B + \frac{C}{T}$	$(\Delta H_{ox})_{50\%}$= 0.3 eV $\gamma_{ef,ox} = B + \frac{C}{T}$

3.2.1 Electrical Overstress

Electrical overstress occurs when a higher-than-rated voltage or current induces a hot-spot temperature beyond specifications for short periods of time [Alexander, 1978; Canali, 1981; Smith, 1978]. Hot-spot development typically occurs at a semiconductor junction as the current flow increases to accommodate the additional stress in the device. As the junction heats up, the increased temperature encourages even greater current flow, as the silicon resistance lowers at higher temperature; this in turn further heats the junction. The fundamental cause of failure is joule heating due to power dissipation along current-conducting paths. If the material reaches its melting temperature or its eutectic temperature, permanent damage may result. In a typical device, the greatest power dissipated per unit volume is either in the depletion region of the junction or in the lightest doped material. The temperature rises due to power dissipation, resulting in decreased silicon resistivity. Figures 2.28 to 2.31 show the various manifestations of EOS damage.

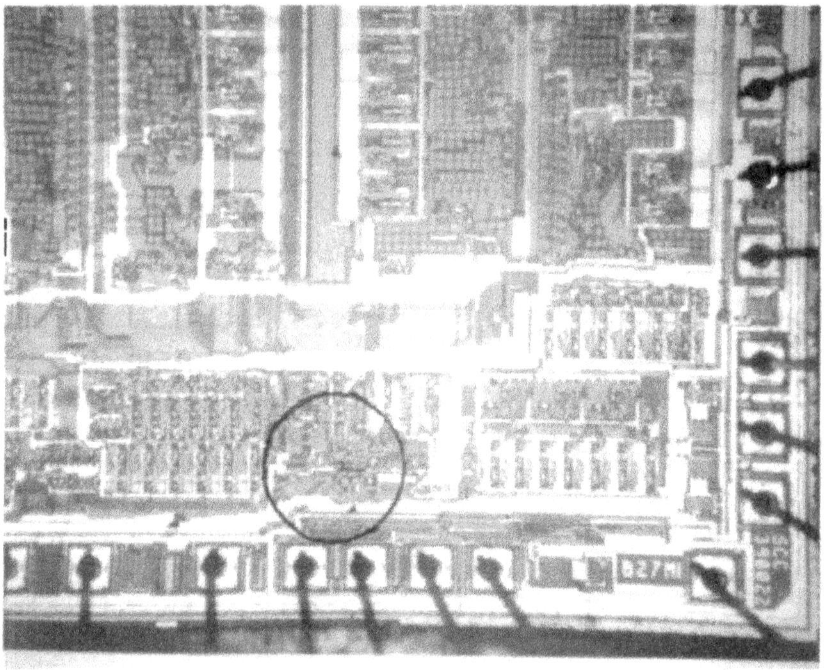

Figure 2.28a Photomicrograph locates an area of electrical overstress on an IC which was probably initiated by an ESD event. (Courtesy of Motorola, Inc.)

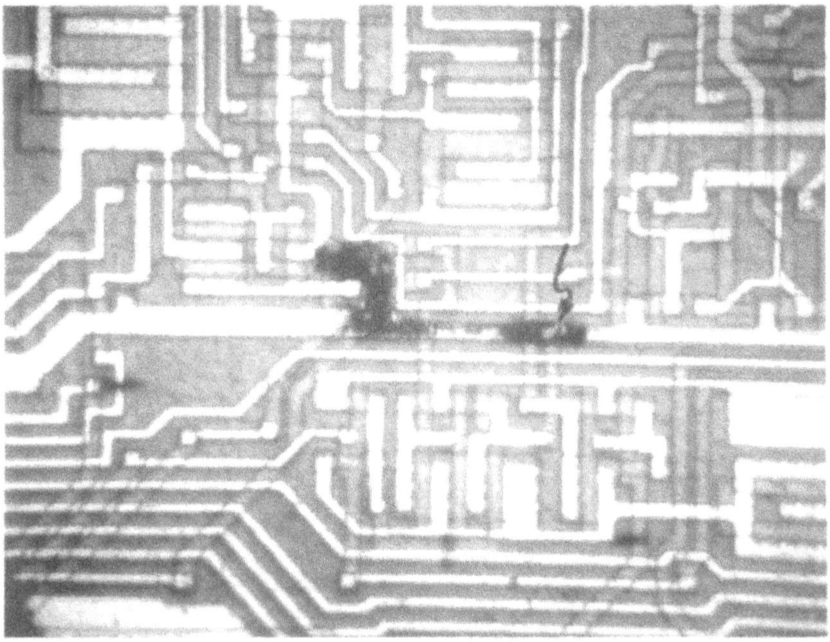

Figure 2.28b Photomicrograph shows the EOS-charred metallization with a probable ESD crater indicated by an arrow. (Courtesy of Motorola, Inc.)

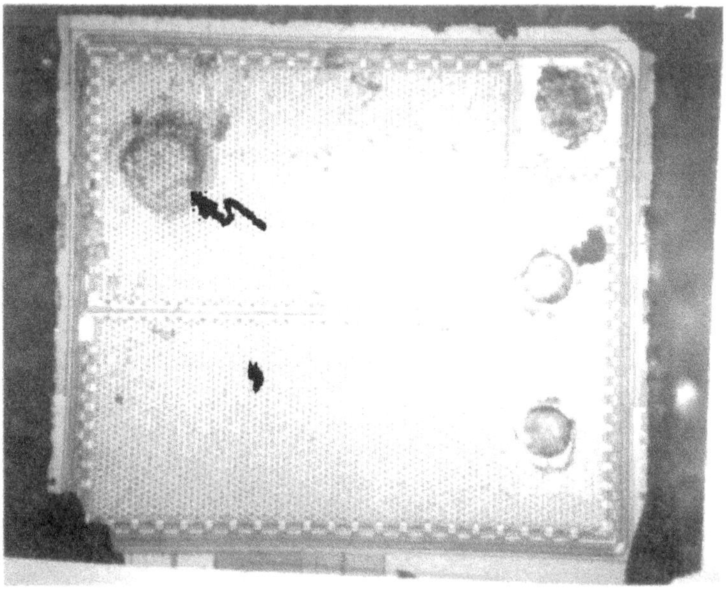

Figure 2.29 Photomicrograph shows the site of an electrical overstress artifact on a load switching integrated circuit marked with an arrow. The circular area of EOS-charred metallization surrounds a circular surface crack in the IC. The size and shape of the crack strongly suggests damage due to a disturbance of a wirebonding tool, which probably predisposed the failure. (Courtesy of Motorola, Inc.)

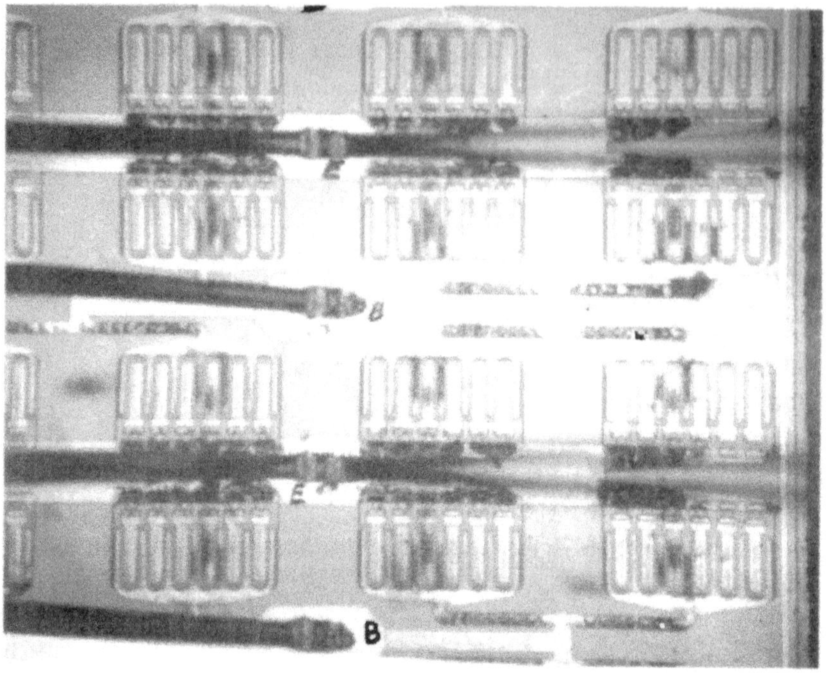

Figure 2.30 Photomicrograph shows effects of electrical/thermal overstress on parallel segments of a radio frequency power transistor. Charred, melted, or premolten emitter and base metallization resulted when the device was unable to dissipate heat generated by the transistor. This was due to a conchoidal fracture of the ceramic attachment between the transistor and heat sink. (Courtesy of Motorola, Inc.)

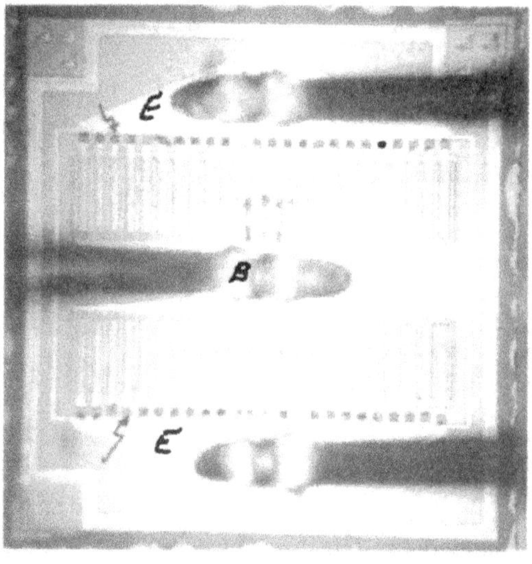

Figure 2.31 Photomicrograph shows electrical/thermal overstress of a radio frequency power transistor evidenced by all emitter fusing links (arrows) being blown. Some melted emitter and base metallization is also apparent. Fusing links may sometimes limit the effects of transient electrical overstress and enable the transistor to function at reduced power. (Courtesy of Motorola, Inc.)

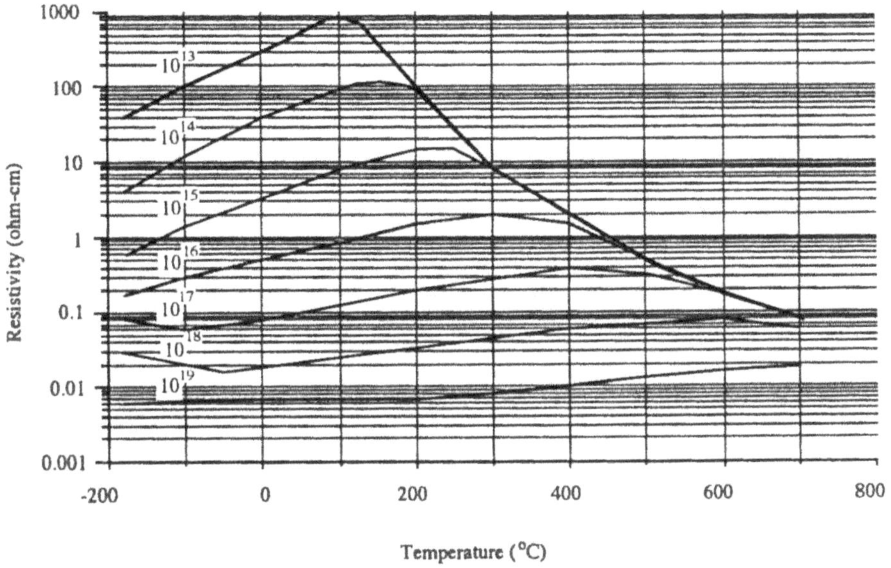

Figure 2.32 Silicon resistivity vs temperature [Runayan, 1965].

Runayan [1965] plotted the resistivity of silicon as a function of temperature for a variety of doping concentrations. Silicon initially exhibits a positive temperature coefficient of resistivity, but reaches a peak value in the neighborhood of 160°C, and thereafter exhibits a large negative temperature coefficient of resistivity. As the temperature of a region rises, the initial increase in the resistance of the region tends to spread the current to the cooler region (Figure 2.32). If the situation continues, the hot-spot temperature may exceed the intrinsic temperature of silicon, beyond which the resistivity of silicon decreases greatly. This allows more current to pass through the hot spot, further increasing the temperature and resulting in thermal runaway - that is, the temperature of the hot spot rises suddenly, while the resistance of the device drops. The process continues until the silicon in the hot spot melts, destroying the silicon crystal structure. If the electrical transient continues, the interconnects melt as well. If the transient is of high voltage, an electrical arc may occur when the metallization opens. The temperature in the vicinity of the arc can cause the metallization to vaporize. The energy required to raise the junction to an unstable temperature can be calculated using a power model [Pancholy, 1978].

The temperature rises due to power dissipation, resulting in decreased silicon resistivity. EOS failures typically manifest themselves as craters and cracks in the silicon and the plastic encapsulant. Figure 2.33 shows the effect of massive electrical overstress on a PNP transistor, resulting in a large crater and cracks in the phenolic plastic encapsulant. Figure 2.34 shows electrical overstress in three parallel bond wires which carried unfused B+ to the collector pedestal of the power transistor of a radio frequency power amplifier. Figure 2.35 shows electrical overstress in a zener diode chip, where the anode metallization melting and discolored semicircle around the remnant wirebond indicates that failure actually may have been accentuated by wire bond cratering. Black spots on the anode are remnant bits of plastic irreversibly polymerized by localized heating due to electrical overstress. Figure 2.36 shows electrical overstress failure in an Avalanche Diode, which hailed in a transient suppressor application. Figure 2.36 shows the large crater extending from the anode metallization well into the silicon substrate.

Figure 2.33 Photomicrograph shows effect of massive electrical/thermal overstress on a PNP transistor. This is evidenced by a large crater and cracks in the phenolic plastic encapsulant. A short circuit of a related printed circuit board runner caused this secondary failure. (Courtesy of Motorola, Inc.)

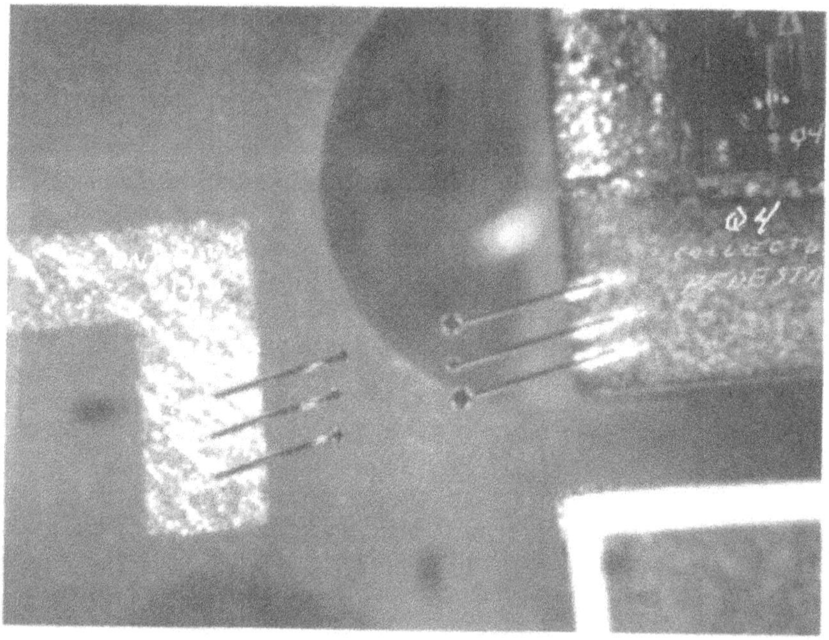

Figure 2.34 Photomicrograph shows electrical overstress indicated by fusing of three parallel wirebonds that carried unfused B+ to the collector pedestal of a power transistor, part of a radio frequency power amplifier. (Courtesy of Motorola, Inc.)

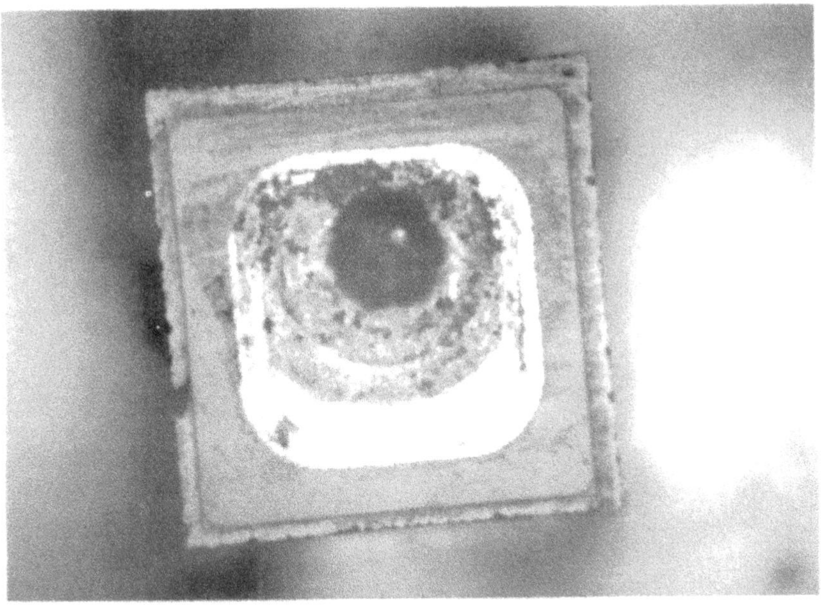

Figure 2.35 Photomicrograph shows effects of massive electrical overstress on Zener diode chip. The anode metallization was melted and discolored in a semicircle around the remnant wirebond, suggesting the failure might have been predisposed by wirebonding tool damage. Black spots on the anode are remnant bits of plastic irreversibly polymerized by localized heating due to the overstress. (Courtesy of Motorola, Inc.)

4. TEMPERATURE DEPENDENCIES OF FAILURE MECHANISMS IN THE DEVICE

In this section, the temperature dependence of failure mechanisms in the device structure of MOS and bipolar devices are discussed. The failure mechanisms discussed in this section include ionic contamination, second breakdown, and surface charge spreading. While ionic contamination and surface charge spreading are failure processes for MOSFETs, second breakdown is a failure mechanism typically found in bipolar devices.

4.1 Ionic Contamination

Ionic contamination causes reversible degradation phenomena, such as threshold voltage shift and gain reduction, due to the presence of mobile charge ions within the oxide or at the device-oxide interface, typically at temperatures above 100°C [Schnable, 1988]. A survey of published data shows that about 30% of microelectronic MOS device failures result from the ionic contamination caused by mobile ions in semiconductor devices [Brambilla, 1981; Johnson, 1976]. Contamination can arise during packaging and interconnect processing, assembly, testing, screening, and operation. Sodium, chloride, and potassium ions are the most common contaminants. The most dangerous is sodium, which can cause failures in extremely small quantities (10^{11} per cm^2).

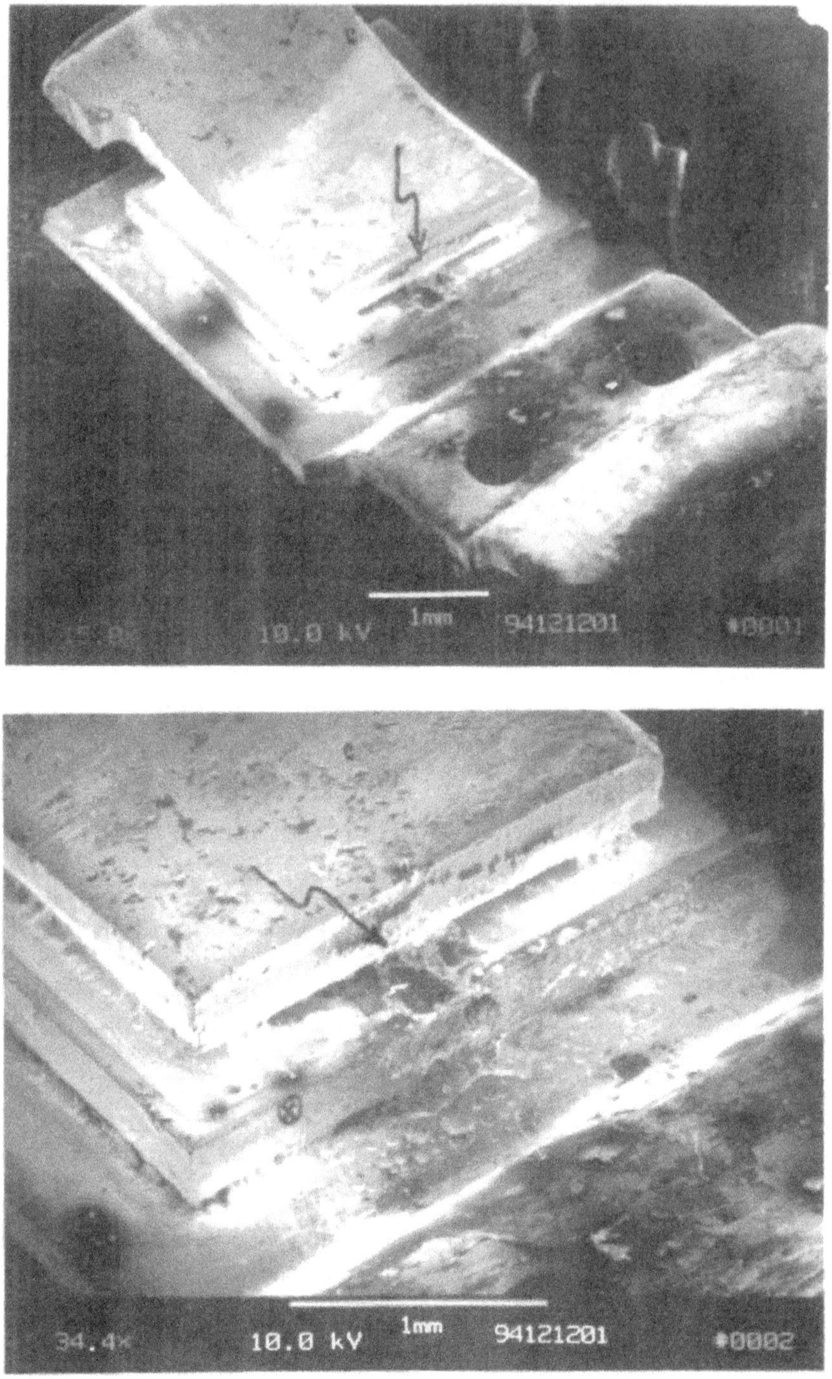

Figure 2.36 SEM micrograph (upper) shows decapsulated avalanche diode that failed in a transient suppressor application due to massive electrical overstress. SEM micrograph (lower) details the large crater extending from the anode metallization and well into the silicon substrate. Failure was indicated by an anode to cathode short circuit. (Courtesy of Motorola, Inc.)

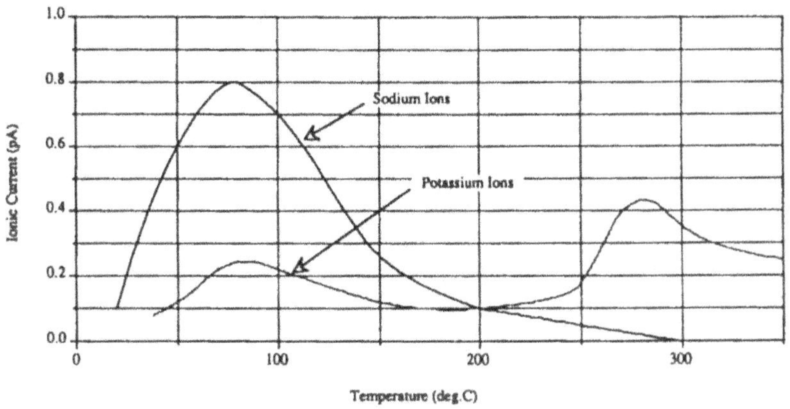

Figure 2.37 Influence of temperature on ionic current due to sodium and potassium ions [Hillen, 1986].

The amount and distribution of alkali ions, sodium ions, lithium ions, and potassium ions (Na^+, Li^+, K^+) in or near the FET gate dielectric region will influence the device threshold voltage by superposing ionic charges on externally applied FET device voltage. An extra positive charge at the silicon/silicon dioxide interface induces extra negative voltage in the n-channel, resulting in a decrease in the threshold voltage of the device. The direction of the field at the gate oxide of a p-channel device (PMOS) repels the positively charged ions away from the gate area; p-channel devices are, therefore, less sensitive to ionic contamination than n-channel devices. Figure 2.37 shows the variation of ionic current vs temperature in devices with Na^+ and K^+ ion implants (the curves have been generated using the TSIC technique[1]). The ionic current increases to 100°C for Na^+ ions, and decreases to a negligibly low value for temperatures higher than 200°C. The K^+ ion, on the other hand, shows two peaks in ionic current at temperatures in the neighborhood of 100°C and 300°C (Figure 25). The ionic current due to mobile ions is thus a complex function of steady-state temperature [Hillen, 1986; Boudry, 1979; Nauta, 1978; Derbenwick, 1977].

The mobility of the ions is temperature-dependent [Hemmert, 1980; 1981]. Consequently, high- temperature storage bake and exposure to high temperature during burn-in screen out ionic contamination failures [Hemmert, 1980; 1981; Bell, 1980]. In particular, the position of ionic charges relative to the silicon-silicon dioxide (Si-SiO_2) interface greatly influences the effectiveness of altering the threshold voltage, V_T. A uniformly distributed contaminant (Na^+) within the oxide is redistributed to the silicon (Si) channel surface due to the influence of an electric field, E_f, and thermal activation energy. The change in threshold voltage as a function of steady-state temperature, T, has been modeled by Wager [1984]:

$$\Delta V_{th} \propto E_{f,ch}^{\frac{1}{2}} \, t_{bs}^{\frac{1}{2}} \, e^{-\frac{\Delta H}{K_B T}} \tag{144}$$

where $E_{f,ch}$ is electric field across dielectric/oxide air drain-source channel, t_{bs} is the time under bias, ΔH is the activation energy, K_B is Boltzmann's constant, and ΔV_{th} is threshold voltage shift.

[1]Three techniques are generally used to measure the ionic current due to mobile charges versus temperature. These include BTS (Bais Temperature Stress), TSIC (Thermally Stimulated Ionic Current), and TVS (Triangular Voltage Sweep).

N-channel MOSFETs are covered with phosphosilicate glass films to stabilize the threshold voltage against changes resulting from ionic contaminants (sodium). Hemmert investigated the temperature bias kinetics for sodium ion drift in phosphosilicate glass [Hemmert, 1980; 1981]. Microelectronic devices may be subject to a variety of defects; one such defect is exposed gate oxide resulting from extraneous etched holes contiguous to the gate that subject the phosphosilicate glass film to either a high local ionic concentration or physical damage. The temperature at which ionic drift occurs in defective devices is not as high as in non-defective devices. Hemmert conducted experiments to measure ionic stability on a 1.5-mm MOS consisting of gate oxide, phosphosilicate glass film, and aluminum-copper metallurgy. He found that below 250°C, the plot of voltage vs temperature shows a polarization plateau that represents saturation PSG polarization voltage. The activation energy, ΔH, was 2.0 eV. This value is uncharacteristically high, compared to the value usually reported for sodium-related failure mechanisms (ΔH=1.0 eV); the cumulative percent defect predictions were found to be more accurate assuming a temperature-dependent defect level. In the temperature range of 250 to 350°C, flatband voltage shifts occurred. This temperature, called the break temperature, occurs when the contributions of flatband voltage shift to ionic drift and to polarization are equal. The maximum variation in the flat-band voltage is related to the mobile charge by the relation

$$Q_{m,ox} = C_{ox}\, \Delta V_{fb} \tag{145}$$

where c_{ox} is the oxide capacitance, $Q_{m,ox}$ is mobile charge per unit area in oxide, and the threshold voltage V_{th} and the flatband voltage, V_{fb} are related by the equation

$$V_{th} = V_{fb} + 2\phi_{fi} + \gamma\sqrt{2\phi_{fi}} \tag{146}$$

where ϕ_{fi} is the difference in Fermi-level and the intrinsic level in the bulk of the semiconductor, C_{ox} is oxide capacitance and V_{fb} is flatband voltage. It is evident that the threshold voltage depends on the flatband voltage and the factor γ is represented by

$$\gamma = \sqrt{\frac{2\, E_s\, q\, N_{d,a}}{C_{ox}^{\ 2}}} \tag{147}$$

where E_s is the oxide permittivity, q is the election/hole charge, $N_{d,a}$ is the donor/acceptor atomic density, and C_{ox} is the oxide capacitance. Thus, the usual assumption that the defect level is temperature-independent was found to be invalid.

A common characteristic of contamination failures is reversibility under high temperatures (in the neighborhood of 150° to 200°C), which partially or fully restores device characteristics [Hemmert, 1980; 1981; Bell, 1980]. However, this simply reflects a disordering of charge accumulations resulting from ion mobility and applied bias. The problem of ionic contamination has been solved to a large extent by the use of high-purity materials and chemicals for processing, use of HCl during oxidation [Robinson and Heiman, 1971], and use of phosphosilicate glass (PSG) or borophosphosilicate glass (BPSG) over the polysilicon to getter the ions [Schnable, 1988]. Passivation layers can provide additional protection against the ingress of alkali ion contamination in completed devices [Schnable, 1988].

4.2 Second Breakdown

Second breakdown is the transition to a state of higher conductance in a reverse-biased avalanching semiconductor junction. Second breakdown occurs predominantly in bipolar devices when the bias current density reaches a threshold value; in this case, the operating temperature in some region of the pn junction becomes high enough that thermally generated carriers can take over the conduction process. The avalanche is thus thermally quenched, and second breakdown occurs. A negative temperature coefficient of resistance, associated with thermally generated current, results in the current constricting and flowing through a narrow region in the junction, where the temperature rises significantly. If the temperature of the constriction is higher than T_p (the temperature at which the resistivity of semiconductor materials is the highest), the high-power density region at the perimeter of the constriction will heat up, becoming intrinsic, and the constriction will elongate into the ohmic region. If the constriction temperature is less than T_p, the constriction will not extend itself into the ohmic region, and the current will flow through the constriction in the junction and fan out. Second breakdown is a limiting phenomenon in power transistors that are switched in the presence of inductive loads, and typically involves three stages [Budenstein, 1972; Sunshine, 1970; Schafft and French, 1966; Reich and Hakim, 1966; Chiang and Lauritzen, 1970; Thornton and Simmons, 1958]:

- nucleation of filament;
- growth of a relatively broad filament across the high resistivity region;
- growth of a second filament interior to the first, in which material is in molten state.

While nucleation and filament growth are non-destructive processes, formation of melt is damaging and irreversible. Typically, nucleation is accompanied by a negligible voltage drop. In reverse-biased diodes, the p-n junction is the hottest region and the place where the filaments nucleate. The increase in the temperature of the p-n junction raises the avalanche voltage, due to a decrease in the collision ionization coefficients. However, as the temperature rises, the reverse saturation current decreases rapidly. The total saturation current is insensitive to the size of the reverse saturation current until the latter is eight-tenths of the former; then the avalanche voltage drops to zero, and the reverse saturation current density equals the total current density, resulting in the initiation of filamentation [Budenstein, 1972; Sunshine, 1970; Chiang and Lauritzen, 1970].

Filament growth involves the development of a moderately hot filament across the high-resistivity region. When the broad filament completely bridges the high resistivity region, melt initiates. Formation of melt involves the development of a melt channel through the hot central portion of the broad filament. The melt channel alters the device irreversibly, the degree of degradation depending on the size and location of melt. Broad filaments typically range from 25 to 50 μm across, while the melt channel, at currents near its formation threshold, is about 1 μm in diameter. Growth and formation of melt, however, are accompanied by an appreciable voltage drop. Junction nucleation occurs when the reverse saturation current density at the local site on the junction becomes equal to the total current density at that site [Budenstein, 1972; Sunshine, 1970, Chiang and Lauritzen, 1970]. Typically there is a time difference between the application of the high current pulse and second breakdown, called the delay time.

Second breakdown is characterized by two different failure mechanisms: forward second breakdown, and reverse second breakdown. Both forward and reverse bias second breakdown of diodes is accompanied by current filamentation and a high current transition. In reverse bias, the filament typically starts at the junction and reaches into the high-resistance side of the junction to the electrode. In forward bias, the filament typically starts well within the interior of the high-resistance region. The transition from avalanche to second breakdown of the base-collector junction occurs at a lower V_{CE} when the emitter is forward-biased than when it is reverse-biased. Second breakdown damage manifests itself in the form of changes in I-V characteristics due to resistive shorts of one or both transistor junctions. The critical condition for filament formation occurs when the temperature in the high-resistance side of the junction reaches the value at which the resistivity of the semiconductor material is at maximum. Filaments are formed during forward and reverse bias at the same current levels. The onset of negative resistance occurs before the filament is clearly distinguishable from its surroundings. The failure mechanism of second breakdown is characterized by an abrupt reduction in voltage and increased localized current density. The consequence is high temperature and permanent device damage or degradation [Budenstein, 1972; Sunshine, 1970; Chiang and Lauritzen, 1970; Beatty, 1976; Chen, 1983; Hower, 1970; Hu, 1982].

4.3 Forward Second Breakdown

Forward second breakdown is typically a result of thermal runaway at a point in the transistor. Current density and temperature increase jointly until thermal runaway occurs. A power transistor can be considered an aggregate of many elementary transistors that operate under different conditions. When a power pulse is applied to the transistor, the junction temperature rises. Since the resistivity of the silicon increases with temperature, the temperature rise causes the current to increase:

$$dP_{sb,1} = V_{ce} dI_c \tag{148}$$

where V_{ce} is collector-emitter voltage, I_c is the collector current, and $P_{sb,1}$is the power dissipated due to a change in the resistivity of silicon with temperature. The increase in temperature causes an increase in the power dissipation:

$$dP_{sb,2} = \frac{dT}{R_{th}} \tag{149}$$

where $P_{sb,2}$ is the power dissipated due to changed resistance, R_{th} is the resistance as a function of temperature, and T is the temperature. If $dP_1 < dP_2$, a thermal equilibrium is reached. If, on the contrary, $dP_{sb,1} > dP_{sb,2}$, the temperature and current increases enhance each other and thermal runaway occurs, causing transistor destruction.

Usually temperatures above 160°C are required to cause device failure due to second breakdown. The major cause of device destruction is localized heating. Second breakdown has also been found to occur in vertical power MOSFETs.

Reverse second breakdown. Reverse second breakdown is an internal phenomenon that occurs in power transistors during switching operations on an inductive load, and is characterized by a precipitous drop in voltage and a rapid increase in current. The second breakdown voltage increases mildly with an increase in temperature in the range of 25°C to 150°C. The breakdown voltage increases from 650V to 680V when the temperature increases from 25°C to 150°C [Beatty, 1976].

Mathematical model for second breakdown: Budenstein's model [1972] assumes that the formation of a melt channel within the broad filament that spans the high-resistivity region marks the initiation of second breakdown. The model assumes the following:

- The central portion of the filament is at the melting point of the silicon and the filament grows radially, while keeping a constant length. This allows the use of a cylindrical geometry for the heat transfer problem.
- The filament is about 1 μm in diameter at the end of transition for current amplitudes close to the threshold for filamentation.
- The filament is in parallel with a fixed resistance, R_h, of the remainder of the device.

The power dissipated in the filament is equal to the rate that the energy is absorbed in the latent heat of fusion plus the rate at which the energy is lost by conduction through the boundary surface of the filament:

$$P_{df} = \frac{d(L_{f,s} M_{mf})}{dt} + C_{hf} A_{sf} \quad (150)$$

where P_{df} is the electrical power dissipated in the filament, $L_{f,s}$ is the latent heat of fusion for the semiconducting material, M_{mf} is the molten filament mass (Equation 151), C_{hf} is the heat transfer coefficient (heat flow per unit time per unit area), A_{sf} is the lateral surface area of the filament, and

$$M_{mf} = d_{sm} \cdot l_{flt} \, \pi \, r_{flt}^2 \quad (151)$$

$$R_{flt} = \frac{\rho_{flt} \, l_{flt}}{\pi r_{flt}^2} \quad (152)$$

where ρ_{flt} is the resistivity of the filament, l_{flt} is the length of filament, r_{flt} is the filament radius, and R_{flt} filament resistance.

$$A_{sf} = 2\pi r_{flt} \, l_{flt} \quad (153)$$

$$R_{frd} = \frac{\rho_{sm} \, l_{flt}}{A_{j,sm}} \quad (154)$$

where R_{frd} is the fixed resistance of the remained of device, ρ_{sm} is the resistivity of semiconductor material, l_{flt} is the filament length, $A_{j,sm}$ is the semiconductor function area.

$$P_{df} = I_{flt}^2 \, R_{flt} = \frac{I_{t,fs}^2 R_{frd} R_{flt}}{(R_{frd} + R_{flt})^2} \quad (155)$$

where $I_{t,fs}$ is the total current through filament-semiconductor parallel resistance, R_{frd} is the fixed resistance of remainder of device - in parallel with filament resistance, R_{flt} is the filament resistance and I_{flt} is the current through filament. Substituting Equations 150 to 154 into Equation 149,

$$\frac{dr_{flt}}{dt} = \frac{C_{hf}}{L_{f,s}d_{sm}}\left(\frac{I_{t,fs}R_{frd}\rho_{flt}}{2C_{hf}} \frac{r_{flt}}{(R_{frd}\pi r_{flt}^2 + \rho_f l_{flt})^2} - 1\right) \tag{156}$$

where r_{flt} is the radius of the filament, C_{hf} is the heat transfer coefficient, $L_{f,s}$ is the latent heat for fusion of the semiconducting material, d_{sm} is the mass density of the semiconducting material, $I_{t,fs}$ is the total current (sum of the currents through the filament and the semiconductor device) and ρ_f is the resistivity of the filament. The steady-state condition ($dr_{fl}/dt = 0$) is assumed to be reached when power dissipated in the filament is at a maximum - that is, when $R_{flt} = R_{frd}$ (obtained by differentiating Equation 154 with respect to R_{flt}). The power dissipated in the filament in a steady-state condition is given by

$$P_{df,ss} = C_{hf}A_{sf,ss} = \frac{V_{flt,ss}^{\ 2}}{R_{flt}} \tag{157}$$

where $P_{df,ss}$ is the power dissipated in filament in steady state, $A_{sf,ss}$ is surface area of filamentation steady state, R_{flt} is the resistance of filament, and $I_{t,fs}$ is the total current (sum of currents through filament and semiconductor device). The primes represent the steady-state values of the respective quantities. At steady-state ($dr_{flt}/dt = 0$), the total threshold value of current for filamentation. Substituting Equation 152 and 153 into Equation 156 and using the steady-state condition, $R_{frd} = R_{flt}$.

$$I_{t,fs} = 2\sqrt{2C_{hf}}\left(\frac{\rho_{flt}\, l_{flt}^3}{R_{frd}^3}\right)^{1/4} \tag{158}$$

$$J_{th,fl} = K_{pr,fl}\,\rho_{sm}^{-\frac{3}{4}} \tag{159}$$

$$K_{pr,fl} = 2\sqrt{2C_{hf}}\;\rho_{flt}^{-\frac{1}{4}}\;A_{j,sm}^{-\frac{1}{4}} \tag{160}$$

where $J_{th,fl}$ is the threshold current density for filamentation, $K_{pr,fl}$ is the constant proportionality for filamentation, ρ_{sm} is resistivity of the semiconducting material. $A_{j,sm}$ is semiconductor junction area of remainder of the device.

4.4 Surface-charge Spreading

Surface-charge spreading, largely observed in MOS and memory devices, involves the lateral spreading of ionic charge, from the biased metal conductors along the oxide layer or through moisture on the device surface [Edwards, 1982; Blanks, 1980; Stojadinovic, 1983]. An inversion layer outside the active region of the transistor is formed due to the charge, creating a conduction path between the two diffused regions or extending the p-n junction through a high leakage region, resulting in leakage currents between neighboring conductors. The rate of charge spread increases with temperature. Surface-charge spreading failure, a wear-out mechanism, is usually observed at temperatures around 150°C to 250°C [Lycoudes, 1980]. There are no existing physics-of-failure models to predict surface-charge spreading failures.

5. TEMPERATURE DEPENDENCIES OF FAILURE MECHANISMS IN THE DEVICE OXIDE INTERFACE

In this section, the failure mechanisms at the device oxide interface are examined. Hot electrons, the only known mechanism at this failure site, occur typically in MOS devices with small gate lengths. This mechanism is unique because of its inverse dependence on steady-state temperature.

5.1 Hot Electrons

Hot electrons, or holes, are charge carriers that acquire energies from very high electric fields in excess of those that would be indicated by lattice or ambient temperature. Hot electrons are a phenomenon largely prevalent in high-density, small geometry, MOS memory devices because of small conduction channels and high voltages. It is accelerated at temperatures below 0°C [Stojadinovic, 1983; Woods, 1980]. Threshold voltage shifts or transconductance degradation can occur as renegade charge concentrations build up over time [Ning, 1979]. Penetration of the oxide by hot electrons can lead to excess gate and substrate currents [Takeda, 1983]. There are three possible mechanisms by which the electrons or holes are injected from silicon into the silicon dioxide [Garrigues and Ballan, 1986]:

- Direct tunnel emission: Injection of electrons by direct tunneling is possible in thin oxides (< 100 Å) and large electric fields ($\epsilon_{ox} > 5 \times 10^5$ V cm^{-1}).

- Field-assisted tunneling: Electrons of energy close to but less than the energy barrier at the silicon-silicon dioxide interface may tunnel through the triangular energy barrier resulting from the field in the oxide. For larger electric fields, the energy bands of silicon are so slanted that direct tunneling can occur, regardless of oxide thickness (field emission or Fowler-Nordheim tunneling).

- Injection over the barrier: Electrons may gain high potential energy due to high electric fields present in the conduction channel, and cross the potential barrier at the substrate-oxide interface. An electron in a conduction band of silicon can be injected over the energy barrier into the conduction band of silicon dioxide if its energy, ΔE, is greater than the difference in the conduction band energies of the conduction bands in the silicon and silicon dioxide.

Electron injection through field-assisted tunneling or injection over the barrier is possible by stimulating the electron thermally, optically, or with an electric field; field stimulation is the most common form of electron stimulation [Garrigues and Balland, 1986]. Figure 2.38 shows the effect of ambient temperature on the hot carrier mechanism [Matsumoto, 1981]. The temperature dependence of the substrate current at fixed bias can have three origins: temperature-dependent channel current, temperature-dependent local electric field, and temperature-dependent ionization rate. The low field mobility follows -3/2 power dependence on the temperature for typical n-channel MOSFETs [Sze, 1981]. The reduced mobility increases the saturation voltage, which lowers the maximum-channel electric field at fixed-drain bias and results in a lower substrate current [Ko, 1980; Hu, 1983]. Tam et al. [1983] proposed that the device degradation rate can be related to the substrate current by

$$\Delta V_{th} = C_{st,fac}\left[C_{mt,fac}\left[I_{sub}\right]^{\alpha_{ef}}\right]^{\beta_{gk}} \tag{161}$$

where $C_{st,fac}$ is a structure-related factor, $C_{mt,fac}$ is a material-related factor, α_{ef} is the energy factor typical value 2.9), and β_{gk} is the general kinetics factor ≈0.6 (typical value). The device degradation due to hot electrons very mildly decreases with the increase in temperature for temperatures between 20°C and 100°C, as shown in Figures 2.39 and 2.40 [Hsu, 1984]. A plot of device lifetime as a function of temperature and substrate current is shown in Figure 2.41 [Hsu, 1984]. A calculation of energy and proportionality factors in Tam's equation [1983] indicate that they are constant over the temperature range, suggesting that the physical mechanism for device degradation is the same over the temperature range 20°C to 100°C. The variation of lifetime vs substrate current for temperatures 20°C and 100°C, shown in Figure 29, indicates that hot-electron degradation is temperature-independent in this range. The increase in the ambient temperature decreases the observed ΔV_{th} shift from the channel hot- electron effect. For a given device design, Matsumoto [1981] has described this behavior:

$$\Delta V_{th} = A_{he,cf}\, N_{ox,eff}\left(1 - e^{B_{le,ex}\, N_{inj,e}}\right) \tag{162}$$

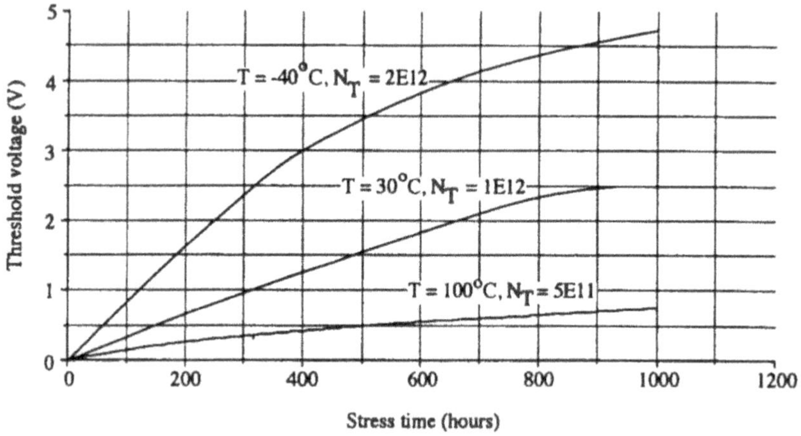

Figure 2.38 Change in threshold voltage due to hot electrons is much greater at lower temperature than at higher temperature [Matsumoto, 1981].

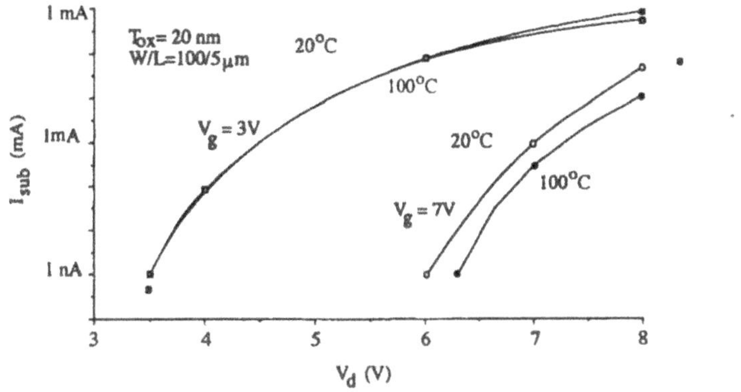

Figure 2.39 Substrate characteristics for a 5μm device at 20°C and 100°C V_g=3V and 7V, V_T = 0.9V [Hsu, 1984]. The variation in device characteristics between 20°C and 100°C indicates that degraduation due to hot electrons is almost temperature independent in this range.

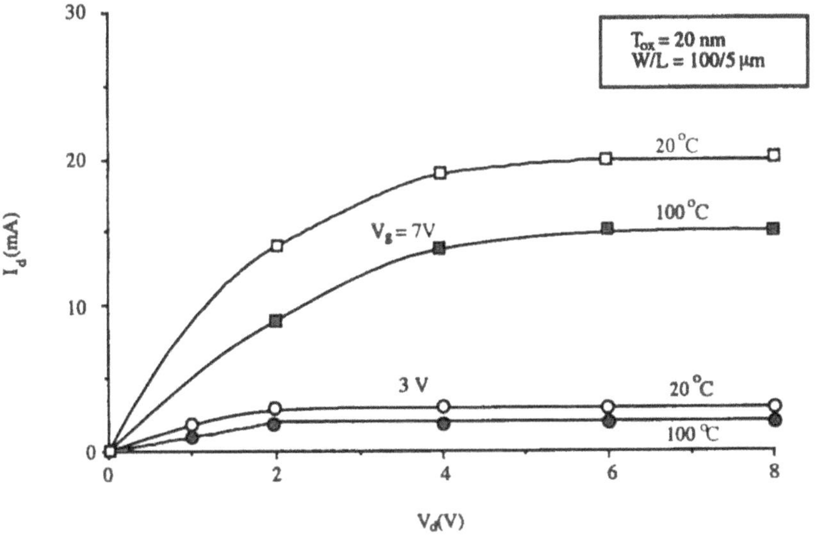

Figure 2.40 Drain characteristics for a 5μm device at 20°C and 100°C V_g = 3V and 7V, V_t = 0.9V [Hsu, 1984]

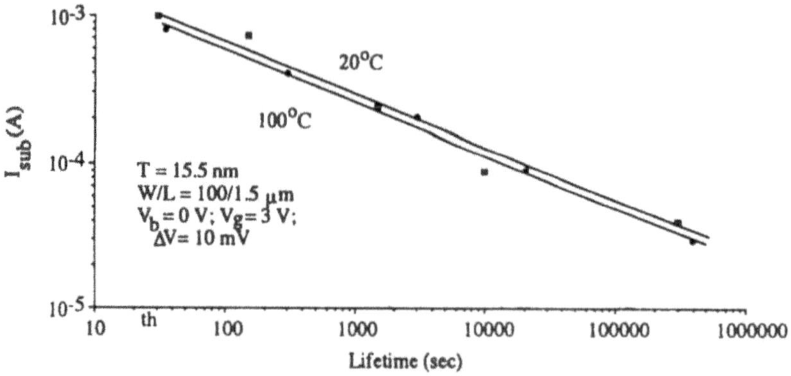

Figure 2.41 Substrate current dependence of device lifetime at 20°C and 100°C. [Hsu, 1984] The small variation in device lifetime for temperature variation between 20°C and 100°C clearly indicates that the degraduation due to hot electrons is temperature independent in this range.

where $N_{ot,eff}$ is the effective oxide-trap density, $N_{inj,e}$ is the density of injected electrons, and $A_{he,cf}$ is the hot electrons coefficient and $B_{le,ex}$ is the hot electrons exponent. From the equation, it is evident that the hot-electron threshold voltage shift increases with both the electrons injected into the oxide and the availability of traps to capture the electrons. Both these quantities decrease with increasing temperature, giving rise to reduced threshold shift due to hot electron trapping at elevated temperatures [Matsumoto, 1981].

The lucky electron model. This model ignores the overall description of the hot-carrier energy distribution and focuses on the distribution tail corresponding to those electrons with sufficient energy to overcome the potential barrier. The lucky electron model accounts for the influence of the electric field profile in silicon, the oxide field, and the lattice temperature. A free electron is injected into silicon dioxide if it reaches the interface with a component of its momentum, normal to the surface, greater than a critical value, P_c, corresponding to the barrier energy, ϕ_{BS}. The probability of tunnel injection through the top of the barrier for an electron with momentum slightly lower than $P_{pr,eme}$ is momentarily neglected. The electron can reach the interface with momentum P_c after traveling various possible trajectories. The trajectory consists of a series of free flights interrupted by various interactions that can be elastic or inelastic, during which the electron energy is modified by the electric field. The lucky electron model assumes that the most probable trajectories are those for which energy, ϕ_{BS}, is gained by the electron in the last free flight without any collisions before reaching the interface. The emission probability is thus close to the probability of the electron making such a trajectory.

The minimum path length required to gain energy before reaching the interface equals the horizontal distance between the top of the barrier and the lower edge of the conduction band. It is assumed that the emission probability for an electron that has collisions within a distance, d, is negligible, irrespective of its previous energy. According to the lucky electron model, the emission probability of an electron is thus equal to its probability of traveling a distance greater than d without a collision [Ning, 1977; Garrigues, 1981]:

$$P_{pr,eme} = P_{pr,ref}\, e^{-\frac{d_{mpl,e}}{\lambda_{mfp,l}\tanh\left(\frac{E_a}{2K_B T}\right)}} \tag{163}$$

where $P_{pr,eme}$ is the emission probability of an electron, $d_{mpl,e}$ is the minimum path length for required for an electron to attain critical energy, $\lambda_{mfp,l}$ is the electron mean free path between lattice interaction, E_a is the activation energy and $P_{pr,ref}$ is the coefficient probability of emission of an electron (≈ 2.9 @ 300^{0}K; ≈ 4.3 @77) °K, K_B is Boltzmann's constant (8.617 x 10^{-5} eV/K or 1.38 x 10^{-23} J/K).

$$d_{mpr,e} = \sqrt{\frac{2\epsilon_{sm}\epsilon_o}{eN_{a,dp}}}\left(\sqrt{\Psi_S} - \sqrt{\Psi_S - \frac{\Phi_{BS}}{e}}\right) \tag{164}$$

$N_{a,dp}$ is the concentration of acceptor doping atoms, e is the electronic charge, ϵ_0 is the free - space permittivity (8.85 x 10^{-14} F cm^{-1}), ϵ_{sm} is dielectric constant of the semiconductor (F cm^{-1}), Ψ_S is the surface potential (volts), and Φ_{BS} represents the silicon-silicon dioxide electrons, taking into account the lowering due to the Schottky effect.

Chapter 3

TEMPERATURE DEPENDENCE OF MICROELECTRONIC PACKAGE FAILURE MECHANISMS

1. TEMPERATURE DEPENDENCIES OF FAILURE MECHANISMS IN THE DIE AND DIE/SUBSTRATE ATTACH

In this section the temperature dependence of failure mechanisms in the die and the die/substrate attach are discussed including die fracture, die thermal breakdown, and die substrate adhesion fatigue. The failure mechanism by which the die or attachment will fail is a function of the material behavior. While several models characterizing the temperature dependence of each of the failure mechanisms have been discussed, each model is valid for some assumptions, which outlines its domain of applicability. All the models characterize die fracture and die/substrate adhesion fatigue with a dominant dependence on temperature cycle magnitude.

1.1 Die Fracture

The die, the substrate, the leadframe, and the case of a microelectronic package typically have different thermal expansion coefficients. For example, die are usually made of silicon, gallium arsenide, or indium phosphide, while the substrate is typically alumina, berylia, or copper, with a coefficient of thermal expansion different from the die material. Microcracks develop in the die during manufacture through crystal growth, wafer scrubbing and slicing, and die separation [Lim, 1989]. As the temperature cycle magnitude rises during temperature and power cycling, tensile stresses develop in the central portion of the die, while shear stresses develop at the edges. A pre-existing defect may develop into a crack under the influence of thermal cycling in the die. This crack may not be of critical size at the applied service stress, but may grow to critical size gradually by stable fatigue propagation. Thus, the rate of fatigue-crack propagation per cycle is determined by the cyclic change in the stress intensity factor, a measure of the stress at or around the crack tip. In some cases, these microcracks may be large enough to cause brittle failure of the die. Ultimate brittle fracture of the die can occur suddenly, without any plastic deformation, when surface cracks at the center of the die or at the edge of the die reach their critical size and propagate during thermal cycling to the critical crack size. The fracture criterion can be quantified by the size of a critical crack on the external die surface. In some cases, these cracks may be large enough to cause brittle fracture of the die. Figure 3.1 shows the cross-section of die fracture in the chip scale package. The top-layer in Figure 3.1 is the silicon chip, the middle layer is the underfill and the bottom is a FR4-substrate. A defect in the back side of the die has caused the die to fracture during thermal cycling. Typically, pre-existing defects are a manifestation of the package manufacturing process. Figures 3.2 and 3.3 show damage at the backside of a package die (observed de-encapsulating the package) caused

Figure 3.1 Chip cracking from surface flows in a chip seal package. (Courtesy of Motorola, Inc.)

by ejector pins used to remove the die from the tape after wafer saw. The flaw caused by the ejector pins is in the form of a crater at the back side of the die which propagated into a crack during thermal shock. Figure 3.3 shows the horizontal and vertical dimensions in the flaw using a Laser Scanning Microscope.

Vertical die cracks propagate under tensile stresses and horizontal die cracks propagate under shear stresses at the edges. Horizontal edge cracks, developed from die-cutting damage, may propagate from the corner of the die to active chip elements and induce device failure, or may propagate horizontally, causing the die to lift. Often, devices with such horizontal cracks will pass a full functional test when the wirebonds are still connected. Although die fracture is mainly governed by the sizes, shapes, and locations of defects in the die, voids in the attachment material or in the die-attach interface may also affect die fracture by perturbing the thermal and stress transfer mechanisms.

Westergaard-Bolger-Paris equation-based formulation. Die fracture has been described using Bolger's equations for stresses at the die-attachment interface during temperature cycling, coupled with fatigue-life estimates from Paris's power law, based on surface and edge crack sizes [Bolger, 1982[a], 1984[b], Paris 1961]. The Paris power law used in this formulation assumes that the crack is larger than the subthreshold crack size, i.e. the rate of crack growth per cycle is linearly proportional to the stress intensity factor amplitude. Paris's power law predicts the fatigue-crack propagation based on the stress intensity factor, at or near the crack tip in the plane of propagation. As the stress varies during thermomechanical cycling, K_f will proportionately vary as follows:

$$\Delta K_f = K_{f,max} - K_{f,min} \tag{1}$$

Then the rate of propagation, *da/dN*, will be given by Paris's power law:

$$\frac{da}{dN} = A_{paris} \, (\Delta K_f)^{n_{paris}} \tag{2}$$

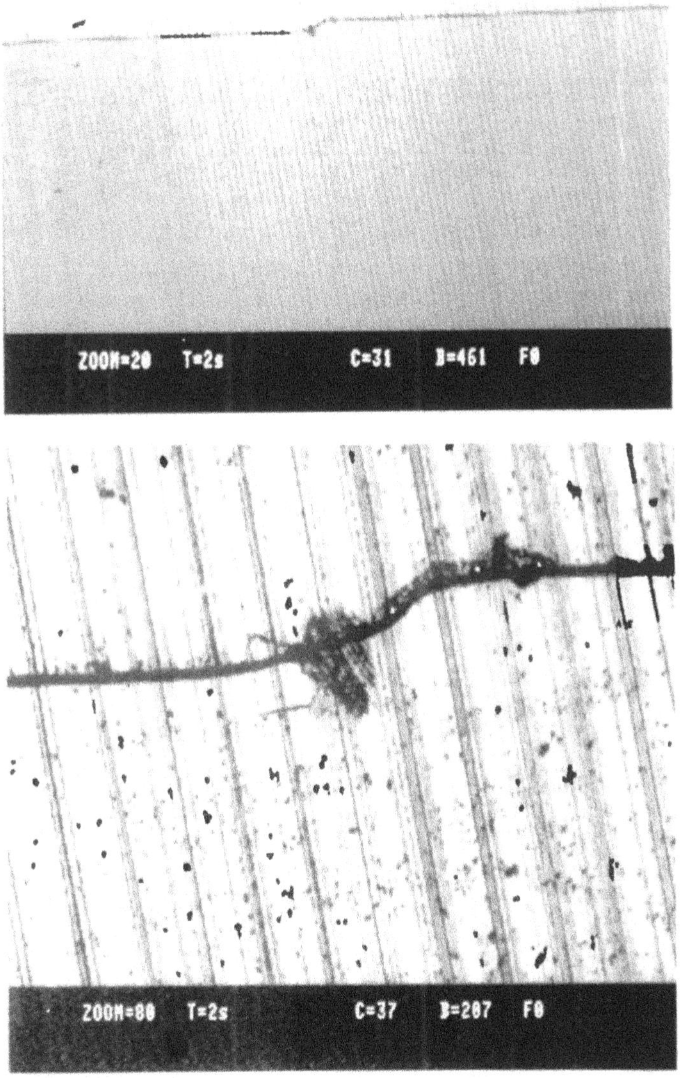

Figure 3.2 Die fracture caused by cracks initiation from an indentation on backside of die, caused by eject pins. (Courtesy of Motorola, Inc.)

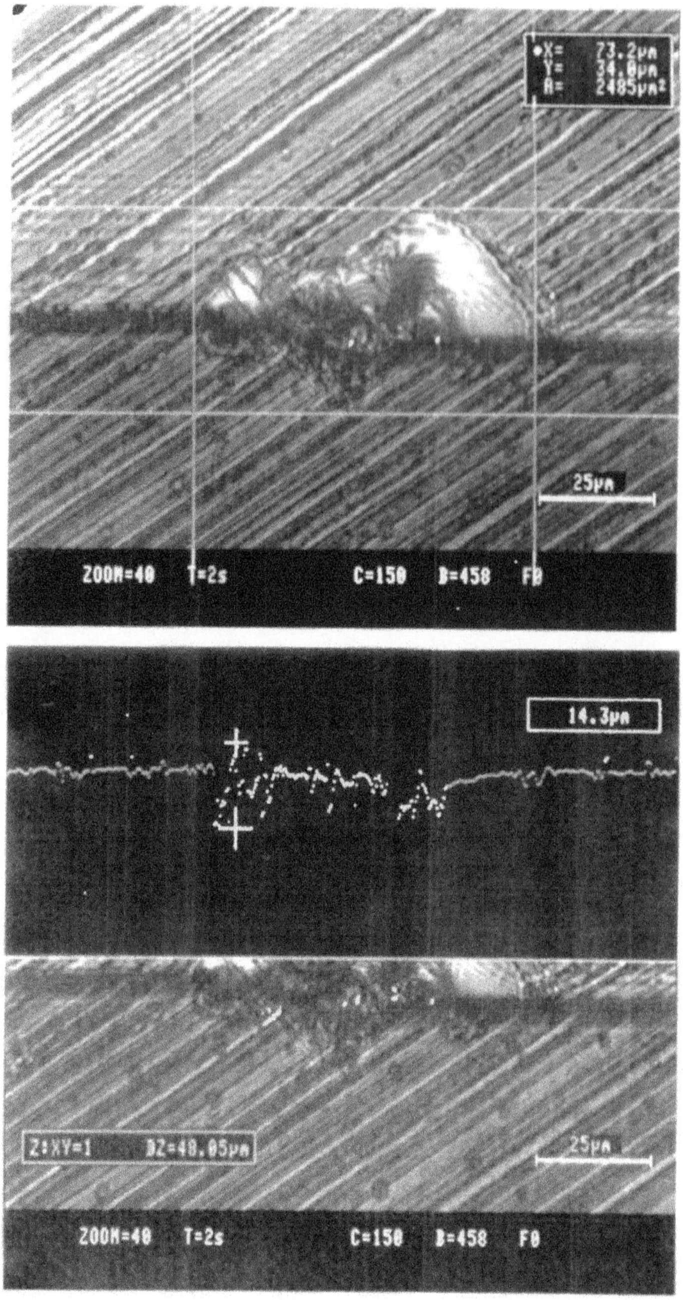

Figure 3.3 Vertical depth of indentation measured using an ESM. (Courtesy of Motorola, Inc.)

where A_{paris} is Paris's coefficient, n_{paris} is Paris's exponent, ΔK_f is the stress intensity factor amplitude, and da/dN is the crack growth per load cycle. The stress intensity factor for an infinite solid was given by Westergaard [1939] as

$$K_{f,mdI} = \sigma_{mdI}\sqrt{\pi a} \tag{3}$$

where σ_{mdI} is the mode I far-field stress, $2a$ is the crack size. $K_{f,mdI}$ is the mode I stress intensity factor, The thermomechanical stress level in the die has been investigated by various authors, including Bolger, who represented the stress level by

$$\Delta\sigma_{mdI} = 10^{-6}\left|\alpha_{sub}-\alpha_{die}\right|\Delta T\sqrt{\frac{E_{sub}E_{att}L_{die}}{X_{da}}} \tag{4}$$

where $\Delta\sigma_{mdI}$ is the thernomechanical stress amplitude, α_{sub} is the substrate CTE, α_{die} is the die CTE, ΔT is temperature cycle amplitude, E_{sub} is elastic modulus of substrate, E_{att} is the elastic modulus of die attach, L_{die} is the diagonal length of die, and X_{da} is the thickness of die attach. Substituting for the stress level and the stress intensity factor in Paris's power law gives

$$\frac{da}{dN} = A_{paris}\left(\Delta\sigma_{mdI}\sqrt{\pi a}\right)^{n_{paris}} \tag{5}$$

Integrating the above equation, the number of cycles to failure is given by

$$N_f = \frac{2}{(n_{paris,die}-2)A_{paris,die}}(\Delta\sigma_{mdI})^{n_{paris,die}}\,\pi^{\frac{n_{paris}}{2}}\left(\frac{1}{a_i^{\frac{(n_{paris,die}-2)}{2}}}-\frac{1}{a_f^{\frac{(n_{paris,die}-2)}{2}}}\right) \tag{6}$$

where $n_{paris,die}$ is Paris's exponent for die crack propagation, $A_{paris,die}$ is Paris's coefficient for die crack propagation, $\Delta\sigma_{mdI}$ is the stress amplitude in thermomechanical fatigue, a_i is the initial crack size, and a_f is the final crack length in die, and N_f is number of cycles to failure.

Improper die-to-header bonds or unsuitable packaging conditions can increase the incidence of die cracks [Tan, 1987]. Wafer backprocessing, such as lapping or thinning, can result in silicon flaws that are susceptible to fracture during temperature and power cycling [Hawkins, 1987]. Die fracture and die-adhesion fatigue are dependent primarily on the magnitude of the temperature cycle.

Suhir-Paris equation based formulation [Hu, Pecht and Dasgupta, 1991]. Die fracture has been described using Suhir's equations for stresses at the die-attachment interface during temperature cycling [Broek, 1986], coupled with fatigue-life estimates based on surface and edge crack sizes using Paris's power law, [Suhir, 1986, 1987, 1989, 1990, Paris 1961]. Suhir model assumes that the die attachment is very thin and compliant compared to the mating surfaces (die and substrate). The Paris power law used in this formulation assumes that the crack is larger than the subthreshold crack size, i.e., the rate of crack growth per cycle is linearly proportional to the stress intensity factor amplitude. The stress intensity factors for vertical die cracks, $K_{f,v\text{-}die}$, and for horizontal die cracks, $K_{f,h\text{-}die}$, are determined by

$$K_{f,v\text{-}die} = \sigma_{da}\sqrt{\pi\, a_{v\text{-}die}}\,F_g \tag{7}$$

$$K_{f,h\text{-}die} = p_{da}\sqrt{\pi\, a_{h\text{-}die}}\,F_g \tag{8}$$

where σ_{da} is static tensile stress perpendicular to crack, $a_{v\text{-}die}$ is the depth of the vertical crack on the top surface of the die, and $K_{f,\,v\text{-}die}$ is the stress intensity factor for a vertical crack, p_{da} is peeling stress along plane of crack, and $K_{f,h\text{-}die}$ is stress intensity factor for horizontal crack, F_g is the geometric correction factor expressed by equation 18 in chapter 3. The static tensile, peeling, and shear stresses at the die-attachment interface, σ_{da}, p_{da}, and τ_{da}, are

$$\sigma_{da} = \frac{\dfrac{1}{t_{die}} + 3\,(t_{die}+t_{sub})\,\dfrac{E_{die}\,t_{die}}{12\,(1-\nu_{die}^2)\,D_f}}{\dfrac{1-\nu_{die}}{E_{die}\,t_{die}} + \dfrac{1-\nu_{sub}}{E_{sub}\,t_{sub}} + \dfrac{(t_{die}+t_{sub})^2}{4\,D_f}} \left[1 - \frac{1}{\cosh(A_{da}L_{die})}\right] (\alpha_{sub} - \alpha_{die})\,\Delta T \tag{9}$$

$$p_{datt} = -\left[\frac{t_{sub}\,t_{die}^3\,E_{die}}{12\,(1-\nu_{die}^2)} - \frac{t_{die}\,t_{sub}^3\,E_{sub}}{12\,(1-\nu_{sub}^2)}\right] \left[\frac{t_{die}}{3\,G_{die}} + \frac{2\,t_{att}}{3\,G_{att}} + \frac{t_{sub}}{3\,G_{sub}}\right]^{-1} \frac{1}{2\,D_f}\,(\alpha_{sub} - \alpha_{die})\,\Delta T \tag{10}$$

$$\tau_{da} = \left[\frac{1-\nu_{sub}}{E_{sub}\,t_{sub}} + \frac{1-\nu_{die}}{E_{die}\,t_{die}} + \frac{(t_{die}+t_{sub})^2}{4\,D_f}\right]^{-\frac{1}{2}} \left[\frac{t_{die}}{3\,G_{die}} + \frac{2t_{att}}{3G_{att}} + \frac{t_{sub}}{3G_{sub}}\right]^{-\frac{1}{2}} \tanh(A_{da}L_{die})\,(\alpha_{sub} - \alpha_{die})\,\Delta T \tag{11}$$

where E, G, ν, and α are the modulus of elasticity, the shear modulus, Poisson's ratio, and the coefficients of thermal expansion (CTE), respectively, and t is the thickness [Suhir, 1986, 1987, 1989, 1990]. The subscripts *die*, *a*, and *sub* denote the die, attachment, and substrate, respectively. ΔT is the temperature change, L is the half-diagonal length of the die, and the constant A is

$$A_{da} = \sqrt{\frac{\dfrac{1-\nu_{sub}}{E_{sub}\,t_{sub}} + \dfrac{1-\nu_{die}}{E_{die}\,t_{die}} + \dfrac{(t_{die}+t_{sub})^2}{4\,D_f}}{\dfrac{t_{die}}{3\,G_{die}} + \dfrac{2t_a}{3G_a} + \dfrac{t_{sub}}{3G_{sub}}}} \tag{12}$$

The flexural rigidity, D_f, in Equations 9, 10, and 11 is defined as

$$D_f = \frac{E_{die}\, t_{die}^3}{12\,(1-\nu_{die}^2)} + \frac{E_{att}\, t_{att}^3}{12\,(1-\nu_{att}^2)} + \frac{E_{sub}\, t_{sub}^3}{12\,(1-\nu_{sub}^2)} \tag{13}$$

The principle stress at the die-attachment interface, σ_1, is given by

$$\sigma_1 = \frac{p_{da}}{2} + \sqrt{\frac{p_{da}^2}{4} + \tau_{da}^2} \tag{14}$$

where p_{da} is the peeling stress given by Equation 10, and τ_{da} is the shear stress at the die-attachment interface given by Equation 14. The die will fail in the first temperature cycle if the tensile stress or the principal stress is greater than the modulus of rupture of the die.

$$\begin{aligned} \sigma &< \sigma_{Rupture-die} \quad @\ die\ middle \\ \sigma_1 &< \sigma_{Rupture-die} \quad @\ die\ edge \end{aligned} \tag{15}$$

where σ is the tensile stress in the middle of the die (Equation 1.9), σ_l is the principal stress at the die edge (Equation 1.14), and $\sigma_{Rupture-die}$ is the modulus of rupture of the die. The die will also fracture in the first temperature cycle if the stress intensity factor of the surface or edge crack is larger than the fracture toughness of the die material.

$$K_{v,die} < K_{C,die} \tag{16}$$

$$K_{h,die} < K_{C,die} \tag{17}$$

where $K_{C,die}$ is the fracture toughness of the die, $K_{v,die}$ and $K_{h,die}$ are the stress intensity factors for vertical cracks on the top surface of the die and for horizontal cracks at the edge of the die (Equations 7 and 8). The term F_g in Equations 7 and 8 is a geometric correction factor[1]:

---

1 Sneddon [1946] showed that for a circular surface crack in an infinite three dimensional solid, the stress intensity factor is expressed by

$$K_I = \frac{2}{\pi}\,\sigma\,\sqrt{\pi\, a}$$

where a is the crack penetration depth. Based on the stress field around an ellipsoidal cavity, Irwin [1962] extended Sneddon's results to the surface flaw of 2c length along the surface, with a semi-elliptical shape penetrating to depth a below the surface, and gave the following expression of the stress intensity factor:

$$K_I = \frac{1.12}{\sqrt{Q_d}}\,\sigma\,\sqrt{\pi\, a}$$

where 12 is the correction for the front free surface, and

$$Q_d = \Phi^2 - 0.212\,(\frac{\sigma}{\sigma_Y})^2$$

where Φ is an elliptical integral of the second kind, given by

$$F_K = \frac{2.85\ [0.953 - 2.369(\frac{a}{t_d}) + 2.74\tan(\frac{a}{t_d})]}{3 + \frac{a^2}{c^2}} \tag{18}$$

where t_d is the thickness of the die, a is the crack depth, and c is the half-length of the crack. The phenomenon of fatigue-crack propagation under environmental temperature cycles and operational power on-off cycles can be described by Paris' power law. By integrating Paris' law, the number of cycles to failure can be estimated. $N_{f(v\text{-}die)}$ and $N_{f(h\text{-}die)}$ are the calculated number of cycles to failure for vertical and horizontal crack propagation in the die, and are determined from the following equations:

$$\Phi = \int_0^{\frac{\pi}{2}} (1 - \frac{c^2 - a^2}{c^2} \sin^2\phi)^{1/2}\, d\phi$$

For a brittle solid such as a semiconductor, $Q=\Phi^2$. Values for Φ can be found in mathematical tables or in graphs of elliptic functions. When a=c, Q=2.4, and when a=0.4c, Q=1.41. Taking the first two or three terms of a series expression for Φ gives the following approximation [Broek 1986]:

$$\Phi = \frac{3\pi}{8} + \frac{\pi}{8}\frac{a^2}{c^2}$$

or

$$\Phi = \frac{\pi}{2}\left\{1 - \frac{1}{4}\frac{c^2 - a^2}{c^2} - \frac{3}{64}\left(\frac{c^2 - a^2}{c^2}\right)^2\right\}$$

To account for the effect of back free-surface in the thickness direction, Kobayashi et al. [1965] introduced a M_K factor, and obtained the following expression for the stress intensity factor:

$$K_I = \frac{1.12\ M_K}{\sqrt{Q_d}}\ \sigma\sqrt{\pi a}$$

where

$$M_K = 0.953 - 2.369(\frac{a}{t_b}) + 2.74\tan(\frac{a}{t_b})$$

and t_b is the thickness of the cracked body. For a surface crack on the top or bottom of a die, the stress intensity factor can be estimated by substituting for M_K and Q:

$$K_I = \frac{2.85\ [0.953 - 2.369(\frac{a}{t_d}) + 2.74\tan(\frac{a}{t_d})]}{3 + \frac{a^2}{c^2}}\ \sigma\sqrt{\pi a}$$

$$N_{f,v} = \frac{1}{(1-\frac{n_{paris,die}}{2}) A_{paris,die}^{n_{paris,die}} \sigma^{n_{paris,die}} \pi^{\frac{n_{paris,die}}{2}}} [a_{f,die}^{1-\frac{n_{paris,die}}{2}} - a_{i,die}^{1-\frac{n_{paris,die}}{2}}] \quad (19)$$

$$N_{f,h} = \frac{1}{(1-\frac{n_{paris,die}}{2}) A_{paris,die}^{n_{paris,die}} p^{n_{paris,die}} \pi^{\frac{n_{paris,die}}{2}}} [a_{f,die}^{1-\frac{n_{paris,die}}{2}} - a_{i,die}^{1-\frac{n_{paris,die}}{2}}] \quad (20)$$

where $a_{i,die}$ is the observed or measured initial crack size, $a_{f,die}$ is the final crack depth or critical crack size, $A_{paris,die}$ and $n_{paris,die}$ are fatigue properties of the die, and σ and p are the cyclic tensile and peeling stresses.

1.2 Die Thermal Breakdown

The thermal performance of a package depends on the geometry of the assembly and the bulk thermal conductivities of the die, attachment, and substrate materials. However, in many types of packages, the size, shape, and total area of the voids play important roles in determining the package thermal performance [Mahalingam, 1989]. In particular, increased void area in the attachment leads to a dramatic increase in the thermal resistance and creates a large temperature gradient in the device. With low-power active elements, small concentrations of random voids have little effect on the peak junction temperature. When a relatively large contiguous void is present, the heat transfer path has to circumvent the void, creating a large temperature gradient in the die and severely degrading the package's thermal performance. If a large void is instead subdivided into many smaller voids, the perturbation of heat flow is lower, with a smaller temperature gradient. In summary, for a given void area, many randomly distributed small voids cause a smaller increase in thermal resistance than a few large, coalesced voids. Kessel [1983] benefit of a small controlled number of voids formed due to solvents in the die-attach adhesive.

The phenomenon of thermal breakdown can be analyzed through a thermal conduction model. In this model, the effective thermal resistance of the die-attach assembly depends on the geometry and thermal conductivities of the materials the thermal boundary conditions. The failure criterion for die failure due to increased thermal conductivity is based on the comparison of the allowable junction temperature and the calculated junction temperature, which depend on power dissipation, effective thermal resistance, and ambient temperature. Die thermal breakdown occurs when the calculated junction temperature is greater than the allowable junction temperature of the device:

$$T_{junc} > T_{junc,allow} \quad (21)$$

where $T_{junc,allow}$ is the allowable junction temperature. T_{junc} is the effective junction temperature calculated from

$$T_{junc} = T_{amb,max} + Q_d \sum_{i=1}^{3} R_{th,i} \quad (22)$$

where $T_{amb,max}$ is the mean value of the maximum ambient temperature, and Q_d is the power dissipated by the devices. The effective thermal resistance of each layer, $R_{th,i}$, (die, attachment, and substrate) is determined from

$$R_{th,i} = \frac{1}{2K_i(l_i - w_i)}\left[\ln\frac{l_i(w_i+2t_i)}{w_i(l_i+2t_i)}\right] \tag{23}$$

where l_i, w_i, t_i are the lengths, widths, and thicknesses of each layer, and K_i is the thermal conductivity of each layer.

1.3 Die and Substrate Adhesion Fatigue

In a typical power or environmental temperature cycle, the die, the die attach, the die substrate, and the package experience temperature and temperature gradient differences. Because the die and the substrate, also have different coefficients of thermal expansion, the bond between them can experience fatigue failure.

Voids are common defects in the attachment layer. Voids can form in the attachment from melting anomalies associated with oxides or organic films on the bonding surfaces, trapped air in the attachment, local non-wetting, outgassing, and attachment shrinkage during solidification. Insufficient plating, improper storage, lack of cleaning, or even diffusion of oxidation-prone elements from an underlying layer can generate voids during attachment melting [Mittal et al., 1984]. In other instances, solder dewetting results in large voids, especially when poorly solderable underlying metal or excess soldering time results in an intermetallic compound not readily wetted by the solder [Estes, 1991, Okikawa et al., 1984]. Even under ideal production conditions, voids are often produced by solvent evaporation or normal outgassing during cooling of organic adhesives.

Stress concentrations near voids can cause local stresses more than three times the nominal stress, depending on the shape of the voids. The presence of edge voids in the die attach induces high longitudinal stresses during power and environmental temperature cycling. These voids can act as microcracks, which propagate during power cycling, resulting in debonding of the die from the substrate or of the substrate from the case. Voids are responsible for weak adhesion, die lifting, increased thermal resistance, and poor power-cycling performance. Solder, and some epoxies, have also been shown to crack and detach during temperature cycling. Edge voids are most likely to produce die cracking, due to high longitudinal stresses; the compressive nature of the stress near a center void greatly reduces the possibility for the crack to propagate. Die-cracking statistics for two samples of equal size with edge and center voids subject to ten cycles of thermal shock are summarized by Chiang [1984]. Devices with center voids show no cracks, while devices with edge voids show nearly a 50% failure rate due to die cracking.

A large void may reduce the thermal performance of the device by creating a large temperature gradient [Mahalingham, 1984], but in general, small concentrations of random voids have little effect on the peak junction temperature. However, when a relatively large contiguous void is present, the heat must flow around the void, creating a large temperature gradient in the silicon. Again, if a large void is instead broken up into many smaller voids, the perturbation to heat flow is reduced.

Bolger-Coffin-Manson equation-based formulation. Attachment fracture for ductile materials has been described using Bolger's equations, for stresses at the die-attachment interface during temperature cycling, coupled with fatigue-life estimates from the Coffin-Manson power relation. Bolger model used in this formulation assumes that the mating surface are very rigid. Estimates given by the Bolger model are typically very conservative [Dasgupta, 1993]. The number of cycles to failure of the die attach as a result of plastic strain cycling is:

$$N_f = 0.5\left(\Delta\gamma_{p,att}/\epsilon_{cf}\right)^{1/c_{cf}} \tag{24}$$

where $\Delta\gamma_{p,att}$ is the plastic straining amplitude in die attach, C_{cf} is a fatigue ductility exponent (Coffin-Manson), ϵ_{cf} is the fatigue ductility coefficient (Coffin-Manson) and N_f is the number of cycles to failure, $\gamma_{p,att}$ is the plastic strain amplitude given by [Smith, 1963, Manson, 1964, Coffin, 1954, Tavernelli, 1959]. Then,

$$\gamma_{p,att} = \frac{L_{die}|\alpha_{sub} - \alpha_{die}|\Delta T}{x_{da}} \tag{25}$$

where x_{da} is the thickness of die attach, α_{sub} is the CTE substrate, α_{die} is the CTE of the die, L_{die} is the hay diagonal length of die. The fatigue failure of the substrate attach is similar to that of the die attach, except that the dimensions of the materials are different. The number of cycles to failure of the fatigued substrate attach is given by the Coffin-Manson relation as

$$N_f = 0.5\left[\frac{L_s(\alpha_{case} - \alpha_{sub})\Delta T}{\epsilon_{cf}x_{sa}}\right]^{\frac{1}{C_{cf}}} \tag{26}$$

where x_{sa} is the thickness of substrate attach, L_s diagonal length of substrate, ϵ_{cf}, C_{cf} the same as in equation 24, α_{case} is CTE of case material, and α_{sub} is CTE of substrate material.

Suhir-(Coffin-Manson/Paris) based formulation [Hu, Pecht and Dasgupta, 1993]. For brittle attach materials: Attachment fracture for brittle attach materials has been described [Broek, 1986] using Suhir's equations for stresses at the die-attachment interface during temperature cycling, coupled with fatigue-life estimates based on surface and edge crack sizes using Paris's power law, [Suhir, 1986, 1987, 1989, 1990, Paris, 1961]. Suhir model assumes that the die attachment is very thin and compliant compared to the mating surfaces (die and substrate). The Paris power law used in this formulation assumes that the crack is larger than the subthreshold crack size, i.e., the rate of crack growth per cycle is linearly proportional to the stress intensity factor amplitude. Brittle attachment material may also cause die fracture when stresses generated by thermal mismatches between the die and the substrate cause the voids at the edge of the attachment to propagate into the die. The stress intensity factors for horizontal attach cracks, $K_{f,hatt}$, are determined by

$$K_{f,hatt} = p_{att}\sqrt{\pi a_{hatt}}\,F_g \tag{27}$$

where F_g is a geometric correction factor (Equation 18, Chapter 3), $2a_{hatt}$ is the crack size in attachment, p_{att} is the peeling stress in attachment (Equation 10, Chapter 3). The attach will fail in the first temperature cycle if the stress intensity factor for horizontal cracks in the attachment is greater than the critical stress intensity factor for the attachment material.

$$K_{hatt} > K_{c,att} \tag{28}$$

where $K_{c,att}$ is the fracture toughness of die/substrate attach, and K_{hatt} is the intensity factor to horizontal cracks in die/substrate attach.

The phenomenon of fatigue-crack propagation for brittle attachment materials can be described by Paris' power law. The number of cycles to failure in temperature cycling is given by

$$N_{f,hatt} = \frac{1}{\left(1 - \frac{n_{paris,att}}{2}\right) A_{paris,att}^{n_{paris,att}} P_{da}^{n_{paris,att}} \pi^{\frac{n_{paris,att}}{2}}}\left[a_{f,att}^{1 - \frac{n_{paris,att}}{2}} - a_{i,att}^{1 - \frac{n_{paris,att}}{2}}\right] \tag{29}$$

where $a_{i,att}$ is the initial attach crack size, and $a_{f,att}$ is the final attach crack depth. $A_{paris,att}$ and $n_{paris,att}$ are fatigue properties of the brittle attachment material, and p_{da} is the cyclic peeling stress determined in Equation 10.

For ductile attach materials. Attachment failure for ductile attach materials has been described by Broek [1986] using Suhir's equations for stresses at the die-attachment interface during temperature cycling, coupled with fatigue-life estimates based on principal and von Mises' stress cycle magnitudes using the Coffin-Manson relation [Suhir, 1986, 1987, 1989, 1990]. The attach will fail in the first temperature cycle if the principal and von Mises stresses in the attachment are greater than the ultimate tensile and shear strength for the attachment material.

$$\begin{aligned} \sigma_{1,att} &> \sigma_{ult,att} \\ \tau_{m,att} &> \tau_{ult,att} \end{aligned} \tag{30}$$

where $\sigma_{ult,att}$ and $\tau_{ult,att}$ are the tensile strength and shear strength of the attachment material. $\sigma_{1,att}$ and $\tau_{m,att}$ are the local principal stress and the local von Mises' stress:

$$\begin{aligned} \sigma_{1,att} &= \frac{3p_{da}}{2} + \sqrt{\left(\frac{3p_{da}}{2}\right)^2 + (4\tau_{da})^2} \\ \tau_{m,att} &= \sqrt{(3p_{da})^2 + 3(4\tau_{da})^2} \end{aligned} \tag{31}$$

where p_{da} and τ_{da} are the peeling and shear stresses given by Equations 10 and 11. The number of cycles to failure of the ductile attachment material due to tensile fatigue and shear fatigue are determined from

$$\begin{aligned} N_{f,att} &= C_{cf,att1} (\sigma_{1,att})^{m_{cf,att1}} \\ N_{f,att} &= C_{cf,att2} (\tau_{m,att})^{m_{cf,att2}} \end{aligned} \tag{32}$$

where $C_{cf,att1}$ and $C_{cf,att2}$ are the Coffin-Manson fatigue coefficients for attachment. This Coffin-Manson fatigue coefficients for attachment indicate tensile and shear, respectively. $m_{cf,att1}$, and $m_{cf,att2}$ are the Coffin Manson exponent for attachment. The Coffin Manson exponent for attachment indicate tensile and shear, respectively.

2. TEMPERATURE DEPENDENCIES OF FAILURE MECHANISMS IN FIRST-LEVEL INTERCONNECTIONS

In this section, the temperature dependence of failure mechanisms in first-level interconnects in microelectronic devices are examined. The interconnects include wirebonds, tape-automated bonds (TAB), and flip-chip bonds. The failure mechanisms discussed for wirebond interconnects include wire fatigue and wirebond fatigue (includes excessive intermetallics). The failure mechanisms discussed for TAB include lead fracture, thermally activated interdiffusion-induced bond fracture, thermally activated solder-joint fatigue, and corrosion. The failure mechanism discussed for flip-chip bonds include flip-chip solder low cycle fatigue. While various models have been proposed to characterize the temperature dependence of these failure mechanisms, most of the mechanisms with the exception of thermally activated solder-joint fatigue and corrosion have a dominant dependence on temperature cycle magnitude. Assumptions stating the domain of applicability of each of the models have also been discussed.

2.1 Wirebonded Interconnections

Wire fatigue. In wirebond interconnections between wire, bond pad and substrate as a result of differential thermal expansion [Philosky, 1973, Pecht 1989]. Wirebond fatigue failures are accentuated by bonding parameter drifts which cause underbonding of the bonding wire. Figure 3.4a shows an underbonded wedge bond caused by insufficient bond energy either due to bonding force or bonding time drift. Underbonded wedge bonds when subjected to accelerated tests such as thermal cycling would typically fail due to wirebond lifts (Figure 3.4a). Figure 3.4b shows the normal wedge bond shape. Wirebond fatigue failures may also be accentuated by overbonding of thermocompression bonds. Overbonding typically results in the pad metallization being squeezed out from under the bond or chip craters. Figure 3.5 shows an overbonded gold ball bond. Thermal fatigue failures of overbonded wires may also result in chip craters. Further, improper adhesion of the pad metallization to the substrate may cause pad peel from the substrate after thermal cycling. Figure 3.6 shows copper metallization peel from a NdTi substrate after a few thermal shock cycles.

The wires used to connect die bond pads to leads, or die bond pads to other die bond pads in the case of hybrid packages, can fail due to cyclic temperature changes resulting from repeated flexing of the wire. The most prevalent failure site is the heel of the wire [Gaffeny, 1968]. Villela [1970], Ravi [1972], and Phillips [1974] conducted extensive studies of wirebond failures and found failures to be due to the differential in the coefficients of thermal expansion of the wire and the package as the device heats and cools during temperature and power cycling. Failures in small-diameter wires may be inhibited by high loops, though the small dimensions of present-day packages impose stringent requirements on the loop dimensions, making this solution largely impractical. Such a solution is only helpful for large-diameter wires, because wire stiffness prevents easy flexing. Wire fatigue failure may be accentuated by defects during the bonding process. Figure 3.7a shows a wire bonded to a thick film. Improper bonding parameters including bonding head force or bond time has caused a severe wire wire neckdown at the wedge bond. This wirebond morphology is more susceptible to thermal fatigue failure at the weakened bond heel. Figure 3.7b shows cracks in the heel region developed after a few thermal cycles. Thermocompression ball bonds typically fail above the ball during thermal fatigue. Thermal fatigue failures in thermocompression bonds are further accentuated due to wire neckdown caused by improper bonding parameters or damaged capillary. Figure 3.5 shows a neckdown defect in the bonding wire above the thermocompression ball bond.

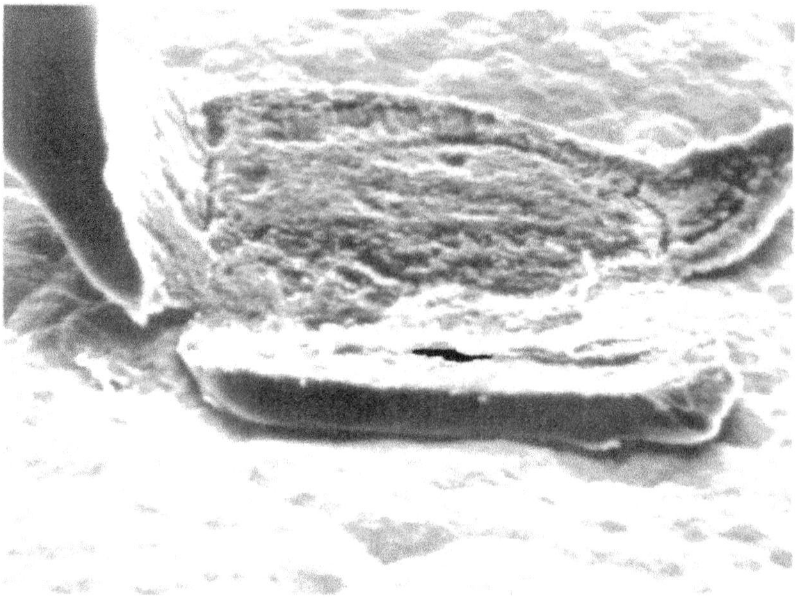

Figure 3.4a SEM micrograph - wirebond to thick film, overbonded. (Courtesy of Motorola, Inc.)

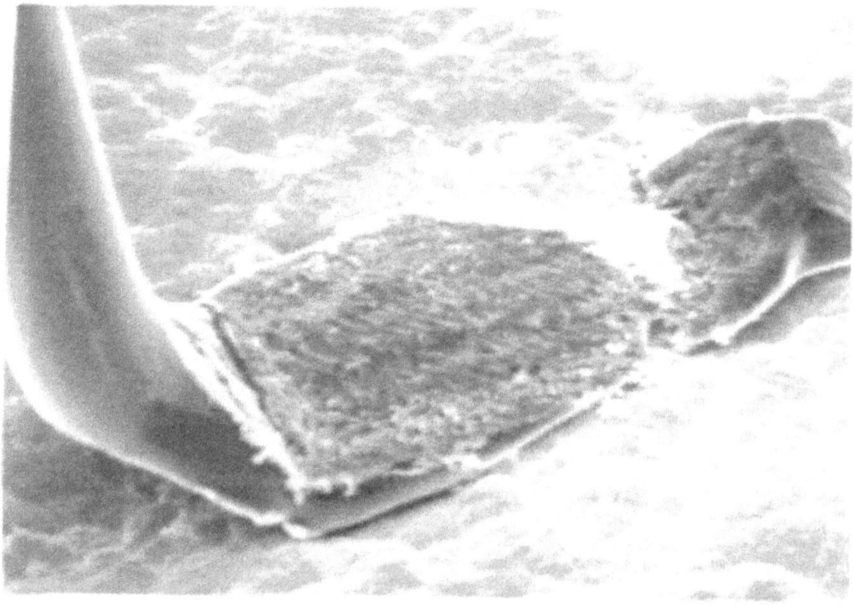

Figure 3.4b SEM micrograph - wirebond heel crack, overbonded. (Courtesy of Motorola, Inc.)

Figure 3.5a SEM micrograph - normal aluminum wirebond. (Courtesy of Motorola, Inc.)

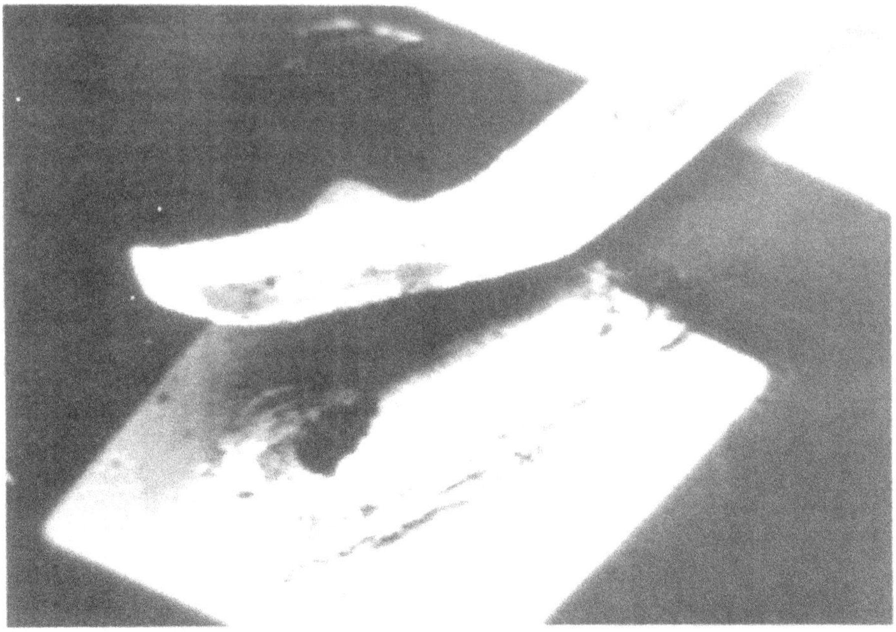

Figure 3.5b SEM micrograph - lifted wirebond, underbonded. (Courtesy of Motorola, Inc.)

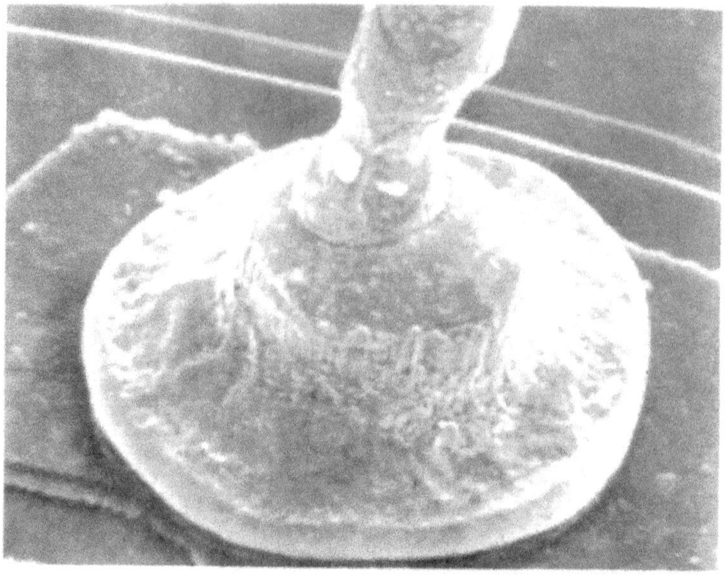

Figure 3.6 SEM micrograph - neckdown of overbonded gold ball bond. (Courtesy of Motorola, Inc.)

The stress in a wirebond interconnect has been modeled as a function of wire orientation. The cycles to failure estimate, $N_{f,wb}$ is

$$N_{f,wb} = A_{cf,wb}(\Delta\epsilon_f)^{n_{cf,wb}} \tag{33}$$

where

$$\Delta\epsilon_f = \frac{r_{wr}}{\rho_{o,wb}}\left[\frac{Cos^{-1}((Cos\lambda_{o,wb})(1 - (\alpha_{wr} - \alpha_{sub})\Delta T))}{\lambda_{o,wb}} - 1\right] \tag{34}$$

where r_{wr} is the wire radius, $\rho_{o,wb}$ is the initial radius of curvature of the wire, $\lambda_{0,wb}$ is the angle of the wire with the substrate, α_{wr} is CTE of the wire, α_{sub} is the CTE of the substrate, $\Delta\epsilon_f$ is the strain amplitude. The time to failure of a wire undergoing a cyclic temperature is thus dependent on the temperature-cycle magnitude, the geometry of the wire, and its material properties, including the CTE. However, wire flexing is, for the most part, directly dependent on the magnitude of the temperature change, and is steady-state temperature independent.

The wirebond interconnect has also been modeled as a function of the aspect ratio of a beam under pure bending due to thermal expansion. The variation of bending stress at the heel of the wire due to thermal expansion during temperature cycles (ΔT) is given by

$$\Delta\sigma_{wr} = 6E_{wr}\frac{r_{wr}}{D_{wr}}\left(\frac{L_{wr}}{D_{wr}} - 1\right)^{\frac{1}{2}}\left(2\alpha_{sub} + \frac{\alpha_{sub} - \alpha_{wr}}{1 - \frac{D_{wr}}{L_{wr}}}\right)\Delta T \tag{35}$$

where $2L$ is the wire length, $2D$ is the wire span, E_{wr} is the elastic modulus of the wire, α_{wr} and α_{sub} are the CTEs of the wire and the substrate materials, respectively, and ΔT is the temperature change encountered during operation [Hu, Pecht, Dasgupta, 1991]. The number of cycles to failure in flexure, $N_{f,wb}$ is related to the stress change in fatigue (Basquin's relation) and is

$$N_{f,wb} = C_{cf,wr}(\Delta\sigma_{wr})^{-m_{cf,wr}} \tag{36}$$

where $C_{cf,wr}$ is the Coffin-Manson coefficient for wirebond, E_{wr} is the elastic modulus wire, r_{wr} is radius of wire, $2D_{wr}$ is the span of wirebond, $2L_{wr}$ is the length of wire, α_{wr} is CTE of wire, α_{sub} is CTE of substrate, and $m_{cf,wr}$ is Coffin-Manson exponent for wirebond.

Thallium has been identified as a major source of wire flexure and contamination-induced wirebond failures [McDonald, 1971, 1973]. Thallium can be transferred to gold wires from gold-plated leadframes during crescent-bond breakoff [James, 1980]. It diffuses rapidly during bond formation and concentrates over grain boundaries above the neck of the ball, where it forms a low-melting eutectic. The forces and temperature applied during plastic encapsulation, or during temperature cycling in operational life, lead to wire breaks and device failures [James, 1980]. Thallium and lead can cause premature wirebond failures during burn-in by accelerating cracks or Kirkendall-like void formation under the bond [Wakabayashi, 1982, Evans, 1984].

Axial fatigue during temperature and power cycling can also contribute to wire failure, especially in plastic-encapsulated packages in which the wire is encapsulated by a material with a higher coefficient of thermal expansion than the wire. This can result in the partial neckdown and eventual breakage of the wire. However, wires commonly used (i.e., gold for smaller packages and aluminum for larger packages that require high current density) are ductile enough to absorb large deformations. The number of cycles to failure is typically over 200,000 power cycles for 0.002-in.-diameter aluminum/1% silicon ultrasonic bonds with loop heights greater than 25% of the bond-to-bond spacing, for temperatures ranging from 38° C to 170° C [Harman, 1974]. The number of cycles to failure was around 18,000 cycles for 0.008-in.-diameter 99.99% pure aluminum bonds on a 2N4863 power transistor subjected to temperatures from 25° C to 125° C [Harman, 1974].

Wirebond fatigue. A wirebond subjected to temperature change experiences shear stresses between the wire, the bond pad, and the substrate as a result of differential thermal expansion [Philosky, 1973, Pecht 1989]. The number of cycles to failure, $N_{f,wb}$, as a result of the shear between the bondpad and the substrate due to a temperature change, ΔT, is typically given by the closed form fatigue relation:

$$N_{f,wb} = C_{cf,bp}\,\tau_{max}^{\;-m_{cf,bp}} \tag{37}$$

Using the shear lag theory, assuming the bond pad to be a thin interlayer between the wire and the substrate, ignoring the bending deformation in the thickness direction of the bond pad, and equating the in-plane thermal deformation of the wire and the substrate, the maximum value of shear stress between the bond pad and substrate can be estimated by the relation

$$\Delta\tau_{max} = C_{intl,wr}\,\Delta T \tag{38}$$

and

$$C_{intl,wr} = \left(\frac{G_{bp}}{t_{bp}Z_{wb}}\right)\left((\alpha_{wr}-\alpha_{sub}) - \frac{(\alpha_{sub}-\alpha_{bp})}{\left(1+\dfrac{E_{sub}A_{cr,sub}}{\dfrac{E_{bp}A_{cr,bp}}{(1-\nu_{sub})}}\right)}\right) \tag{39}$$

where G_{bp} is the shear modulus for bondpad, t_{bp} is the thickness of bond pad, $C_{intl,wr}$ is the intermediate term, α_{bp} is CTE of bond pad, and Z is given by Equation 43.

The life prediction estimate for the number of cycles to failure as a result of shear between the wire and the bondpad under thermal cycling, $N_{f,wb}$, can be empirically related to the shear stress amplitude by the following Coffin-Manson relation:

$$N_{f,wb} = C_{cf,bp}(\Delta\tau_{max,wr})^{-m_{cf,bp}} \tag{40}$$

Assuming that the wire and substrate near the bonding layer are subjected to longitudinal normal stress, the normal traction in the wire and substrate near the bondline can be determined by integrating the equations of equilibrium at the wire substrate interface. The shear stress under thermal expansion mismatch is then given by

$$N_{f,wb} = C_{cf,bp}(\Delta\tau_{max,wr-sub})^{-m_{cf,bp}} \tag{41}$$

where

$$\Delta\tau_{max,wr} = \left(\frac{r_{wr}^2}{4Z_{wb}^2 A_{cr,wr}{}^2}\left(\frac{Cosh(z_{wb}\,x_w)}{Cosh(z_{wb}\,l_w)} - 1\right)^2 + \frac{Sinh^2(Z_{wb}\,x_w)}{Cosh^2(Z_{wb}\,l_w)}\right)^{\frac{1}{2}} C_{intl,wr}\Delta T \tag{42}$$

and

$$\Delta\tau_{max,wr-sub} = \left(\left(\frac{w_{bp}C_{intl,wr}}{2Z_{wb}A_{cr,sub}}\left(1 - \frac{Cosh(Z_{wb}x_{sub})}{Cosh(Z_{wb}l_{sub})}\right) + \frac{(\alpha_{sub}-\alpha_{bp})}{\frac{(1-\nu_{sub})}{E_{sub}A_{sub}} + \frac{1}{(E_{bp}A_{bp})}}\right)^2 + C_{intl,wr}^2\frac{Sinh^2(Z_{wb}x_{sub})}{Cosh^2(Z_{wb}l_{sub})}\right)^{\frac{1}{2}}\Delta T \tag{43}$$

where

$$Z_{wb}^2 = \frac{G_{bp}}{t_{bp}}\left(\frac{r_{wr}}{E_{wr}A_{cr,wr}} + \frac{(1-\nu_{sub})W_{bp}}{E_{sub}A_{cr,sub}}\right) \tag{44}$$

where E_{bp} and E_{sub} are the modulus of elasticity of the pad and the substrate materials, respectively; ν_{sub} is Poisson's ratio for the substrate materials; A_{bp} is the cross-sectional area of the pad; and $A_{cr,sub}$ is the effective cross-sectional area of the substrate, equal to $b_{sub}(w_{bp}+w_{sub})/2$, where w_{bp} is the width of the bond pad; x_{wr} gives the location of the maximum stress in the wire; r_{wr} is the radius of the wire; ΔT is the temperature cycle; $C_{intl,wr}$ is defined by Equation 39.

Intermetallic brittleness and growth is enhanced by temperature cycling. Bonded together, gold and aluminum (Au and Al) intermetallics are more susceptible to flexure damage when subjected to temperature cycling than pure gold (Au) and aluminum (Al) wires. When the contacting materials are gold and aluminum, Kirkendall voids occur at the interface between the wire and the bond pad as one element diffuses out of a region faster than the other can diffuse in from the opposite side of that region. Vacancies pile up and condense to form voids, normally on the gold-rich side of the gold-to-aluminum interface.

Rates of diffusion vary with temperature and are dependent upon adjacent phases, as well as on the number of vacancies in the original metals. The compounds formed in the intermetallic during thermocompression or thermosonic bonding are often called purple plague, because of the purple appearance of the gold-aluminum compound. Five intermetallic compounds can arise: Au_5Al_2, Au_2Al, $AuAl_2$, AuAl, and Au_4Al. Prolonged exposure to high temperatures results in continued diffusion until all the gold or aluminum is consumed. The initial growth rate is indicated by a parabolic relationship:

$$t_{int} = k_{rc,int}\, t^{1/2} \tag{45}$$

where t_{int} is the intermetallic layer thickness, t is the time, and $k_{rc,int}$ is the rate constant, given:

$$k_{rc,int} = c_x\, e^{-\frac{E_a}{K_B T}}, \quad \textit{for } T > 150°C \tag{46}$$

where c_x is a constant, E_a is the activation energy for layer growth, and T is the steady-state temperature. The value of $k_{rc,int}$ changes for each intermetallic phase. Figure 3.8 shows that the formation of significant amounts of purple plague can take years at 150°C and below.

Intermetallic compounds are mechanically strong, brittle, and electrically conductive. Temperature and power cycling can cause failure as a result of differential thermal expansion between the intermetallic and the surrounding metal, and can reduce mechanical bond strength as a result of the voiding of the surrounding metal, which usually accompanies intermetallic formation. Gold-aluminum intermetallics are stronger than pure metals and their growth is enhanced by increased temperature and temperature cycling [Philosky, 1973].

Newsome [1976] conducted experiments on the effect of temperature on the intermetallic formation in two thick-film systems (Owens-Illinois and EMCA 212B) and in conventional

Figure 3.7a SEM micrograph - gold ball bond, overbonded. (Courtesy of Motorola, Inc.)

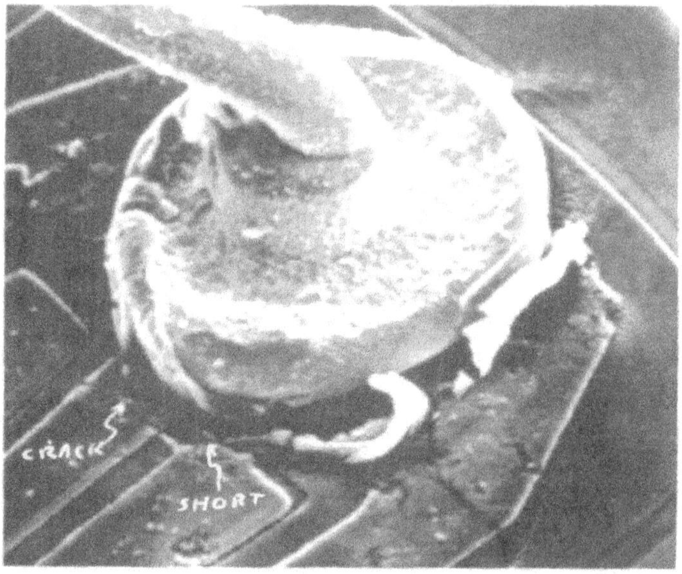

Figure 3.7b SEM micrograph - gold overbond, magnified. (Courtesy of Motorola, Inc.)

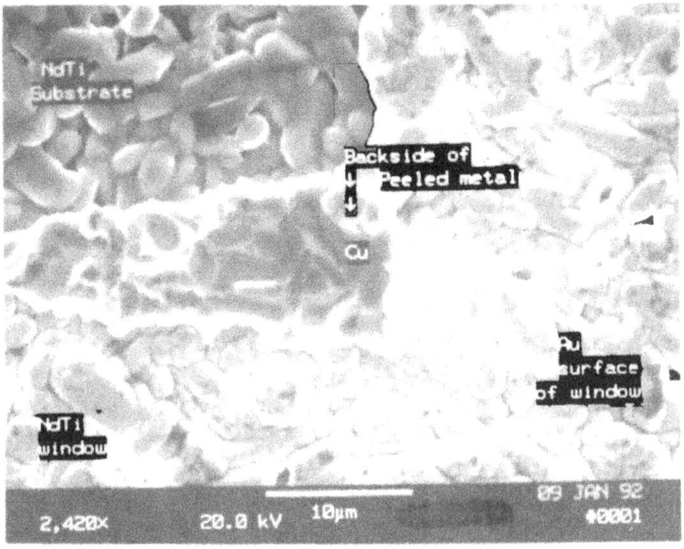

Figure 3.8 De-adhesion of a copper lead from a Neodymium-titanate substrate. (Courtesy of Motorola, Inc.)

nichrome, nickel, and gold thin-film systems, conditioning them at 25, 75, 125, 150, 160, 175, 200°C for 168 hours in an air circulation oven. The data from energy-dispersive X-ray spectrometer (EDS) analysis, shown in Figure 3.9, indicates that the purple phase of the intermetallic did not appear until 175°C, even though a significant thickness of Au_4Al was formed in the EMCA 212B at 75°C. The change in bond resistance vs temperature is shown in Figure 3.10; Table 1 lists the observations in the various phases. The process of intermetallic growth was accompanied by volumetric expansion of the intermetallics formed at temperatures in the neighborhood of 160 to 175°C, creating severe mechanical stress cracks at the vertical interface with the gold. Kirkendall voiding and mechanical stress due to the volumetric expansion accompanying intermetallic formation was found to increase with a larger volume of gold under the bonds. Philosky [1970, 1971] listed nine different rate constants for five gold-aluminum compounds; Gerling [Gerling, 1984] published an extensive compilation of reported activation energies for various types of gold-aluminum rich couples. Aluminum-rich gold-aluminum compounds have a high melting point and are relatively stable.

Newsome [1976] observed Kirkendall voiding in the neighborhood of 150°C to 200°C (150°C for EMCA 212B and Honeywell thin-film and around 175°C to 200°C for the Owens-Illinois system). Lateral voiding at the aluminum interface appeared at around 200°C. Kirkendall voiding in present-day packages are impurity - or corrosion-driven reactions, and are temperature-independent below temperatures around 150°C [Newsome, 1976]. However, in the presence of halogenated species, this process is accelerated at a lower temperature [Khan, 1986].

	EMCA 212B							Owens-Illinois 99+							Honeywell thin film						
	25	75	125	150	160	175	200	25	75	125	150	160	175	200	25	75	125	150	160	175	200
Purple ($AuAl_2$)					■	▒	■					■	■	▒					▒	■	■
White (AuAl)						■	■						■	■						■	■
Tan (Au_2Al)																					
Tan (Au_5Al_2)			■	■	■	■	■			▒	■	■	▒	▒			■	■	■	▒	■
Tan (Au_4Al)		▒	■	■	■	■	■														

Figure 3.9 Phase identification vs temperature for EMCA 212B, Owen-Illinois 99+, and Honeywell thin film system [Newsome, 1976].

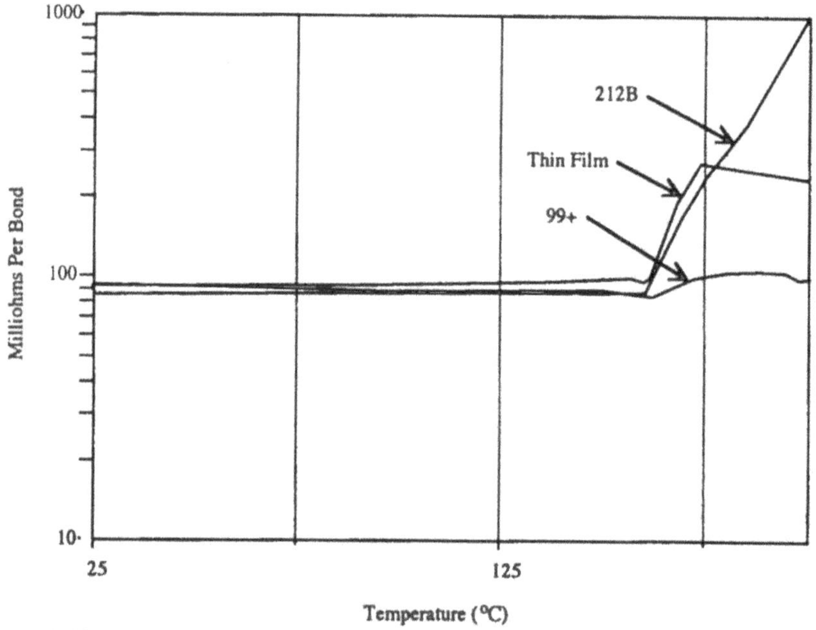

Figure 3.10 Change in bond resistance vs temperature [Newsome, 1976]

Table 1 Summary of optical and SEM findings for three gold systems [Newsome 1976]

	EMCA 212B	Owens-Illinois 99	Honeywell Thin-film
Physical characteristics	Rough, pitted gold surface. Irregular interface between gold and aluminum causes non-uniform intermetallic growth.	Rough gold surface, but not as pitted as EMCA 212B. Intermetallic growth is more uniform than for EMCA 212B.	Uniform surface; gold 1/3 the thickness of thick film conductors. Interface between gold and aluminum is uniform as is intermetallic growth.
Constituents and energy dispersive spectrometer (EDS) analysis	Thick-film ink with frit for mechanical adhesion to substrate. The frit contains silicon, bismuth, and cadmium as major constituents. The concentration of bismuth was found to be higher at the intermetallic diffusion front than in either the intermetallic or the gold. Au_4Al, Au_5Al_2, AuAl, and $AuAl_2$ intermetallic phases identified with the AuAl and $AuAl_2$ not detectable for low temperatures.	Fritless thick-film ink with copper additive for wetting the substrate. No other detectable impurities. The concentration of copper follows the same pattern as the bismuth in EMCA 212B. Au_5Al_2, AuAl, and $AuAl_2$, intermetallic phases present, but no Au_4Al. AuAl and $AuAl_2$ appear at the higher temperatures.	Gold over nickel and nichrome. No detectable impurities. Au_5Al_2, AuAl, and $AuAl_2$ intermetallic phases present in detectable quantities with AuAl and $AuAl_2$ appearing at the higher temperatures.
Temperature: 25°C	No detectable intermetallics	No detectable intermetallics	No detectable intermetallics
75°C	Significant thickness of gold-rich intermetallic; visible Au_4Al (tan)	No detectable intermetallics	No detectable intermetallics

125°C	Visible lateral voiding and annular cracks at gold interface. Visible elevation of bond above surface of gold due to volumetric expansion of intermetallic. Second gold-rich phase appears: Au_5Al_2 (tan).	First gold-rich intermetallic phase appears. Identified as same phase appearing at this temperature for EMCA 212B.	First gold-rich intermetallic phase appears. Identified as same phase appearing at this temperature for EMCA 212B.
150°C	Consumption of gold continues, with associated increase in lateral voiding, annular cracks, and bond elevation. Au_4Al narrowed to thin line by growth of Au_5Al_2 when supply of gold to intermetallic has been limited by voiding.	Intermetallic continues to consume gold.	Gold under the bond consumed if lateral voiding has not occurred. Some annular cracking and lateral voiding observed.
160°C	The gold under the bond is consumed if lateral voiding has not occurred. The width of the Au_4Al and Au_5Al_2 is approximately the same. A fine line of a third intermetallic phase appears at the aluminum interface, identified by coloration as $AuAl_2$ (purple).	Annular cracking becomes visible with elevation of the bond above the surface of the gold. $AuAl_2$ appears to be present as a fine line at the aluminum interface.	Visible elevation of the bond above the gold surface with an increase in the annular cracking. $AuAl_2$ appears to be present as a fine line at the aluminum interface.

175°C	AuAl (white) phase appears and width of $AuAl_2$ increases to a clearly defined band. Extensive annular cracking and lateral voiding cause slowdown in the growth of the Au_4Al and Au_5Al_2. Au Al reduced to a narrow band adjacent to the diffusion front on all samples.	AuAl phase appears and width of $AuAl_2$ increases. Some lateral voiding found at gold interface where initial intermetallic diffusion front was irregular. Annular cracking continues to increase.	AuAl phase appears and width of $AuAl_2$ increases. The increase in annular cracking with growth of the intermetallics continues.
200°C	Width of aluminum-rich phases increases, growing into both the bond and the gold-rich phases. Lateral voiding appears at the aluminum interface. The width of $AuAl_2$ is approximately three times that of AuAl.	Width of aluminum-rich phases increases, growing into both the bond and gold-rich phases. Lateral voiding appears at aluminum interface. Gold under bond consumed if lateral voiding has not occurred. The width of the $AuAl_2$ is approximately twice that of the AuAl and one-half the width of the $AuAl_2$ in EMCA 212B samples.	Width of aluminum-rich phases increases, growing into both the bond and gold-rich phases. The AuAl and $AuAl_2$ phases are each about one-half the width of the $AuAl_2$ phase in EMCA 212B.

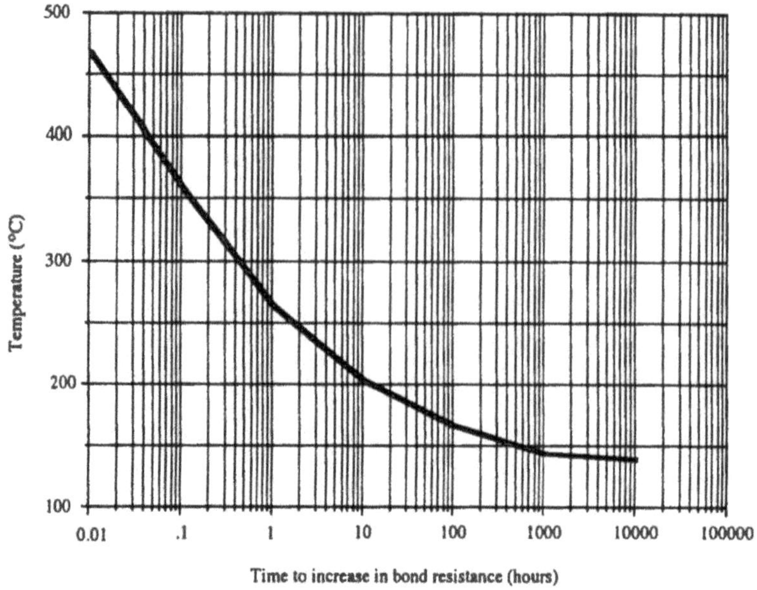

Figure 3.11 The formation of significant amounts of purple plague can take years in equipment operating range of -55 to 125°C [DM Data, 1990].

Gold wires bonded to bare copper leadframes and to copper thick-films in hybrids react, forming three intermetallic phases: Cu_3Au, AuCu, Au_3 Cu. Temperatures above 300°C accelerate the rate of intermetallic compound formation. The time to decrease bond strength decreases with an increase in temperature [Hall, 1975]. Moreover, temperature time studies on thermocompression leadframe bonds indicate a decrease in strength as a result of void formation [Pinnel, 1972, Feinstein, 1979[a], Feinstein, 1979[b]]. Figure 3.11 shows the temperature dependence of the time needed to decrease the copper-to-gold bond strength 40% below the as-bonded strength for temperatures above 300 °C. Pitt et al. [1982] studied gold thermosonic bonds to thick-film copper and found little strength degradation at 150 °C for up to 3,000 hours, and no failures at 250 °C for over 3,000 hours. This suggests that the copper-gold bond is not advisable for microelectronic devices operating at high temperatures (above 300°C). The difference in hardness and material properties, including the coefficients of thermal expansion of the intermetallic and surrounding metal, make the interface a potential site for failure. Generally, the bond strength is more a function of temperature cycling than of steady-state temperature in the range between -55 °C and 125°C, although the bond strength decreases as a function of temperature above 150°C for gold-aluminum bonds [Newsome, 1976] and above 300 °C for gold-copper bonds [Hall, 1975]. Table 2 addresses the temperature dependence of various metal systems in microelectronic packages.

2.2 Tape Automated Bonds

Tape automated bonding is another alternative for chip-to-leadframe interconnection. This method requires a tape that is usually gang-bonded to a bumped chip. Copper is used extensively as the tape metallization because of the ease of the lamination processes for copper-to-plastic carriers. Copper is readily etched and has good electrical and thermal conductivity. The two types of copper used are electrodeposited and rolled-annealed [Oswald, 1977]. The tape metallization is mounted on a plastic carrier made of polyimide, polyester, polyethersulfone (PES), or polyparabanic acid (PPA) [Lyman, 1975]. Among them, polyimide is most widely used because it can survive 365°C. Applications involving polyester are limited to only 160°C. PES and PPA are new materials, with maximum short-term temperatures of 220°C and 275°C, respectively. The tape and the plastic carrier are adhesively bonded; many kinds of adhesives have been used in three-layer tape, including polyimides, polyesters, epoxies, acrylics, and phenolic-butyrals [Tummala, 1989]. These are rated for only 20 to 30 sec. at 200°C and, therefore, limit the temperature at which the tape can be processed. Tapes are normally gold- or tin-plated to optimize inner-lead (ILB) or outer-lead bonding (OLB) operations, as well as to provide required shelf life and corrosion resistance. Gold-plated copper leads are used for T/C and S/R bonding, tin-plated copper leads for eutectic solder bonding [Speerschneider, 1989].

Lead fracture. Liljestrand [1986], while conducting temperature-cycling experiments between -55°C and 125°C on TAB on ceramic substrates with copper (Cu) thick film, glass-fibre reinforced epoxy, copper-Invar-copper (Cu-Invar-Cu) with polyimide coating, and ceramic with polyimide coating, found that the dominant mode of failure was lead fracture near the Kapton ring. Failure was initiated at this site due to the hinged action of the TAB leads at the Kapton ring and the solder joint during thermal cycling [Lau, 1989]. Lau's predictions, based on a nonlinear three-dimensional finite element method, agreed with the bond-pull tests performed by Liljestrand [1985]. The failure mechanism is enhanced by the difference in the coefficients of thermal expansion between the copper leads and the Kapton ring. The number of cycles to failure corresponded to a plastic strain of 1 to 2%, according to Hagge [1982].

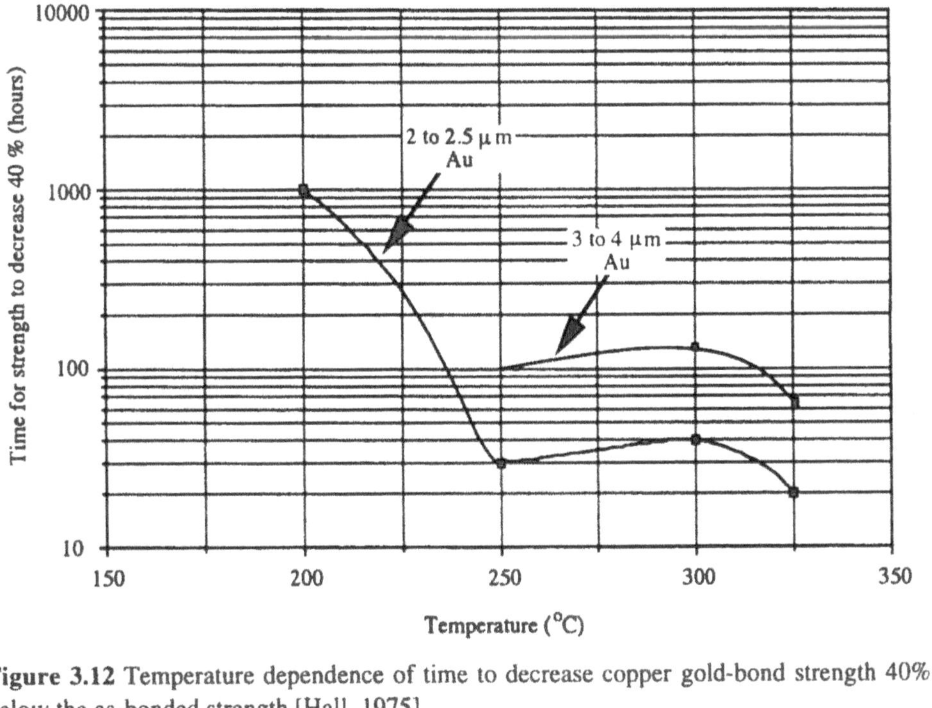

Figure 3.12 Temperature dependence of time to decrease copper gold-bond strength 40% below the as-bonded strength [Hall, 1975].

Table 2 Temperature dependence of interdiffusion in various metal systems

Metal system	Temperature of intermetallic formation	References
Au-Al interdiffusion	Independent of steady state temperature below 150°C; independent of steady state temperature above lower temperatures (T < 150°C) in presence of halogenated compounds	[Newsome, 1976], [Philosky, 1970, 1971], [Gerling, 1984], [Khan, 1986], [Pinnel, 1972], [Feinstein, 1979], [Feinstein, 1979], [Pitt, 1982] [Khan, 1986], [Villela, 1971]
Ti/Pt/Au metal system on gallium arsenide; Ti-Ga-As interdiffusion results. Ga diffuses into Ti and Pt diffuses into GaAs	250°C-275 °C	[Morgan, 1981; Davey, 1981]
Ti metallization on GaAs; TiAs formation	180°C-220 °C	[Christou, 1981]
Au-Cr interdiffusion	100°C-275 °C	[Bresse, 1985]
Al-GaAs interdiffusion	210°C-250 °C	[Kashiwagi, 1987]
	380 °C	[Kim, Kniffins, Sinclair, and Helma, 1988]
Gate-channel interdiffusion resulting in NiCr resistor degradation (MMIC device)	175°C-310 °C	[Fraser and Ogbounah, 1985]
Electromigration in Ohmic contacts; Ti oxidation and As outdiffusion	180°C-220 °C	[Christou, 1981]
Au-Al ohmic contact degradation	180°C-337 °C	[Mizugashira and Sakaguchi 1985]
Ti/W/Au	240°C-300 °C (Au interdiffusion) 425 °C (Ga outdiffusion on Au)	[Drukier and Silcox, 1979] [Allen, 1986]

Thermally activated interdiffusion-induced bond fracture. Problems caused by diffusion in TAB fall roughly into three groups [Hall, 1977]. First, a surface can be contaminated by diffusion of an underlying material to the surface through the protective top layer via grain boundaries. Typically, a base metal diffuses through a gold top layer and an oxide is formed that inhibits a good bond. Second, an interface can redistribute material to form a weakened brittle layer. This interdiffusion may cause either Kirkendall voids or intermetallic compounds, resulting in loss of strength or delamination of an interface. For example, interdiffusion can create a layer of solder depleted of tin, leaving behind a weaker lead-rich layer. Third, the basic material properties of a film may change due to diffusion.

A metal barrier layer is usually used between two easily diffusing metals; for example, nickel may be used between copper and gold. Studying the effects of gold thickness in thermocompression bonds, Feinstein found that 1 μm of gold is better than the usual thickness of 2 μm for gold plating over copper in bump and lead metallurgy [Feinstein, 1979[b]]. Hall suggested that 0.7 μm is the optimal thickness for gold plating. For many applications, a nickel barrier for gold plating over copper can be omitted without significant loss in the reliability of thermocompression bonds [Hall, 1977].

Liljestrand found that temperature cycling produced no significant reduction in the ILB strength, though the bond strengths for ILB were insignificantly changed after 1,000 cycles from -55°C to 125°C. The OLB strengths were changed by values in the neighborhood of 25g [1986].

Extended storage at 125°C for 2,000 hours caused ILB leads to fracture, while the OLBs showed reduced adhesion between the leads and substrate. Extended storage at 200°C for 2,000 hours reduced the ILB and resulted in delamination of the OLB from the substrate [Liljestrand, 1986]. The mechanisms responsible for these failures were interdiffusion between aluminum (Al) and gold (Au) across the titanium/tungsten (Ti/W) barrier on the aluminum metallization, and diffusion between copper (Cu) and gold (Au), with the subsequent formation of Kirkendall pores in the copper leads.

Harman conducted pull tests and fatigue tests on a 35-mm tape with tin-plated copper leads melt-bonded to solid gold bumps, as well as on a 11-mm tape with bare copper leads thermocompression bonded to gold-plated bumps, and found that the 35-mm tape had a pull strength of around 40g, but failed at very low vibration cycles (around 140 cycles) because of the brittle gold-tin (Au-Sn) intermetallic compound formation. The 11-mm tape withstood a pull strength of around 27g and vibration cycles in the neighborhood of 1,600 cycles. Harman also tested a number of gold-plated copper (Au-plated Cu) bumped tapes with leads bonded to chip aluminum (Al) pads, and found the fatigue life to be comparable to that of 11-mm tape. Therefore, even though the brittle intermetallic compound was not observed to degrade the bond pull strength, it caused poor reliability, since intermetallic cracks propagated rapidly under the kind of low-amplitude repeated flexing stress encountered in thermal cycling.

Corrosion. With improvements in die-passivation technology, metallization corrosion has been greatly reduced. However, the passivation may have defects, such as cracks and pinholes, which lead to corrosion of the underlying metal. Certain metals, such as TAB leads, cannot be passivated because of the assembly process; this exposes these metals to corrosion attack. Observed corrosion failures in TAB include corrosion of positive and negative metallization tracks (type A) and corrosion at the die metallization pad-bump interfaces (type B) [Padmanabhan, 1985]. Corrosion of positive tracks, caused by electrolytic attack, is predominant. Corrosion is primarily intergranular, suggesting that the grain boundaries provide paths for the migration of anions from the bond pads. Cl^- was the major anionic embrittling species and K^+ was a major cationic species. Cl^- ions migrate along the die-metallization pad-bump interfaces and reach potential corrosion sites in the metal tracks by diffusing along the grain boundaries. Corrosion leads to the formation of $Al(OH)^+$ ions at pinhole defects that offer a direct path to the underlying metal through the passivation. Cathodic corrosion occurs due to the migration of K^+ ions to the cathode, increasing pH at these sites and enhancing the subsequent formation of $KAlO_2$ or $Al(OH)_3$. While cathodic corrosion tends to be localized at passivation defect sites, anodic corrosion often extends back to the aluminum pads.

Anodic corrosion of aluminum by Cl^- is usually explained as a series of chemical or electrochemical reactions [Nguyen, 1980, Padmanabhan, 1985]. The first reaction is the dissolution of the protective oxide layer on aluminum:

$$Al(OH)_3 + Cl^- \rightarrow Al(OH)_2 + OH^- + Cl \quad (47)$$

Corrosion of aluminum occurs via the electrochemical reaction

$$Al + 4Cl^- \rightarrow AlCl_4^- + 3e^- \quad (48)$$

The transient complex $AlCl_4^-$, produced in Reaction 48, hydrolyses, producing the aggressive anion, Cl^-:

$$AlCl_4^- + 3H_2O \rightarrow Al(OH)_3 + 3H^+ + 4Cl^- \tag{49}$$

Reaction 49 provides Cl^- ions to continue the corrosion process cycle, so a small number of Cl^- ions can cause considerable corrosion damage.

Type B failure is due to anodic corrosion of the pads by Cl^- and poor adhesion at the metallization pad-bump interface. Type B failure also occurs because the galvanic cells formed by different metals are in contact. In either case, adhesion between the various metal layers in the bump structure is a key factor in the corrosion process. Corrosion resistance can be improved by restricting the entry of ionic species into the system, eliminating moisture condensation sites by improving adhesion between the various metal systems and the passivation, and controlling defects in the passivation.

2.3 Flip-chip Joints

Flip-chip joints are solder interconnections between the chip and the substrate. Flip-chip bond failure mechanisms are a function of BLM and TSM materials and structure, geometric configuration, and solder and substrate materials. The failure mechanisms are also influenced by manufacturing parameters and the mission profile of the package.

Puttlitz, while working on lead-tin and lead-indium solder joints, found that solder-joint fracture modes were a function of the joining temperature of flip-chip bonds [1990]. For lead-tin solder joints joined at a low peak reflow. temperature (approximately 340°C), fracture initiated at the BLM periphery of high neutral point distance (DNP) pads and propagated through the solder. However, the predominant failure initiation location switched to the TSM periphery if joints experienced high reflow temperatures (approximately 365°C) and/or excess post-join reflow cycles (greater than ten). Cracks propagated inward along the brittle gold-tin intermetallic/solder interface and then angled off into the bulk solder until the separation was complete.

Lead-indium solder-terminated chips joined at low peak temperatures (approximately 265°C) and a low number of reflows (less than ten) experienced failure due to a TSM separation at the Cu_7In_4 intermetallic/solder interface when stressed to thermal cyclic conditions. Chips joined at intermediate-to-high (approximately 265°C-325°C) temperatures and more than ten cycles exhibited failure at the BLM interface due to the bond instability caused by the loss of Cu_7In_4 at the BLM interface [Puttlitz, 1990].

Solder exhibits a unique material behavior, which stems from its time- and temperature-dependent stress-relaxation/creep characteristics. Steady-state dwells at the extremes of the functional cycle allow complete stress relaxation in the solder, which prevents the solder from supporting any significant stresses. All the cyclic strains due to thermal expansion mismatch of the solder joints are converted to plastic strains [Engelmaier, 1984]. In high-temperature, low-cycle fatigue, a slower strain application or dwell time introduced at the extremes reduces the number of cycles to failure because of creep. Since either slower strain application or dwell times also prolong each cycle, the lifetime decreases with decreasing thermal cycling frequency.

3. TEMPERATURE DEPENDENCIES OF FAILURE MECHANISMS IN THE PACKAGE CASE

In this section, the temperature dependence of failure mechanisms in the encapsulant of the plastic packages or the case of hermetic packages are discussed. The failure mechanisms discussed include: cracking in plastic packages, reversion or depolymerization of polymeric bonds, whisker and dendritic growth and modular case fatigue failure. While reversion or depolymerization is dependent on steady state temperature, the temperature at which this mechanism becomes active is a function of molding compound composition. Typical temperatures for this mechanism are in the neighborhood of 360°C. None of the other mechanisms has any dependence on steady state temperature — cracking in plastic packages has a dominant dependence on time dependent temperature change, whisker and dendritic growth in hermetic cases is a contamination actuated mechanism, and modular case fatigue failure has a dominant dependence on temperature cycle magnitude.

3.1 Cracking in Plastic Packages

Coefficient of thermal (CTE) mismatches drive the failure mechanisms for cracking of plastic packages. Silicon has a CTE of about 3 ppm/°C, while molding compounds typically have a CTE of about 20 ppm/°C below the glass transition temperature. Most leadframes are made of either alloy 42 (with a CTE of 4.7 ppm/°C) or copper (with a CTE of about 17 ppm/°C).

A typical process involves die attachment with adhesive cure at 270°C (polyimide adhesive) or 170°C (epoxy adhesive), followed by encapsulation and cure of the molding compound at 170°C. Negligible stress is established at these elevated temperatures, but coefficient of thermal expansion differentials cause increasing stress as the ambient temperature is lowered. The die attach produces a bending moment at room temperature that puts the surface of the die into tension and the bottom of the die into compression. The magnitude of the tensile stress at the surface of the die is proportional to the thickness of the die. Encapsulation superimposes a compressive stress on the die, and tends to place the molding compound under tension. Cracks, therefore, have a tendency to propagate in the molding compound. Cracks in packages are a reliability concern because they provide paths for the entry of contaminants. Shear stresses are operative on the surface of the chip due to differential thermal expansion between the plastic encapsulant and the passivation layer, resulting in the fracture of the passivation layer; this forfeits its function as an impurity getter and barrier. It also causes lateral displacement of the interconnection lines on the surface of the chip, and shorts between the overlying and underlying interconnection patterns in devices with two or more layers of interconnection [Schnable, 1969]. The effects of stresses applied by plastic are more pronounced in larger chips, especially at the chip corners. Resilient conformal coating on the die and the use of plastic encapsulants with a lower coefficient of thermal expansion are methods used to decrease thermal stress effects in the passivation layer. Figure 3.15 shows the x-section of a 48 pin PLCC, including the areas of delamination. The typical areas of delamination are die surface, die paddle lead frame finger tips, and lead frame-mold interface. The areas of delamination most detrimental to the life of the component are the die surface and lead finger tips. The main reason is that delamination at these two interface often pulls the wirebonds, leading to an open connection at the ball bond pad, bond cratering, pulled or stressed wires, and degraded bonds.

Figure 3.13, 3.16, and 3.17 show the various delamination failure modes during popcorning. Figure 3.13a shows the CSAM image of the top surface of a die in a 48 pin PLCC, showing that greater than 75% of the die is delaminated. Figure 3.13b shows the CAM image of a 48 pin PLCC with all the lead finger tips delaminated. The wirebonds have been pulled

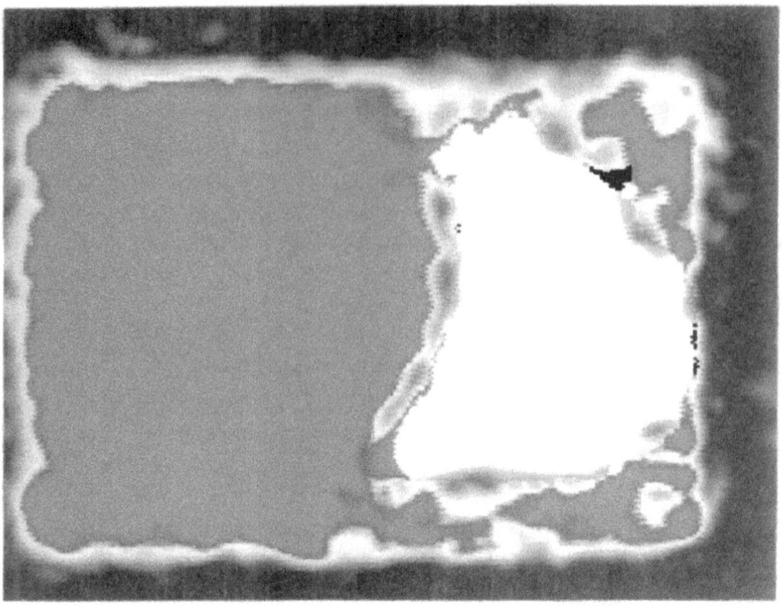

Figure 3.13a C-SAM image of the top of the die, showing that greater than 75% of the interface is delaminated. (Courtesy of Motorola, Inc.)

Figure 3.13b C-SAM image of the lead frame showing that all of the tips of the bonding fingers are delaminated. (Courtesy of Motorola, Inc.)

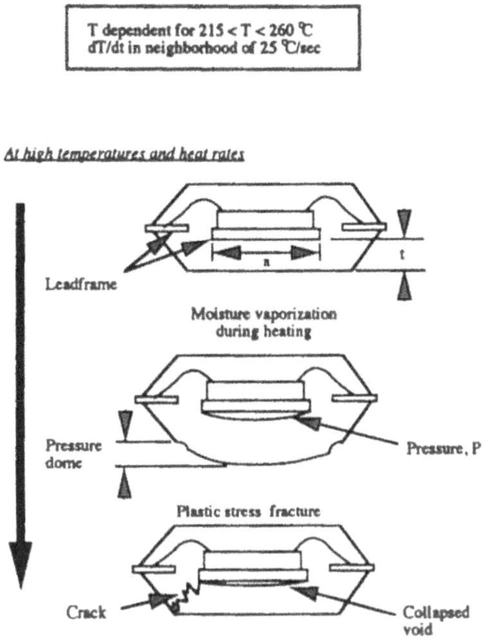

Figure 3.14 Temperature dependencies of plastic package cracking.

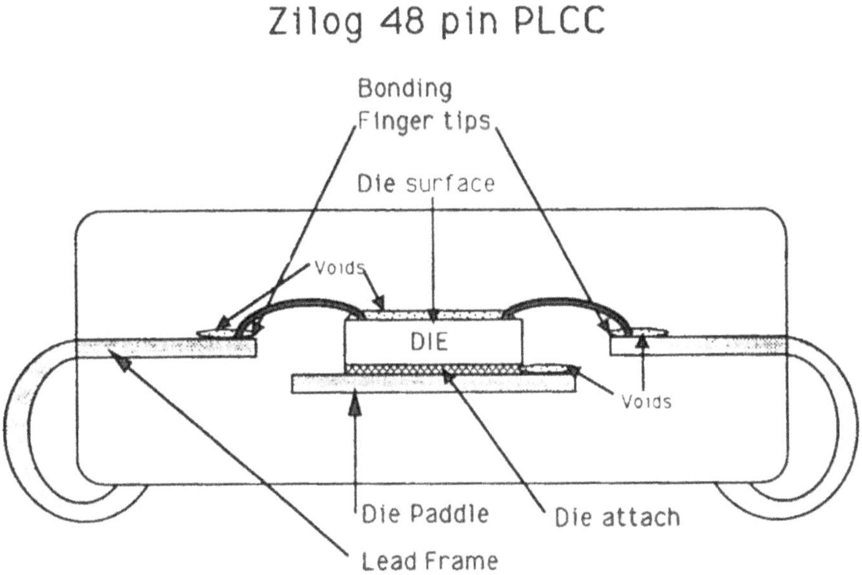

Figure 3.15 Drawing of a 48 pin PLCC showing areas of delamination.

from the bond pads. Figure 3.16a shows the CSAM image of a 48 pin PLCC with no delamination. Figure 3.16b shows the CSAM image of a PLCC with minor delamination at the leadframe. Figure 3.17 shows the CSAM image of a 48 pin PLCC with 100% delamination at the die paddle. Inayoski et al. reported damage in the passivating film of plastic-encapsulated LSI devices due to thermally induced stress at the encapsulant-passivation layer interface [1979]. The location of aluminum corrosion coincided with defects in the passivating film, typically caused by differential thermal stress from the encapsulant. Sim and Lawson found a phosphorus content of 1.8% to be optimum for plastic device reliability [1979].

Cracks also initiate due to stress concentrations at the top edge of the die or at the bottom edge of the leadframe, or due to voids in the molding compound. The cracks propagate under the influence of temperature cycling. Nishimura et al. [1987] predicted crack propagation using the relation

$$\frac{da}{dN} = A_{paris,mc}(\Delta K_f)^{m_{paris,mc}} \tag{50}$$

where $A_{paris,\ mc}$ is Paris's coefficient for molding compound, $m_{paris,mc}$ is Paris's exponent for molding compound, and ΔK_f is the stress intensity factor amplitude. Package cracking can be controlled by downsetting the die below the plane of the leads so that the thickness of the molding compound both above and below the die is roughly equal. Modifications to the molding compound to control crack propagation characteristics include filler coatings to reduce delamination at the epoxy filler interface and control of the filler size distribution.

Delamination is a failure mechanism that occurs between the die and the molding compound, or between the molding compound and the leadframe, due to shrinkage after molding, especially in plastic surface-mount components (PSMCs) which are typically subjected to temperatures in the range of 215-260°C and heating rates of as high as 25°C/sec during board assembly (Figure 3.12).

In addition to CTE driven stresses, two additional factors affect internal stresses inside the package. If the heating or cooling rates are much higher than 10°C/sec, thermal stresses associated with internal temperature gradients can be appreciable. Moisture absorbed by the package molding compound evaporates under high temperatures, creating internal pressure that causes cracking (generally referred to as the popcorn mechanism). PSMCs can absorb moisture during storage in normal ambients, allowing vapor-driven stresses to dominate the mechanics of the package. The main factors influencing the propensity to crack include the peak temperature reached during soldering, the moisture content of the molding compound, the dimensions of the die, the thickness of the molding compound under the die, and the adhesion of the die to the leadframe. Fukuzawa et al. tied together all the factors affecting the failure mechanism of thermal cracking of plastic packages [1985]. The steam vapor pressure in the gap under the die pad causes the molding compound to delaminate from the pad, expand, and form a dome. A crack occurs where the maximum bending stress, σ_{max}, exceeds the fracture stress characteristic of the molding compound at elevated temperature T_{elev}. The relation is represented mathematically as

$$\sigma_{max,mc} > \sigma_{crit,mc}(T_{elev}) \tag{51}$$

Figure 3.16a C-SAM image of the top of the die, showing that 0% delamination is detected at this interface. (Courtesy of Motorola, Inc.)

Figure 3.16b C-SAM image of the lead frame, showing that minor delamination detected on only six bonding fingers. (Courtesy of Motorola, Inc.)

Figure 3.17 C-SAM image of the die paddle from a "New Virgin" part. 100% delamination is detected at this interface surrounding the die. Leadframe and die surfaces are not in focus. Only paddle is in focus and can be interpreted. (Courtesy of Motorola, Inc.)

The maximum bending stress is first reached at the center of the long side of the die pad and is given by

$$\sigma_{max,mc} = 6K_f\left(\frac{a_{die}}{t_{mc}}\right)^2 P_c \tag{52}$$

where $\sigma_{crit,mc}$ is stress elevated temperature prior fracture, $\sigma_{max,mc}$ is the max bending stress in molding, K_s is a dimensionless stress concentration factor that depends on the aspect ratio of the die pad (K_s=0.05 for a square pad), a_{die} is the length of the short side of the die, t_{mc} is the thickness of the molding compound beneath, and P_c is the vapor pressure in the cavity. For packages of different dimensions, saturated at the same temperature and relative humidity, followed by the same temperature shock, P_c equals the value of critical stress $\sigma_{crit,mc}$, and the ratio (a/t) is the major determinant of whether the package will crack. Kitano et al. measured water- vapor absorption and the water diffusivity of the molding compound and calculated the internal vapor pressure as a function of temperature from deflection measurements of the pressure dome [1988]. During solder temperature shock there is not enough water vapor inside the gap under the die pad to account for this pressure; however, because the molding compound is hygroscopic, a larger amount of water vapor is available to generate steam pressure within the diffusion length of the cavity over the duration of the temperature shock. The pressure, P_c, generated inside the package due to thermal shock is represented by

$$P_c = RH_{sat}P_{sat,elev} \tag{53}$$

where $P_{sat\ elev}$ is the saturation pressure @ elevated temperature, RH_{sat} is the relative humidity of the saturation ambient prior to temperature shock. The equations for pressure and stress model for the package cracking sensitivity as a function of temperature shock (temperatures around 215°C-260°C).

3.2 Reversion or Depolymerization of Polymeric Bonds

Reversion or depolymerization is characterized by the breaking of polymeric bonds (say, in the plastic encapsulant) which turns the solid polymer into a gummy liquid comprising monomers, dimers, and low molecular-weight species. Reversion refers to the driving of the condensation reaction in a direction opposite to polymerization [Tummala, 1989]. Elevated temperature is an accelerator of such failures, that is, temperature increases the reaction rate. However, most polymers used in the industry degrade only at temperatures over 300°C.

Thermal degradation may also be encountered in power packages brominated-epoxy flame retardants as encapsulants. Degradation involves scission of highly linked polymers into constituent free radicals, monomers, and so on, at elevated temperatures in the neighborhood of 320°C to 350°C for most electronic materials. Figure 3.18 [Tummala, 1989] shows a differential scanning calorimetry plot of a typical epoxy molding resin for plastic packages, in which the degradation temperature is recognized as a exotherm. Softening and melting temperatures for thermoplastics are a function of molecular weight, with high-temperature properties associated with longer-chain molecules. Thermoplastics with a lower crosslink density generally have a glass transition temperature in the range of 150°C to 200°C [Manzione, 1990]. Generally, the degradation temperature is much higher than semiconductor operational ranges and, therefore, has little impact on the packaging of these chips. The high temperature at which this failure phenomenon occurs makes it a recessive mechanism during normal device operation.

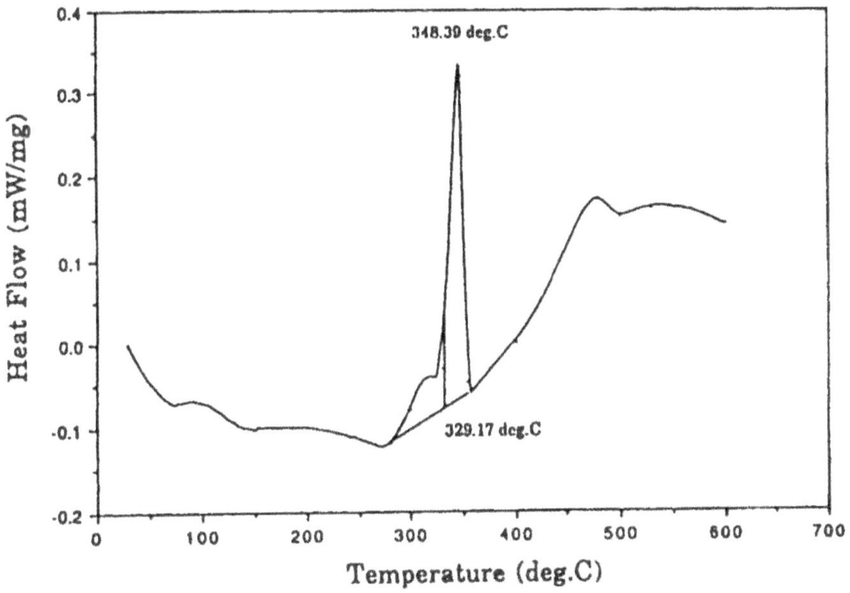

Figure 3.18 Differential scanning calorimetry plot of a typical epoxy resin for plastic packages. The two degradation exotherms: in the neighborhood of 320°C and 350°C indicate degradation of bulk molding compound [Tummala, 1989].

3.3 Whisker and Dendritic Growth

Cases coated with tin are prone to the growth of whiskers and dendrites. Whisker and dendrite formation is well documented and will only be summarized here. Tin dendrites form as a result of electromigration, in which a direct current potential induces metal filaments to grow from the cathode to the anode. Dendrite growth is accelerated in the presence of ionic species; tin whiskers, in the form of single crystals that grow spontaneously, are formed due to compressive stresses in the tin plating. Tin whiskers and dendrites can cause shorts or leakage currents in the component.

Increasing the thickness of the coating reduces the stress, decreasing the chance of whisker growth. Tin coatings of thicknesses exceeding 0.0004" (10 micrometers) prevent the growth of whiskers [NASA, 1992]. To obtain stress-free coating, the method of applying the tin coat should preferably be hot dipping. If electroplating is used, the tin coat should be reflowed to relieve the residual plating stress. Using 2 to 3% lead in conjunction with tin also protects against whisker formation [NASA, 1992, Knott, 1992]. This is a contamination-actuated mechanism and has no temperature dependence.

3.3.1 Modular-case Fatigue Failure

Modular case configurations, often used when the case wall does not play a significant role in heat conduction, use a case wall material different from the case header material. The case header and walls are joined by brazing. Due to differences in the coefficients of thermal expansion of the header and wall materials, stresses are generated in the brazed joint under temperature cycling loads. The stresses in the joint are evaluated by considering the joint as a two-dimensional bi-metal strip. The maximum shear stress in the braze layer can thus be estimated by using models for thermo-mechanical stresses at the interface of a bi-metal strip [Suhir ,1989]. The maximum shear stress in the braze layer for a temperature cycle of magnitude ΔT is given as

$$\Delta\tau_{max,case} = \left[\frac{1-\nu_{hdr}}{E_{hdr}t_{hdr}} + \frac{1-\nu_{wal}}{E_{wal}t_{wal}} + \frac{(t_{wal}+t_{hdr})^2}{4D_{f,cs}}\right]^{-\frac{1}{2}} \left[\frac{t_{wal}}{3G_{wal}} + \frac{2t_{att}}{3G_{att}} + \frac{t_{hdr}}{3G_{hdr}}\right]^{-\frac{1}{2}} \tanh(C^0_{intl,cs}L_{jt})\Delta\alpha\Delta T \tag{54}$$

where ν is Poisson's ratio, E is the modulus of elasticity, t is the thickness, G is the shear modulus, L_{jt} is the length of the joint, and $\Delta\alpha$ is the difference in the CTEs of the case wall and header materials. Subscripts *hdr, wal* and *att*, respectively, denote the header, wall, and attach material. $C^0_{intl,cs}$ is defined as

$$C^{0}_{intl,cs} = \sqrt{\frac{\dfrac{1-\nu_{hdr}}{E_{hdr}t_{hdr}} + \dfrac{1-\nu_{wal}}{E_{wal}t_{wal}} + \dfrac{(t_{wal}+t_{hdr})^2}{4D_{f,cs}}}{\dfrac{t_{wal}}{3G_{wal}} + \dfrac{2t_{att}}{3G_{att}} + \dfrac{t_{hdr}}{3G_{hdr}}}} \qquad (55)$$

The flexural rigidity of the structure, $D_{f,cs}$ is defined as

$$D_{f,cs} = \frac{E_{hdr}t_{hdr}^{3}}{12(1-\nu_{hdr}^{2})} + \frac{E_{att}t_{att}^{3}}{12(1-\nu_{att}^{2})} + \frac{E_{wal}t_{wal}^{3}}{12(1-\nu_{wal}^{2})} \qquad (56)$$

The stress in the seal can cause overstress failure of the case joints or, more likely, result in fatigue damage due to cyclic loads. Equations 54 to 56 give the magnitude of the stress cycle. The fatigue life of the case joints is estimated using Basquin's law [Hertzberg, 1976]:

$$\Delta\tau_{max,case} = 2\sigma_{cf,cs}(2N_{f,cs})^{m_{cf,cs}} \qquad (57)$$

where $\Delta\tau_{max,case}$ is the stress amplitude as calculated from Equations 54 to 56, $2N_{f,cs}$ is the number of cycles to failure ($N_{f,cs}$ is the number of load reversals), $\sigma_{cf,cs}$ is the fatigue strength coefficient for the braze material (defined by the stress to failure for one load reversal - that is, at $N_{f,cs} = 1$ on a graph of elastic stress versus number of cycles to failure), and $m_{cf,cs}$ is the fatigue strength exponent for the braze material.

4. TEMPERATURE DEPENDENCE OF FAILURE MECHANISMS IN LID SEALS

In this section, the temperature dependence of failure mechanisms in lid seals of packages is discussed including thermal fatigue of lid seal. This failure mechanism occurs only in hermetic packages and has a dominant dependence on temperature cycle magnitude and no steady state temperature dependence.

4.1 Thermal Fatigue of Lid Seal

The package lid seal and lid provide a barrier against the ingress of moisture and ionic impurities that can cause corrosion of both active and passive elements and interconnects. The lid seal and lid, along with the case, also serve to protect the contents of the package from mechanical and radiation loads. The seal material can be solder or glass, typically in a preform configuration. In the case of welded seals, the lid and case materials themselves (or the plating) become the seal material. If the case and the lid are made of the same material, there is only a local mismatch of coefficients of thermal expansion between the seal material and the case and lid material. On the other hand, if different materials are used for the case and the lid, the CTE mismatch is global, between the joint parts themselves (the lid and the case), rather than just in the joining material (the seal). Figure 3.20a shows a nickel-silver shield (misnomer - since the composition is copper, nickel, zinc) soldered onto a copper pad on a PC board substrate using 62Sn/36Pb/2Ag solder. Cross-section shows a copper-tin intermetallic at the solder joint interface, which is detrimental to joint strength in thermal fatigue. Figure 3.20b shows a tin plated nickel-silver shield showing similar copper-tin intermetallics at the solder joint interface. Figure 3.19 shows the difference in the wetability characteristics of the nickel-silver and the tin-plated nickel-silver shields. The tin-plated shields show better wetting characteristics as is seen by a toe fillet compared to the unplatted shields and wicking of the solder on the flange.

Case A: Lid and case material are the same

When the lid and the case material are the same, stress in the seal due to temperature excursion is computed using models described by Suhir [1990]. The maximum normal stress in the lid seal occurs at the center cross-section, dropping to zero at the seal corners. On the other hand, the maximum shear stress occurs at the seal corners and diminishes rapidly as one moves away from the corner. The shear stress at any seal cross-section at a distance, x, from the center cross-section (measured from the middle of the long side of the seal) is:

$$\tau_{case}(x) = \frac{\Delta\alpha\Delta T}{\sqrt{C_{c,l}C_{c,jt}}}\,\frac{\sinh\left(\sqrt{\frac{C_{c,l}}{C_{c,jt}}}x\right)}{\sinh\left(\sqrt{\frac{C_{c,l}}{C_{c,jt}}}l_{seal}\right)} \tag{58}$$

where $\Delta\alpha$ is the difference in the coefficients of thermal expansion of the lid and case material and the seal material, ΔT is the difference between the sealant melting point (stress-free temperature) and the operating temperature, and $C_{c,l}$ is the in-plane compliance of the joint $(=[(1-\nu_{case})/E_{case}(h_{cavity}+t_{lid})] + [(1-\nu_{seal})/E_{seal}t_{seal}])$ [Suhir, 1990]. The interfacial compliance for the joint, $C_{c,jt}$, is given as the sum of the individual compliances for the seal and the case or lid material; that is, $C_{c,jt} = C_{c,seal} + C_{c,case}$, while $C_{c,seal} = 2(1+\nu_{seal})t_{seal}/3E_{seal}$ and $C_{c,case} = 2(1+\nu_{case})(h_{cavity}+t_{lid})/3E_{case}$. Half the seal length (or width) is represented by l_{seal}. The symbols E and ν (with subscripts) are, respectively, Young's modulus of elasticity and Poisson's ratio for the element indicated by the subscript. The normal stress at any cross-section at a distance x from the mid-cross-section of the seal is [Suhir, 1990]:

$$\sigma_{case}(x) = \frac{\Delta\alpha\Delta T_{C}}{C_{c,l}l_{seal}}\left[1-\frac{\cosh\left(\sqrt{\frac{C_{c,l}}{C_{c,jt}}}x\right)}{\cosh\left(\sqrt{\frac{C_{c,l}}{C_{c,jt}}}l_{seal}\right)}\right] \tag{59}$$

The maximum principal stress, σ_l, in the seal-end cross-section is found using Mohr's circle. Seal materials in current use (including glasses and gold-tin eutectic solder) are brittle and assumed to fail when the maximum principal stress in the seal exceeds the tensile strength of the seal material. Brittle seal materials also imply that the stresses in the seal will be elastic, and that the possibility of low-cycle plastic strain-dominated fatigue is excluded. High-cycle fatigue failures that typically occur at over 10,000 stress cycles do not pose a threat, since 10,000 cycles far exceed the expected life of most components. Thus, lid-seal failures are due to overstress, rather than fatigue, when subjected to temperature excursions.

Case B: Lid and case material differ

If the lid material is different from the case material, the stresses are determined by the global coefficient of thermal expansion mismatch, between the case and the lid. Whichever material has the greater CTE will have tensile normal stresses, while the other will have compressive stresses. The principal stress in the seal is

$$\sigma_{seal} = \frac{p_p}{2} \pm \sqrt{\frac{p_p^2}{4} + \tau_{seal}^2} \tag{60}$$

where p_p is the peeling stress and τ_{seal} is the shear stress in the seal [Suhir, 1989]. And

$$\tau_{seal} = \left[\frac{1-\nu_{case}}{E_{case} t_{case}} + \frac{1-\nu_l}{E_l t_l} + \frac{(t_{case}+t_l)^2}{4D_f}\right]^{-1/2} (G')^{-1/2} \tanh(A l_{seal})\ \Delta\alpha \Delta T_C \tag{61}$$

$$p_p = -\left[\frac{t_{case} t_{lid}^3 E_{lid}}{12(1-\nu_l^2)} - \frac{t_{lid} t_{case}^3 E_{case}}{12(1-\nu_{case}^2)}\right] \frac{1}{2G' D_f} \Delta\alpha \Delta T_C \tag{62}$$

where the symbols ν, E, l, and t denote Poisson's ratio, Young's modulus of elasticity, length, and thickness, respectively. The subscripts *lid*, *case*, and *seal* following these symbols refer respectively to the lid, case, and seal. Where α is the coefficient of thermal expansion, ΔT_C is the temperature excursion with respect to the stress-free state of the seal, or the sealing temperature. A, D, and G' are defined as

$$A = \sqrt{\frac{\frac{1-\nu_{case}}{E_{case} t_{case}} + \frac{1-\nu_{lid}}{E_{lid} t_{lid}} + \frac{(t_{case}+t_{lid})^2}{4D_f}}{G'}} \tag{63}$$

$$D_f = \frac{E_{case} t_{case}^3}{12(1-\nu_{case}^2)} + \frac{E_{lid} t_{lid}^3}{12(1-\nu_{lid}^2)} + \frac{E_{seal} t_{seal}^3}{12(1-\nu_{seal}^2)} \tag{64}$$

$$G' = \frac{t_{case}}{3G_{case}} + \frac{2t_{seal}}{3G_{seal}} + \frac{t_{lid}}{3G_{lid}} \tag{65}$$

where G is the shear modulus of elasticity.

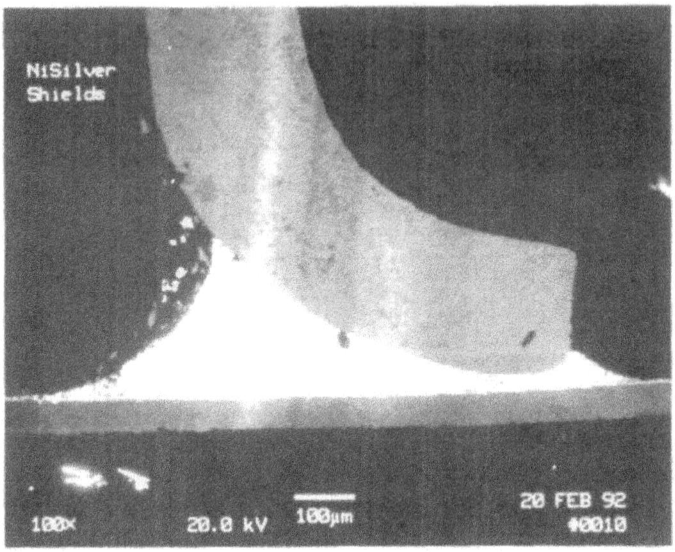

Figure 3.19a Raw nickel silver shield solder joint. (Courtesy of Motorola, Inc.)

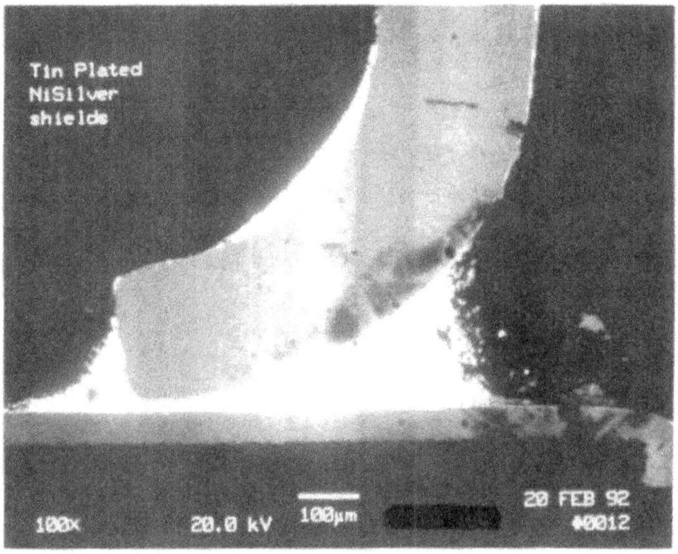

Figure 3.19b Tin plated nickel silver shield solder joint. Note similar geometries. (Courtesy of Motorola, Inc.)

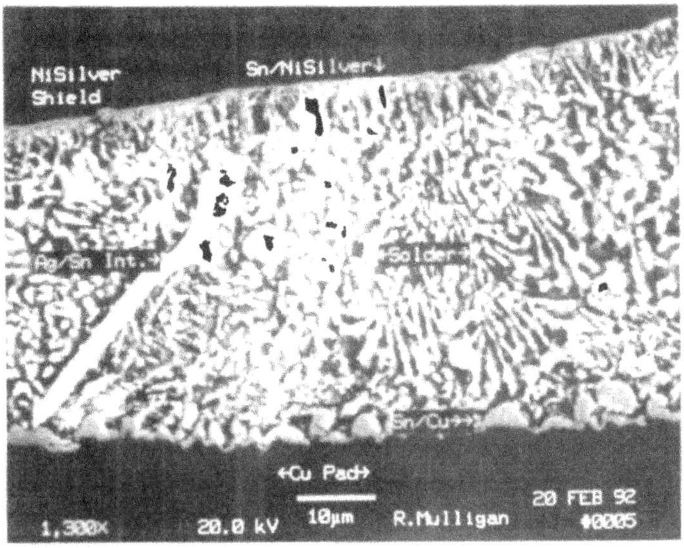

Figure 3.20a Raw nickel silver shield solder joint (Courtesy of Motorola, Inc.)

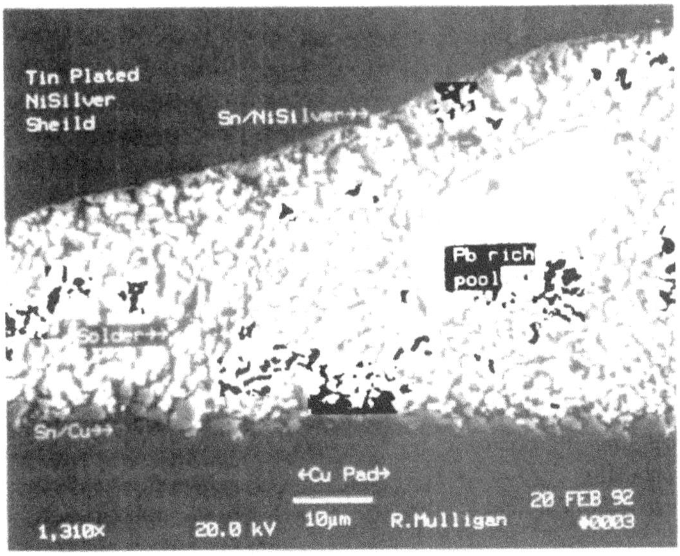

Figure 3.20b Tin plated nickel silver shield solder interface with the Cu pad and shield. (Courtesy of Motorola, Inc.)

Note: similar intermetallics at the solder interface with the Cu pad and shield.

5. TEMPERATURE DEPENDENCIES OF FAILURE MECHANISMS IN LEADS AND LEAD SEALS

In this section, the temperature dependence of failure mechanisms in leads and lead seals is discussed. Mechanisms discussed include mishandling and defect-induced lead-seal failure, post forming defect-localized lead corrosion, stress corrosion at the lead-lead seal interface and lead solder joint fatigue. Most of the lead seal failure mechanisms with the exception of lead solder joint fatigue are defect actuated mechanisms and have no dependence on steady state temperature. Lead solder joint fatigue has a dominant dependence on temperature cycle magnitude and has no steady state temperature dependence.

5.1 Mishandling and Defect-induced Lead-seal Failure

While conducting a study of hermetic packages with glass-to-metal feedthrough seals, metal-to-metal welded lids, and plated leads, that had been clipped, formed, soldered, and tested, Neff found that most lead-seal failures could be tracked to defects that had escaped inspection or were a result of mishandling [1986]. Microscopic examination of the several hundred failed samples revealed that the defect or damage criteria could be classified into nine categories, including handling, radial cracks, poor pin-glass bonds, meniscus chips, eccentricity, poor glass-header bonds, weld problems, inadequate glass, and poor package quality. The relative percentages of failures attributed to each failure cause are documented in Table 3. Defect-induced and mishandling-induced lead-seal failures are thus a function of defect magnitudes, and have no dependence on temperature.

5.2 Post-forming Defect-localized Lead Corrosion

Leads undergo corrosion in the presence of moisture and ionic contaminants, which can result in a change in the electrical properties of the lead and eventual failure. Excessive corrosion can also result in loss of strength at the lead, leading to mechanical failure. Corrosion is usually localized at pinholes in the lead plating, diffusion sites of lead base metal to lead surface, or plating cracks, all of which expose the base metal to the external environment. Cracks can appear in the lead plating, or even in the lead base material, during lead forming.

Typically, thick nickel plating can cause lead cracking during bending or forming. Panousis examined various leads with nickel coatings varying in thickness from 0 to 41 μm, and bent through 90°, under a metallurgical microscope [Panousis, 1976]. A crack was defined as a discontinuity on the lead surface exceeding 10% of lead thickness. It was found that leads did not crack if the nickel plating thickness was less than 1.0 μm; a nickel thickness of about 2.7 μm resulted in cracks with depths of less than 10% of lead thickness. Nickel thickness of greater than 5.7 μm resulted in cracks larger than 10% of the lead thickness. Zakraysek, while evaluating various lead coatings on bent leads with respect to corrosion resistance in salt-spray tests, found that tin and tin-lead finish parts generally passed, while the gold finish parts all failed, when an electroless nickel undercoat was used, because the cracks induced during the bending of the leads exposed the base metal to corrosion [1981]. The presence of gold accelerated corrosion of the base metal. Tin plate over electrolytic nickel was found to be superior to tin over copper, electroless nickel, or bare Kovar after 24 hours of

Table 3 Major causes of lead-seal failure [Neff, 1986]

Defects and failure modes	Percent of packages exhibiting failure mode
Handling damage: Most severe damage in lead seals was attributable to poor handling in the form of bent leads, broken glass, and scratched gold plate.	95%
Radial cracks: The cracks were generated in the seals starting from the pins outward 20 to 60% of the distance to the outer diameter of the glass bead. Cracks resulted from side loading of pins due to improper fixturing, and from lead forming.	75%
Poor pin-glass bonds: Meniscus seal often masks poor pin-glass bonds at the pin-glass junction. The defective packages pass incoming inspection and fail early in product life during lead forming.	60%
Meniscus chips: Meniscus seal is the bond of the glass to the pin, which has flowed or wetted up to the pin to form a very thin circumferential glass-to-metal bond.	55%
Eccentricity: Eccentrically located rectangular pins cause leaks. This type of structural defect seldom withstands thermal shock, such as a solder dip, which forces the sharp corner of the pin against a very small quantity of glass to produce a leak.	20%
Poor glass-header bonds: Poor bonding at the outside diameter of the glass bead results in leaks early in the operational life of the device.	10%
Inadequate glass: Glass seals fabricated with inadequate glass develop leaks in the form of radial cracks.	5%
Poor quality package: Package defects as received from the manufacturer include lack of concentricity with rectangular pins, broken meniscus, broken plating, and glass broken away from pins.	60%

exposure to a salt-fog atmosphere. Tin over electroless nickel performs poorly because highly stressed electroless nickel deposits can develop cracks and expose the base metal to the corrosive environment. Neither 99.7 nor 99.9 gold plate passed the corrosion test when electroless nickel was employed as the undercoat [Zakraysek, 1981].

Although leads can undergo corrosion anywhere on their surfaces, they are particularly susceptible to corrosion at the interface between the lead and the lead seal of glass-sealed leads. Since the lead coating is applied after the lead-sealing process, the part of the lead covered by the sealing glass is not coated. This makes it highly prone to corrosion in the presence of moisture and ionic contaminants. The lead seal does not form a bond at the lead surface, and thus provides a pathway for moisture to come into contact with the bare lead surface. During assembly and handling, the lead is often bent, causing the glass meniscus to crack at the glass-lead interface and exposing the bare lead surface underneath to the external environment.

5.3 Stress Corrosion of Leads at the Lead-lead Seal Interface

Stress corrosion cracking is the interaction between fracture and corrosion because of stress concentrations at corrosion-generated surface flaws (as quantified by the stress intensity factor, K). Electrochemical corrosion, interacting with mechanical stress, is the potential cause of failure, and results in typically transgranular cracks, due to acceleration of the fatigue process by the corrosion of the advancing fatigue crack [Tummala, 1989]. Stress corrosion reduces the fracture strength of the material, such that failure occurs before K_{crit} for the material is reached. The process is synergistic - that is, it is the combined simultaneous interaction of mechanical and chemical forces resulting in crack propagation, whereas neither factor acting independently or alternately would produce the same result [Davis, 1987]. Stress corrosion is a failure phenomenon that occurs around 300°C and above, predominantly in power devices.

Stress corrosion in leads occurs in the form of transgranular or intergranular cracks propagating in a plane normal to the tensile stress and perpendicular to the axis of the leads. In samples exhibiting severe corrosion, base metal attack causes the leads to break as a result of a 90°C bend. Stress-corrosion cracking can occur in package leads even without external loads, because residual stresses from conventional manufacturing practices are sufficient to initiate the attack. Residual stresses originate during such fabrication processes as rapid-quenching heat treatments. Laboratory tests show that Kovar has a tendency toward stress-corrosion cracking, particularly in the presence of chloride. Stress-corrosion failure of a Kovar lead, evidenced by brownish iron rust and intergranular fractures with secondary cracks, results in the separation of the lead at the point of stress. Over-oxidation of Kovar prior to or during glass sealing may contribute to a rapid, localized attack on the lead. Chemical inhomogeneity caused by residual oxide at the grain boundaries increases Kovar's susceptibility to corrosion [Berry, 1987].

Laboratory tests show that failure due to stress-corrosion cracking is significantly more frequent at the lead-glass interface than at other areas of the lead because of the presence of large residual stresses originating with the formation of the glass seal, as well as moisture retention at the lead-glass interface [Berry, 1987]. Adsorbed moisture films caused by relative humidity exposure cause static fatigue of the lead zinc-borate sealing glass; the distinct mechanism is chemical attack on glass, and the time to failure is usually modeled by an Arrhenius equation of temperature that activates above 300°C [Tummala, 1989]. The rate of galvanic corrosion in the presence of a liquid electrolyte increases as temperature is increased, due to a faster rate of electron transfer. Typically, corrosion products in microelectronic devices, such as $Al(OH)_3$ are derived from reaction processes that are monotonically increasing functions of temperature. Although the corrosion rate depends in part on the steady-state temperature, it also depends on the magnitude and polarity of the corrosion galvanic potential, which are functions of the electrolyte concentration, pH, local flow conditions, and aeration effects. As noted previously, elevated temperature acts as a means of slowing the corrosion process.

5.4 Lead Solder-joint Fatigue

Cyclic thermal loads experienced by surface-mount components can cause failure of the component-to-board attachment due to low-cycle fatigue. Typically, the solder joint, rather than the lead, fails under temperature cycling loads. However, the mean fatigue life of lead solder joints for surface-mount technology under temperature cycling loads is affected by the lead material and geometry. The difference in the thermal expansion of the case and board materials under temperature cycling results in cyclic strains in the component attachment to the board, - that is, in the lead and lead-to-board solder joint. Figure 3.21 shows solder joint fatigue cracks in a ball grid array package after liquid-liquid thermal shock. Fatigue cracks in

BGAs are typically characterized by a primary crack at the package-solder interface which propagates from-package outside-to-inside, and a secondary at the solder joint-board interface which propagates from-package inside-to-outside. Complete solder joint failure in BGAs is thus characterized by either of the cracks propagating to failure at the package-solder or the solder-board interfaces. Typically, the primary crack is the first to propagate to failure and thus governs solder joint failure time. Solder joint fatigue cracks first initiate - in the corner or the outmost joints (w.r.t to the neutral point of the package) in CBGAs and in the joints just underneath the die at the die edge (not the outmost) in PBGAs. Figure 3.21b shows the top layer of outermost solder joint, showing fatigue crack at the package-solder interface, and Figure 3.21c shows a fatigue crack at the lower layer outermost joint showing a secondary crack at the solder-board interface. Often, the fatigue damage in surface-mount components is reduced through the use of compliant leads that can withstand high displacements without transferring large stresses across the solder joint. Figure 3.22 shows solder joint fatigue cracks in an Alloy42 TSOP subjected to liquid-liquid thermal shock. TSOP solder joint fatigue failures are characterized by primary crack initiating from the heel of the solder joint and a secondary crack propagating from under the lead at the toe of the solder joint. The failure of the solder joint is thus characterized by the time taken by primary and secondary cracks to propagate across the bottom of the lead.

The steady state creep in solder joints is characterized by a relationship of the form [Garofalo 1965, Murty 1992, Frost 1982]

$$\frac{d\gamma_s}{dt} = C_{mc,sdr}\frac{G_{sdr}}{T}\left(\sinh\left(\alpha_{pr,bk}\frac{\tau}{G_{sdr}}\right)\right)^{n_{sdr}}\exp\left(\frac{-E_{a,sdr}}{kT}\right) \tag{66}$$

where (dr_s/dt) is steady state creep rate, G_{sdr} is the shear modulus, k is the Boltzman constant, T is absolute temperature, τ is the applied stress, $E_{a,sdr}$ is the activation energy, n_{sdr} is the stress exponent, α is the stress level at which the power law dependence breaks down, $C_{mc,sdr}$ is a constant characteristic of the underlying failure mechanism. Based on the above relationship, the steady state creep rate and thus stress relaxation in solder is an exponential function of temperature. The steady state creep rate and stress relaxation in solder during thermal cycling decreases with the decrease in temperature. The lower stress relaxation at lower temperatures results in higher viscoplastic strain energy density and thus higher plastic work in solder joints at lower temperatures than at higher temperatures. The solder fatigue damage which is quantified by the viscoplastic strain energy density amplitude per cycle is thus higher for -40

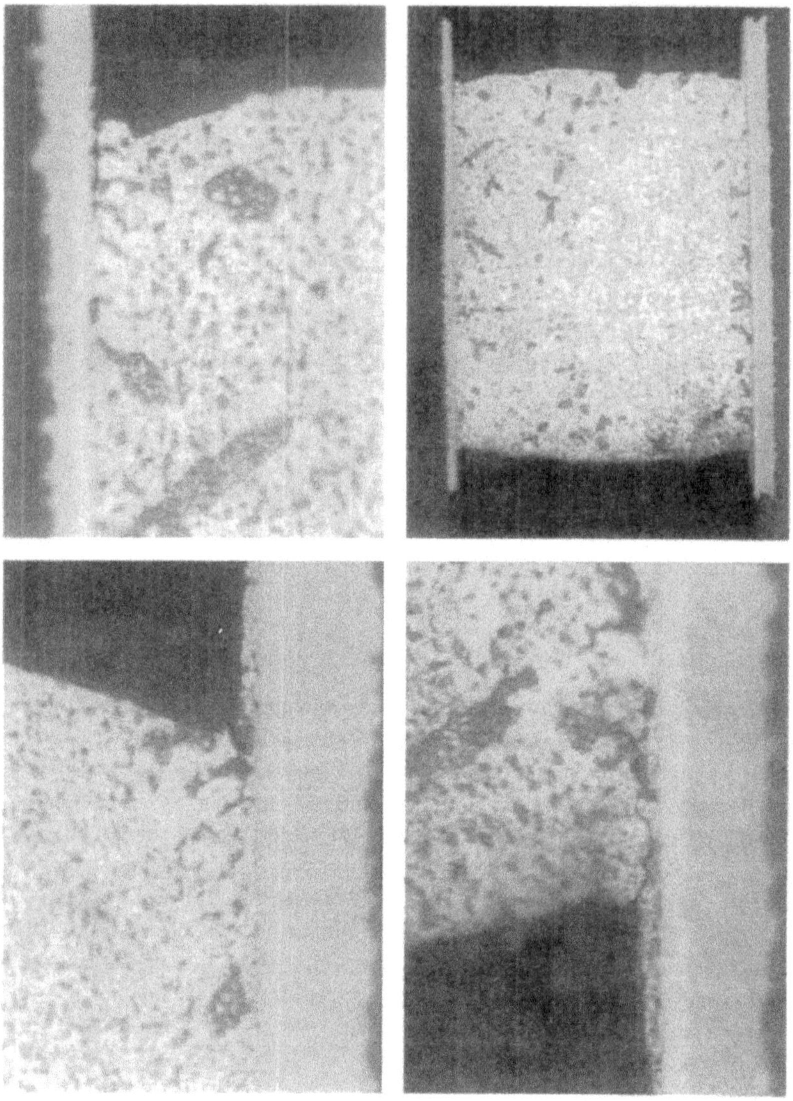

Figure 3.21 Fatigue cracks in a ball grid array package after liquid-liquid thermal shock. (Courtesy of Motorola, Inc.)

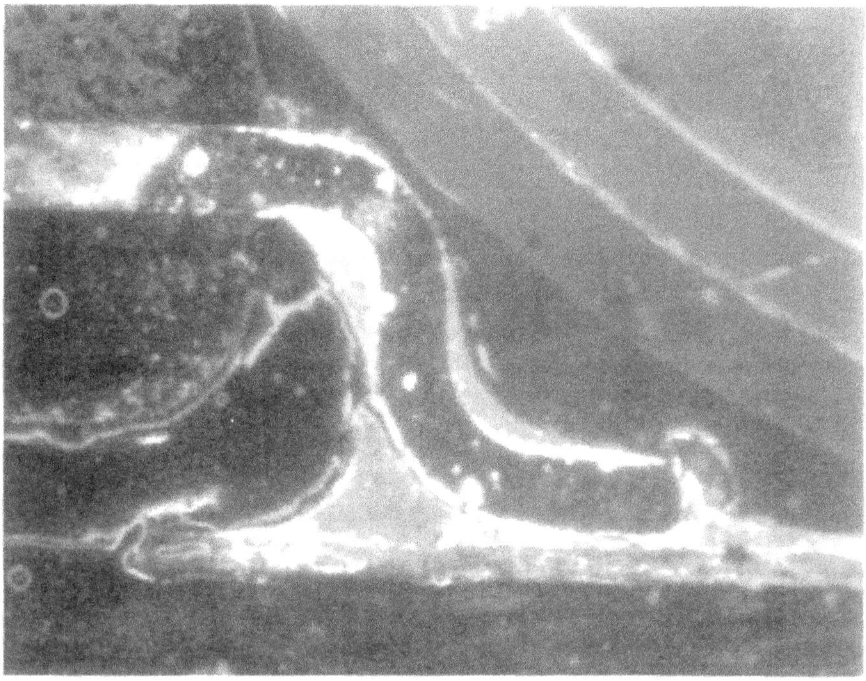

Figure 3.22 Solder joint fatigue cracks in a TSOP package subjected to liquid-liquid thermal shock. (Courtesy of Motorola, Inc.).

to 60°C than a cycle from 0 to 100°C for the same component. The steady state creep rate is generally not achieved immediately after load is applied. A certain amount of transient creep occurs before attaining steady state creep. Typically, the strain rate starts out high, then decelerates and decreases to steady state creep rate as the material work hardens. The transient creep at constant stress and temperature is defined by [Murty et.al. 1992]:

$$\gamma_{ttl,creep} = \frac{d\gamma_s}{dt}t + \gamma_{tr}\left(1 - \exp\left(\beta_{tr,cft}\frac{d\gamma_s}{dt}t\right)\right) \tag{67}$$

where $\gamma_{ttl,creep}$ is the total creep strain, $(d\,r_s/dt)$ is steady state creep rate, γ_{tr} is the transient creep strain, $\beta_{tr,cft}$ is the transient creep coefficient, t is time, and T is absolute temperature. High stresses, $(t/G_{sdr} > 0.001)$, creep is accompanied by time-independent plastic flow which follows a typical strain hardening law:

$$\gamma_{plstc,sdr} = C_{plstc,sdr}\left(\frac{\tau}{G_{sdr}}\right)^{m_{sdr}} \tag{68}$$

where $\gamma_{plstc,sdr}$ is the time independent plastic strain, m_{sdr} is the strain hardening exponent, $C_{plstc,sdr}$ is a constant for time-independent plastic flow. The total strain is given by the sum of the creep strain and strain due to time-independent plastic flow.

$$\gamma_{ttl,sdr} = \gamma_{plstc,sdr} + \gamma_{ttl,creep} \tag{69}$$

where $\gamma_{ttl,sdr}$ is the total strain in solder. The plastic work in the solder during thermal shock can be calculated based on equations (1) to (4). The plastic work or the viscoplastic strain energy density is the area of the stabilized hysteresis loop. The damage-life relationship has been characterized for 62Sn/36Pb/2Ag solder by [Darveaux et al. 1994] using a Paris's power law type equation

$$N_{in,sdr} = 7860(\Delta W)^{-1.0} \tag{70}$$

$$\frac{da}{dN} = 4.96e{-8}\ (\Delta W)^{1.13} \tag{71}$$

where $N_{in,sdr}$ is the number of cycles to crack initiation in the solder, da/dN is the crack propagation rate in solder, and ΔW is the plastic work amplitude per cycle. The variation of shear stress and shear strain versus time are plotted from the FE model of a DCA assembly (-40 to 125°C; 1cph)(Figure 3.23). It is seen that the solder joints completely stress relax at the higher temperature dwell, while they do not at the lower temperature dwell. As the temperature increases, the solder joints become less rigid. The plots of inelastic strain history and shear stress versus inelastic shear strain hysteresis loops show that, up to a critical temperature of 83.75°C (Figure 3.24 & 3.25), unloading of the solder joints is elastic with inelastic strains being almost constant. The critical temperature of 83.75°C, which is a function of creep rate of solder, depends also on the overall stiffness of the assembly and the CTE mismatch between the PCB and the chip carrier. This is so because the stress relaxation rates are a function of the initial stress. Beyond 83.75°C, the solder joints become more and more compliant whereby the inelastic strains increase at a fast rate while the shear stresses decrease. By the time the temperature gets to 125°C, the solder joints are almost fully compliant and the shear stresses are almost constant, and nothing much is happening in terms of stress reduction during the dwell times at 125°C.

Modeling results show that the dwell should be asymmetric with longer dwell at the lower temperature and a shorter dwell at the higher temperature because it takes longer time for solder to creep at the lower temperature than at higher temperature (Figures 3.24 & 3.25). This is evident from plot of the inelastic shear stress and the inelastic shear strain (Figure 3.24 & 3.25), where the strain levels off after a few minutes at the higher extreme and does not level off at the lower extreme in the same time. Figure 3.26 shows that the hysteresis loop stabilizes after the 3-4 cycles. Stabilization of the hysteresis loop indicates that the plastic strain amplitude or the viscoplastic energy amplitude does not change after the 3-4 cycle. Figure 3.27 shows that inspite of the stress relaxation, the dwells do not contribute a whole lot to the damage in the solder joints - this can be seen from the plot of the plastic work done per unit volume versus time where the maximum damage in solder is seen while going from the higher to lower extreme or vice versa. The plastic work per unit volume during dwell is indicated by the flats in the graph indicating little or no plastic work. The dwells should therefore be greater of the time to allow stress relaxation in solder and time to allow thermal equilibrium in the thermal shocked assembly. Lower temperature dwells are more damaging than higher temperature dwells.

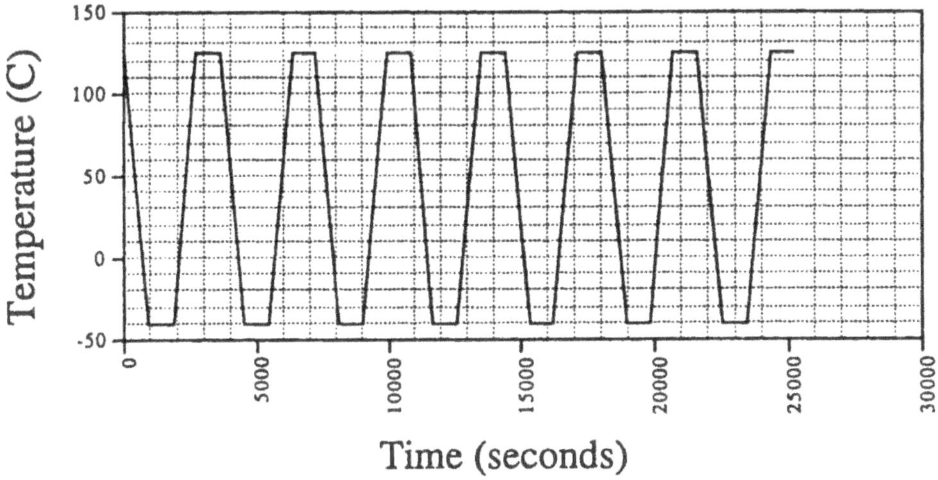

Figure 3.23 Temperature shock profile versus time.

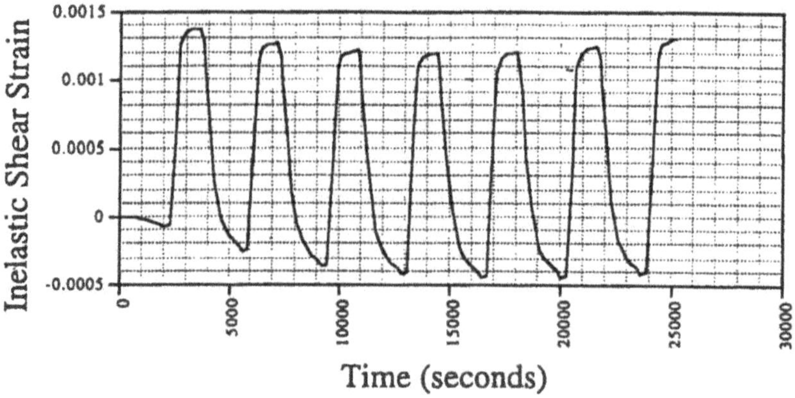

Figure 3.24 Model prediction of inelastic shear strain in solder joints versus time in 7 cycles of temperature shock from -40 to 125°C.

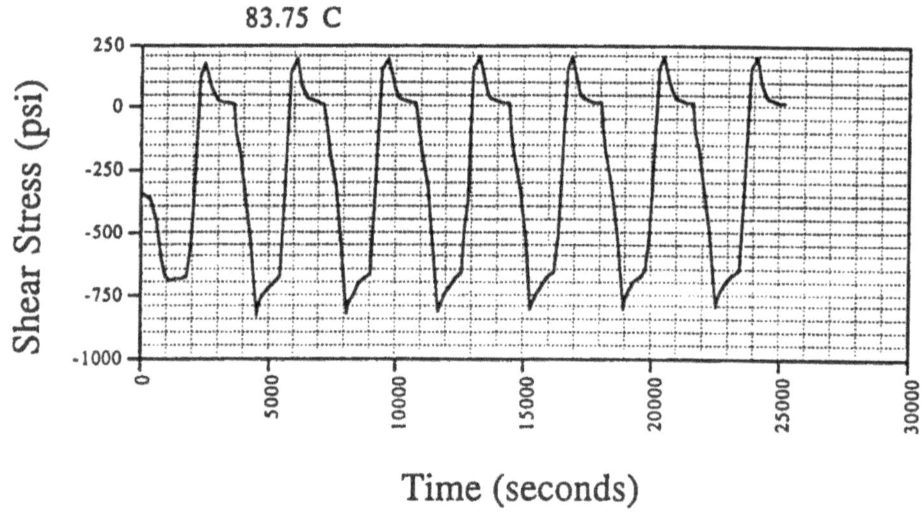

Figure 3.25 Model prediction of inelastic shear stress in solder joints versus time in 7 cycles of temperature shock from -40 125°C.

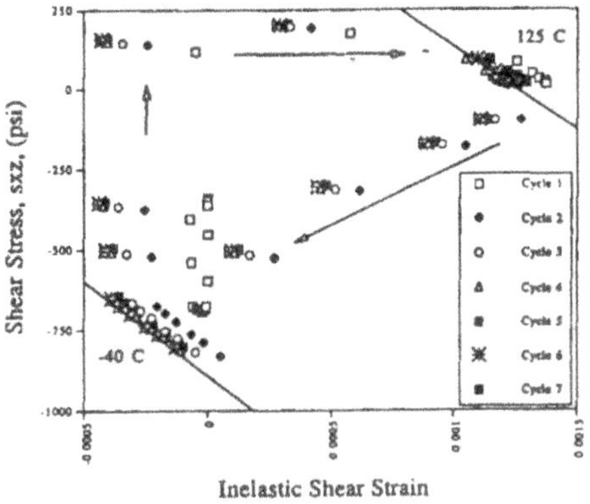

Figure 3.26 Hysteresis loop for solder joint fatigue in 7 cycles of temperature shock from -40 to 125°C.

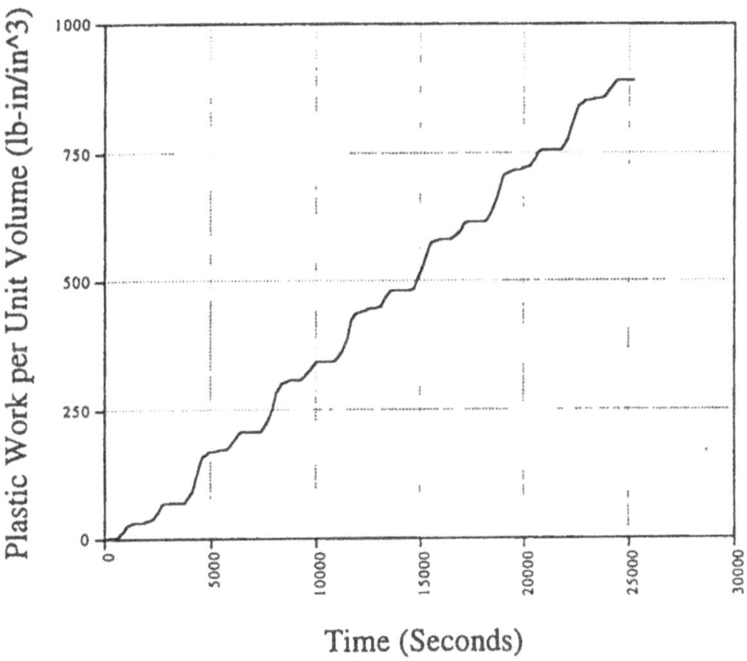

Figure 3.27 Plastic work done per unit volume in the solder joints in 7 cycles of temperature shock from -40 to 125°C.

Chapter 4

ELECTRICAL PARAMETER VARIATIONS IN BIPOLAR DEVICES

This chapter discusses the temperature dependence of electrical parameter variations in bipolar devices. The parameters investigated for bipolar devices include intrinsic carrier concentration, thermal voltage, mobility, current gain, leakage current, collector-emitter saturation voltage, and VTC shift. The parameters investigated for MOSFETs include threshold voltage, mobility, drain current, time delay, strong inversion leakage, subthreshold leakage, and chip availability. The temperature thresholds above which temperature dependence renders the device inoperable have been identified for each of the parameters.

1. TEMPERATURE DEPENDENCE OF BIPOLAR JUNCTION TRANSISTOR PARAMETERS

1.1 Intrinsic Carrier Concentration, Thermal Voltage, and Mobility

The density of the electrons in the conduction band is expressed as

$$n_{ce} = \int_{E_c}^{\infty} P_{pr,occ}(E)\; N_{st}(E)\,dE \tag{1}$$

where $P_{pr,occ}$ is the probability of occupancy, N_{st} is the density of states per unit volume, E_c is the conduction band energy level, n_{ce} is the density of states in the conduction band. *N(E)dE* (in Equation 1) is the density of states per unit volume in the energy range *dE*. The number of electrons per unit volume in the energy range *dE* is the product of the density of states and the probability of occupancy, *f(E)*. The total electron density is the integral over the whole conduction band. The function *N(E)* in Equation 1, can be calculated using quantum mechanics and the Pauli Exclusion Principle. Assuming the electrons in the conduction band to be free, the Schrodinger's Wave Equation for a free electron in three dimensional space can be written as

$$-\frac{h^2}{2m_{elec}}\nabla^2\psi_{wf} = E\psi_{wf} \tag{2}$$

where ψ_{wf} is the wave function of an electron of energy, *E*, M_{elec} is the electron mass, *h* is the Planck's constant. The form of the assumed solution is

$$\psi_{wf} = (const)e^{jk_{c,x}r_v} \tag{3}$$

where $k_{c,x}$ is (Schrodinger equation), a constant in assumed solution, $j = \sqrt{-1}$, and r_v is the radial vector . The electron is described in terms of a set of periodic boundary conditions within the lattice, to measure the electron energies, in a cube of material of side L_{wf}. Periodicity requires that

$$\psi_{wf}(x+L_{wf}, y, z) = \psi_{wf}(x, y, z) \tag{4}$$

where L_{wf} is the periodicity interval of the wave function. Similar boundary conditions need to be satisfied for y and z directions also. The wave function is

$$\psi_{wf} = A_{wf}e^{\left(j\frac{2\pi}{L_{wf}}(n_x x + n_y y + n_z z)\right)} \tag{5}$$

where the factor $2\pi/L$ of Equation 5 is the same as factor k of Equation 3, and guarantees periodicity in the x, y, z direction. Substituting the wave function, ψ in the Schrodinger Wave Equation 2,

$$-\frac{h^2}{2m_{elec}} A_{wf}\nabla^2 e^{j\frac{2\pi}{L_{wf}}(n_x x + n_y y + n_z z)} = EA_{wf}e^{j\frac{2\pi}{L_{wf}}(n_x x + n_y y + n_z z)} \tag{6}$$

where A_{wf} is the coefficient around solution (Schrodinger equations)

$$\begin{aligned} E_n &= \frac{h^2}{2m_{elec}}\left(\frac{2\pi}{L_{wf}}\right)^2 \left(n_x^2 + n_y^2 + n_z^2\right) \\ &= \frac{h^2 n^2}{2m_{elec}V_{cube}^{2/3}} \end{aligned} \tag{7}$$

where n_x, n_y and n_z are directions of cosines of the radial vector r_v and V_{cube} is the volume of the cube of length L_{wf}. According to Pauli's Exclusion Principle, the number of energy states with energy level less than n is twice the volume in the n-space. The total number of energy states in the volume V are

$$N_{st}V_{cube} = \frac{8\pi}{3}n^3 \tag{8}$$

The energy of the nth energy level is

$$
\begin{aligned}
E_n &= \frac{h^2 n^2}{2m_{elec} V_{cube}^{2/3}} \\
&= \frac{h^2}{2m_{elec} V_{cube}^{2/3}} \left(\frac{N_{st} V_{cube}}{\frac{8\pi}{3}} \right)^{2/3} \\
&= \frac{h^2}{2m_{elec}} \left(3\pi^2 N_{st} \right)^{2/3}
\end{aligned} \tag{9}
$$

If the nth energy level is some maximum value such that all the state below this level are filled, and all the states above this are empty, then the energy of this nth level is termed as the Fermi energy at 0°K. The total number of states per unit volume is

$$
\begin{aligned}
N_{st} &= \int N_{st}(E) dE \\
&= \frac{1}{3\pi^2} \left(\frac{2m_{elec} E}{h^2} \right)^{3/2}
\end{aligned} \tag{10}
$$

Differentiating Equation 10,

$$
N_{st}(E) = \frac{3}{2} \left(\frac{1}{3\pi^2} \right) \left(\frac{2m_{elec}}{h^2} \right)^{3/2} E^{1/2} \tag{11}
$$

The probability of occupancy of any one energy level is represented by a Fermi-Dirac distribution.

$$
P_{pr,occ}(E) = \frac{1}{e^{\frac{E - E_f}{K_B T}} + 1} \tag{12}
$$

where E_f is fermi-level energy. The density of electrons in the conduction band at a given temperature is calculated by substituting Equation 10 and 12 in Equation 1.

$$
\begin{aligned}
n_{ce} &= \int_0^{\infty} N_{st}(E)\, P_{pr,occ}(E) dE \\
&= \frac{1}{2\pi^2} \left(\frac{2m_{elec}}{h^2} \right)^{3/2} e^{E_f / K_B T} \int_0^{\infty} E^{1/2} e^{-E/K_B T} dE
\end{aligned} \tag{13}
$$

where the energy in the conduction band has been taken to be zero. The integral has a standard form

$$
\int_0^{\infty} x^{1/2} e^{-ax} dx = \frac{\sqrt{\pi}}{2a\sqrt{a}} \tag{14}
$$

The density of electrons can thus be represented as

$$n_{ce} = 2\left(\frac{2\pi m_{elec} K_B T}{h^2}\right)^{3/2} e^{E_F/K_B T} \tag{15}$$

or by

$$n_{ce} = 2\left(\frac{2\pi m_{elec} K_B T}{h^2}\right)^{3/2} e^{-(E_C-E_f)/K_B T} \tag{16}$$

if the conduction band energy level is assumed to be E_c instead of 0 eV. The effective density of states in the conduction band is thus

$$N_{st,c} = 2\left(\frac{2\pi m_n^* K_B T}{h^2}\right)^{3/2} \tag{17}$$

where $N_{st,c}$ is the density of state in conduction band. Similarly, the density of holes in the valence band is

$$P_{vh} = 2\left(\frac{2\pi m_p^* K_B T}{h^2}\right)^{3/2} e^{-(E_F-E_V)/K_B T} \tag{18}$$

where P_{vh} is the density of hole in valence band, m_p^* is the effective mass of a hole, and E_v is valence band energy level and the effective density of states is

$$N_{st,v} = 2\left(\frac{2\pi m_p^* K_B T}{h^2}\right)^{3/2} \tag{19}$$

where $N_{st,v}$ is the density of states in valence band. For an intrinsic material, the Fermi energy level, E_F, lies in the middle of the band gap and is termed as the intrinsic energy level, E_I. The product of the electron density in the conduction band and the hole density in the valence band is a constant for a particular material and temperature.

$$n_{ce}\, P_{vh} = n_i^2 \tag{20}$$

where n_i is the intrinsic carrier concentration. Variation in temperature, however, varies the intrinsic carrier concentration dramatically.

$$n_i = \sqrt{N_{st,c}\, N_{st,v}}\; e^{-E_a / 2K_B T} \tag{21}$$

The temperature dependence of intrinsic carrier concentration is thus calculated by substituting for $N_{st,c}$ and $N_{st,v}$ from Equations 17 and 19 in Equation 21.

$$n_i(T) = 2\left(\frac{2\pi K_B T}{h^2}\right)^{3/2}\left(m_n^* m_p^*\right)^{3/4} e^{-E_{bg}/2K_B T} \tag{22}$$

The exponential temperature dependence dominates the intrinsic carrier concentration, $n_i(T)$, and the plot of $\ln(n_i)$ versus 1000/T appears almost linear (neglecting the variation in the density of states function due to $T^{3/2}$ dependence and the variation in E_{bg} with temperature. Equations 16 and 18 can be modified as

$$\begin{aligned} n_{cr} &= N_{st,c} e^{-(E_C - E_i)/K_B T} e^{-(E_i - E_F)/K_B T} \\ &= n_i\, e^{-(E_i - E_F)/K_B T} \end{aligned} \tag{23}$$

where E_{bg} is band gap energy.

$$\begin{aligned} p_{vh} &= N_{st,v}\, e^{-(E_f - E_i)/K_B T} e^{-(E_i - E_v)/K_B T} \\ &= n_i\, e^{-(E_f - E_i)/K_B T} \end{aligned} \tag{24}$$

For known intrinsic concentration and temperature, the variables in Equations 23 and 24 include electron carrier density (n_{cr}), hole carrier density (p_{vh}), and Fermi-level position (E_F) relative to the intrinsic carrier concentration. One of the two quantities must be given to find the other. At very low temperatures (temperature < -190°C), negligible intrinsic electron-hole pairs exist, and the donor electrons are bound to donor atoms. As the temperature is raised, these electrons are donated to the conduction band, and at about 100°K all the donor atoms are ionized. The temperature range over which the donor atoms are ionized is called the ionization region. When every available extrinsic electron has been transferred, the electron density in the conduction band is virtually constant with temperature until the concentration of intrinsic carriers becomes comparable to the extrinsic density. At even higher temperatures the intrinsic carrier density becomes greater than the extrinsic carrier density. In most devices it is desirable to control the carrier density by doping rather than thermal electron-hole pair generation. Semiconductor materials are generally doped such that the extrinsic range extends beyond the highest temperature at which the device is to be used.

1.1.1 Thermal Voltage and Mobility

The ideal p-n junction (ideal diode equation) is affected by temperature due to variations in the saturation current, which is function of the intrinsic carrier concentration and thermal voltage. Intrinsic carrier concentration is exponentially dependent on steady-state temperature (Equation 1.21), and thermal voltage is linearly dependent on steady-state temperature ($V_T = K_B T/q$). The assumptions for ideal diode analysis (short base diode) include:

- The lengths of the 'p' and 'n' regions are much shorter than the diffusion lengths in the regions — thus little or no recombination occurs on the bulk of the quasi-neutral region.

- In limit all the injected minority carriers in the 'p' and 'n' regions combine at the ohmic contacts, at either end of the structure.

The forward bias current in the short base pn diode is

$$I_{t,fb} = I_{sat}\left(e^{\frac{V_d}{V_{tv}}} - 1\right) \tag{25}$$

where $I_{t,fb}$ is the total forward bias diode current, V_d is the voltage across the diode, V_{tv} is the thermal voltage ($= K_BT/q$), and I_{sat} is the saturation current given by

$$I_{sat} = eA_{cr,pn}\left(\frac{D_p p_{nh}}{L_{p,df}} + \frac{D_n n_{pc}}{L_{n,df}}\right) \tag{26}$$

where $L_{p,df}$ and $L_{n,df}$ are the diffusion lengths, D_p and D_n are the diffusion coefficients of the band n-regions, respectively. P_{nh} is the minority hole concentration in n region, $A_{cr,pn}$ is the sectional area of p-n junctions,

$$I_{sat} = eA_{cr,pn}\left[\frac{D_p p_{n0}}{L_{p,df}} + \frac{D_n n_{p0}}{L_{n,df}}\right] = eA_{cr,pn}\left(\frac{D_p n_i^2}{L_{p,df}N_{d,dp}} + \frac{D_n n_i^2}{L_{n,df}N_{a,dp}}\right) \tag{27}$$

where $N_{d,dp}$ is the concentration of donor atoms, $N_{a,dp}$ is the concentration of acceptor doping atoms, and n_i is the intrinsic carriers concentration, respectively (one diffusion length is the distance in the neutral region at which the minority carrier concentration is 0.37 of its value at the edge of the depletion, or space-charge, region); p_{n0} is the minority hole concentration in the n-region; and n_{p0} is the minority electron concentration in the p-region. The effect of temperature on saturation current derives its origin from the variation in intrinsic carrier concentration, n_i, which increases dramatically with temperature and is given by Equation 21 N_V and N_C in Equation 21 are the valence and conduction-band effective densities, respectively, given by Equations 17 and 19. Although $N_{st,c}$ and $N_{st,V}$ vary with temperature, the temperature dependence of intrinsic carrier concentration is primarily due to exponential term in Equation 21. In an intrinsic semiconductor, all the carriers result from the excitation of charge carriers across the forbidden energy gap. The intrinsic carrier concentration thus depends on temperature because thermal energy is the source of carrier excitation across the forbidden energy gap. The intrinsic carrier concentration is also a function of the size of the forbidden energy gap since fewer electrons can be excited across a larger gap. For silicon, the energy gap, E_g, is 1.1 eV, thus the intrinsic carrier concentration doubles for every 8°C increase in temperature near room temperature. Furthermore, the temperature dependence of the intrinsic carrier concentration may also be represented by

$$n_i^2(T) = KT^{-3}e^{-\frac{V_{bg}}{V_{tv}}} \tag{28}$$

where K is a constant, T is the absolute temperature (Kelvin), V_{bg} is the band gap voltage (for silicon @ 300° K, V_{bg} is 1.11 V), and V_{tv} is the thermal voltage (@ 300K, V_{tv} ≈ 26 mV) [Hodges and Jackson, 1988].

The temperature rate of change of the saturation current is represented by the temperature coefficient of saturation current, which describes the fractional change for I_{sat} per unit change in temperature, represented by Hodges and Jackson [1988] as

$$\begin{aligned}\frac{1}{I_{sat}}\frac{dI_{sat}}{dT} &= \frac{1}{n_i^2}\frac{dn_i^2}{dT} \\ &= \frac{3}{T} + \frac{1}{T}\frac{V_{bg}}{V_{tv}}\end{aligned} \tag{29}$$

For silicon near-room temperature, the first term in Equation 1.28 is ≈ 1% per Kelvin, and the second term is 14% per Kelvin. In other words, the saturation current approximately doubles for every 5°C rise. Moreover, the band gap energy, E_{bg}, also varies with temperature:

$$E_{bg}(T)=E_{bg}(0)-\frac{\alpha_{bg}T^2}{T+\beta_{bg}} \tag{30}$$

where $E_{bg}(0)$ is the band gap energy @ $T=0$ [Roulston, 1988], α_{bg} is the first band gap energy temperature coefficient =7.02 x 10^{-4}, β_{bg} is the second gap energy temperature coefficient = 1108, and E_{bg} is the band gap energy @ temperature T. The band gap energy decreases with temperature.

Carrier mobility, a function of steady-state temperature, is a ratio of carrier velocity to the electric field causing carrier motion. Mobility decreases with temperature and is represented by Roulston [1990] as

$$\mu_p=54\left(\frac{T}{300}\right)^{-0.57}+\frac{1.36x10^8T^{-2.33}}{1+\left(\frac{N_{d,dp}}{2.35x10^{17}\left(\frac{T}{300}\right)^{2.4}}\right)0.88\left(\frac{T}{300}\right)^{-0.146}} \tag{31}$$

$$\mu_n=0.88\left(\frac{T}{300}\right)^{-0.57}+\frac{7.4x10^8T^{-2.33}}{1+\left(\frac{N_{a,dp}}{1.26x10^{17}\left(\frac{T}{300}\right)^{2.4}}\right)0.88\left(\frac{T}{300}\right)^{-0.146}} \tag{32}$$

The temperature characteristics of these parameters is, however, of minor importance, because signal voltages in most cases are much larger than the variation of the diode voltage over the temperature range -55°C to 125°C [Hodges and Jackson 1988].

1.2 Current Gain

The current gain of the bipolar junction transistor is the collector current divided by the base current, otherwise known as $\beta_{cg,dc}$. The d.c. current gain is expressed as

$$\beta_{cg,dc} = \frac{I_C}{I_B} \tag{33}$$

where I_c is the collector current, I_B is the base current and $\beta_{cg,dc}$ is d.c. current gain. The common emitter d.c. current gain increases appreciably with increasing temperature because of improved emitter efficiency at higher temperatures. Emitter efficiency is exponentially dependent on steady-state temperature Equation 2. The exponential term, represented by $exp(-\Delta E_g/K_BT)$, arises due to the energy-gap reduction of a highly doped emitter [Kauffman and Bergh, 1968; Buhanan, 1969; RCA, 1978].

$$\gamma_{eff,em} = \frac{D_n}{D_p}\left(\frac{\int_0^{W_E} N_{d,dp} e^{-\frac{\Delta E_g}{K_B T}}}{\int_0^{W_B} N_{a,dp} dx}\right) \tag{34}$$

where $\gamma_{eff,em}$ is the emitter efficiency, W_e is the emitter width, W_b is the base width, and ΔE_g is the band gap energy.

For moderately or lightly doped emitters (concentrations below 5 x 10^{19} cm^{-3}), ΔE_g is very small, and the temperature effect on current gain is negligible. For doping concentrations above 10^{19} cm^{-3}, the hole mobility in the emitter is relatively constant with respect to temperature. Einstein's relationship $(D_p/\mu_p = K_BT/q)$ predicts that the diffusion constant must increase with temperature. The minority carrier lifetime in the emitter, τ_p, increases with temperature, which causes the diffusion length, L_p, to also increase with temperature. The increase in the diffusion length of minority carriers adds to the ΔE_g effect by increasing the current gain. The variation in the d.c. current gain is depicted by Figure 4.1. The temperature sensitivity of current gain is a contributing factor to hot-spot formation and affects the second breakdown energy limit. It is therefore advantageous to reduce the current gain temperature dependence by lightly doping the transistor base and limiting the phosphorus-doped emitter surface concentration to about 7 x 10^{19} cm^{-3} [Kauffman and Bergh, 1968; Buhanan, 1969; RCA, 1978].

Ebers-Moll equations for emitter and collector current are represented as [Hodges and Jackson, 1988]

$$I_e = I_{e,sat}\,(e^{V_{be}/V_{tv}} - 1) - \alpha_{r,cb}\, I_{c,sat}\,(e^{V_{bc}/V_{tv}} - 1) \tag{35}$$

where I_e is the emitter current, $I_{e,sat}$ is the emitter saturation current, $I_{c,sat}$ is the collector saturation current, V_{bc} is the base-collector voltage, V_{be} is the base-emitter voltages, and V_{tv} is the thermal voltage $(= K_BT/q)$.

$$I_C = \alpha_{f,cb} I_{e,sat}(e^{V_{be}/V_{tv}} - 1) - I_{c,sat}(e^{V_{bc}/V_{tv}} - 1) \tag{36}$$

where $\alpha_{f,cb}$ is the common base current gain with base-emitter junction forward biased. I_C is collector current.

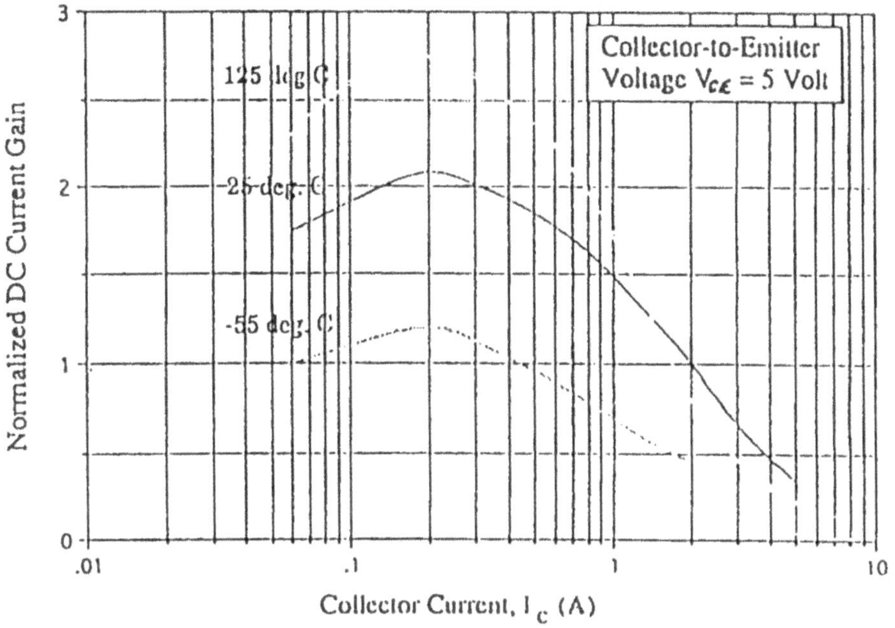

Figure 4.1 Temperature dependence of current gain for a bipolar transistor. The temperature sensitivity of current gain is a contributing factor to hot spot formation and effects the second breakdown energy limit. It is, therefore, advantageous to reduce the current gain temperature dependence by lightly doping the transistor base and limiting the phosphorous doped emitter surface concentration to about 7E19 cu.cm [Kauffman and Bergh, 1968; Buhanan, 1969; RCA, 1978]

$$I_b = I_e - I_c \tag{37}$$

$$I_b = eA_{cr,pn} n_i^2 \left(\frac{D_e}{L_{e,df} N_{e,dp}} \right) \left(e^{\frac{eV_{be}}{K_B T}} - 1 \right) + eA_{cr,pn} n_i^2 \left(\frac{D_c}{L_{c,df} N_{c,dp}} \right) \left(e^{\frac{eV_{bc}}{K_B T}} - 1 \right) \tag{38}$$

where $A_{cr,pn}$ is the cross-sectional area of p-n junction in the base region, n_i is the intrinsic carrier concentration, e is the electronic charge, D_e is the diffusion coefficients of emitter, $L_{e,df}$ is the diffusion length of minority carrier, $N_{c,dp}$ is the special density of dopant in the emitter region, V_{be} is the base-emitter voltage, and V_{tv} is the thermal voltage.

The expression for common emitter d.c. current gain, $\beta_{cg,dc} = I_C/I_B$, holds only for the active region of BJT operation. In the active region, the base-collector junction is reverse-biased, making the exponent << 1. On the assumption that the exponential base-collector term is negligible, the previous equations reduce to

$$I_B = e\, A_{cr,pn}\, n_i^2 \frac{D_e}{L_{e,df}\, N_{e,dp}} e^{V_{be}/V_{tr}} \tag{39}$$

$$I_c = e\, A_{cr,pn}\, n_i^2 \frac{D_B}{W_B\, N_b} e^{V_{be}/V_{tr}} \tag{40}$$

where D_b is the diffusion length of base and W_b is the base width.

Substituting the explicit temperature dependence into the above equations and dividing I_C by I_B, β_{cg} is given as [Marazas, 1992]

$$\beta_{cg} = \frac{N_{e,dp} K_B}{e N_b W_b} \sqrt{T \mu(T) \tau(T)} \tag{41}$$

where $\mu(T)$ is the mobility as a junction of temperature, T is the absolute temperature, $\tau(T)$ is the minority carrier lifetime as a junction of temperature.

1.3 BJT Inverter Voltage Transfer Characteristic (VTC)

Steady-state temperature alters the voltage transfer characteristic of the bipolar junction transistor (BJT). Typically, the transistor moves from the cutoff region to the active region, in the neighborhood of 27°C, when a voltage of approximately 0.7 volts $(V_{be,on})$ is applied across the base-emitter junction (V_{in}). The minimum base-emitter voltage, $V_{be,on}$, required to obtain the same $I_{b,on}$ decreases with an increase in temperature. The temperature dependence of $V_{be,on}$ can be derived from Equation 7. This value, $I_{b,on}$, represents the minimum current necessary to turn on the transistor at 27°C. The equation for $V_{be,on}$ is [Hodges and Jackson, 1988; Muller and Kamins, 1986]

$$V_{be,on}(T) = \frac{K_B T}{e} \log\left[\frac{I_{b,on} L_{e,df} N_{e,dp}\, e}{e n_i^2(T) A_{cr,pn} \mu_e(T) K_B T}\right] \tag{42}$$

where $N_{e,dp}$ is the dopant density in emitter region.

There are three regions of operation for the BJT inverter, known as the cut-off, active, and saturation regions. In the cut-off region, the input voltage, V_{in}, is less than the BJT turn-on voltage, $V_{be,on}$. The base current, I, is zero, and thus the collector current and the emitter current are also zero. The output voltage is therefore equal to V_{CC}:

$$V_{out} = V_{CE} = V_{CC} \tag{43}$$

where V_{out} is the output voltage, V_{CE} is the collector-emitter voltage, and V_{CC} is the collector supply voltage. In the active region, V_{in} is greater than $V_{be,on}$. The base current can be found from the voltage drop across the base resistor

$$I_b = \frac{V_{in} - V_{be,on}}{R_{b,ld}} \tag{44}$$

where $R_{b,ld}$ is the resistance of base-load resistor, V_{in} is the input voltage (bipolar transistor)

$$I_C=\beta_{cg,bjt}I_B \tag{45}$$

The output voltage, V_{ce}, is given by

$$V_{ce}=V_{CC}-R_C I_C \tag{46}$$

where V_{cc} is the collector supply voltage.

The temperature dependence of the output voltage is calculated from Equations 12 through.

$$V_{ce}(T)=V_{CC}-\left(\frac{R_C}{R_{b,ld}}\right)\beta_{cg,bjt}(T)\left(V_{in}-V_{be,on}(T)\right) \tag{47}$$

The third region of operation for the BJT is the saturation region. The collector current is given by

$$I_c=\frac{V_{cc}-V_{ce}}{R_c} \tag{48}$$

where V_{cc} is the collector supply voltage and V_{ce} is the collector voltage. However, when the transistor is not actually in the saturation region but on the edge of saturation, $V_{CB}=0$, so

$$I_c=\frac{V_{cc}-V_{be,on}}{R_c} \tag{49}$$

Equation 49 represents the maximum amount of current allowed in the collector before the transistor actually goes into saturation. Additionally, the maximum allowed base current is

$$I_B=\frac{I_c}{\beta_{cg,bjt}} \tag{50}$$

The input voltage can be derived from Equation 44:

$$V_{in}=I_B R_{b,ld}+V_{be,on} \tag{51}$$

The temperature dependence of the critical input voltage that would push a BJT into saturation is represented as [Marazas, 1992]

$$V_{in,cr}(T)=\frac{R_{b,ld}}{R_c}\left(\frac{V_{cc}-V_{be,on}(T)}{\beta_{cg,bjt}(T)}\right)+V_{be,on}(T) \tag{52}$$

where $V_{in,cr}$ is the critical input voltage. In bipolar devices operating high-voltage current-switching applications, the collector current plays a major role in determining transistor losses; Bromstead represents this by I_{cr} in the following equation [1991]:

$$I_{ce} = eA_{cr,pn}\left(\frac{D_n}{L_{n,df}}\frac{n_i^2}{N_{a,dp}} + \frac{D_p}{L_{p,df}}\frac{n_i^2}{N_{d,dp}}\right) \tag{53}$$

The approximate fractional change in leakage current versus temperature is represented by

$$\frac{1}{I_{ce}}\left(\frac{\partial I_{ce}}{\partial T}\right) = \frac{1}{2}\left(\frac{3}{T} + \frac{E_a}{\sqrt{N_{st,c}N_{st,v}}}\right) \quad @\ V_R\text{=constant} \tag{54}$$

where $N_{st,c}$ is the density of states in conduction band and $N_{st,v}$ is the density of states in valence band.

The change in the collector leakage current is approximately 8 °C for every 12 °C rise in steady-state temperature at the junction for the temperature range 150 °C to 200 °C. Below steady-state temperatures of 150 °C, the leakage current exhibits a plateau - i.e., it does not change at temperatures of up to 150 °C ($I_{CE}(150\ °C)$= 1.2 x 10^{-4} amp, and $I_{CE}(200\ °C)$= 1.5 x 10^{-3} amp)[Bromstead, 1991]. Emitter-base leakage current exhibits saturation in the lower temperature region (under 150°C). For temperatures higher than 150°C, the emitter base leakage current exhibits doubling for every 12°C rise in steady-state temperature,($I_{EB}(150\ °C)$= 1.3 x 10^{-4} Amp, and $I_{EB}(200\ °C)$= 2.3 x 10^{-3} Amp)[Bromstead, 1991].

1.4 Collector-Emitter Saturation Voltage

In high-voltage and current-switching applications, the saturation voltage, $V_{ce,sat}$, is vital for determining transistor losses. Under normal switch or inverter conditions, the major part of the voltage drop across the transistor occurs in the non-conductivity-modulated region close to the n^+ epitaxial layer. Assuming that both the base and the collector have entered high-level injection, the total value of the C-E saturation voltage can be written as [Roulston 1990]

$$V_{ce,sat} = \frac{I_c\rho_{epi}W_R}{A} - 2V_t\ln\left[\frac{p_h(W_b)}{n_i}\right] + 2V_t\ln\left[\frac{p_n(o)}{n_i}\right] \tag{55}$$

where $W_R = W_b - t_{ox}$ (W_b - base width, t_{ox} - oxide thickness), ρ_{epi} is the resistivity of epitaxial layer, $P_n(w_b)$ is the hole concentration @ collector-base junction, and $P_n\ (o)$ is the electron concentration @ base-emitter junction. The resistivity of the epitaxial layer is given as

$$\rho_{epi} \approx \frac{1}{e\mu_n N_{epi}} \tag{56}$$

where μ_n is the electron mobility and N_{epi} is the epitaxial layer density. The temperature dependence of electron mobility is represented as

$$\mu_n = 0.88\left(\frac{T}{300}\right)^{-0.57} + \frac{7.4x10^8T^{-2.33}}{1+\left(\frac{n_{cr}}{1.26x10^{17}\left(\frac{T}{300}\right)^{2.4}}\right)0.88\left(\frac{T}{300}\right)^{-0.146}} \tag{57}$$

where n_{cr} is the electron density, and T is the steady-state temperature in degrees Kelvin [Roulston, 1990]. The temperature dependence of $V_{ce,sat}$ is related to the intrinsic carrier concentration, n_i by

$$n_i = \sqrt{N_{st,c}N_{st,v}}e^{-\frac{E_a}{2K_BT}} \tag{58}$$

where $N_{st,c}$ is the effective density of the conduction band, $N_{st,v}$ is the effective density of the valence band, E_a is the activation energy in eV, K_B is Boltzmann's constant, and T is the steady-state temperature in degrees Kelvin [Pierret, 1987]. Table 1 represents the variation of intrinsic carrier concentration, epitaxial resistivity, and collector-emitter saturation voltage for temperatures between 20 °C and 200 °C [Bromstead, 1991].

Table 1 Variation in intrinsic carrier concentration, epitaxial resistivity, and collector-emitter saturation voltage for temperatures between 20 and 200 °C [Bromstead, 1991]

Temperature (°C)	Intrinsic carrier concentration (cm^{-3})	Epitaxial resistivity (ohm-cm)	$V_{ce,sat}$ (T)	
			Measured value	Calculated value
20	4.85×10^{9}	0.1103	0.55	0.2371
80	3.68×10^{11}	0.1486	0.6	0.3396
140	8.42×10^{12}	0.1952	0.72	0.4709
200	9.07×10^{13}	0.2496	0.85	0.6287

Chapter 5

ELECTRICAL PARAMETER VARIATIONS IN MOSFET DEVICES

1. TEMPERATURE DEPENDENCE OF MOSFET PARAMETERS

This chapter discusses the temperature dependence of electrical parameter variations in MOSFET devices. The parameters investigated for MOSFETs include threshold voltage, mobility, drain current, time delay, strong inversion leakage, subthreshold leakage, and chip availability. The temperature thresholds above which temperature dependence renders the device inoperable have been identified for each of the parameters.

1.1 Threshold Voltage

The threshold voltage is the minimum gate voltage at which the channel starts conducting. The closed-form value of the threshold voltage is derived from the current voltage characteristics. MOSFET behavior with respect to temperature can be derived by finding the incremental voltage drop along the channel, as a function of the channel current. At a distance $y_{channel}$ along the channel, the voltage with respect to the source is $V(y)$ and the gate-to-channel voltage is $V_{GS} - V(y_{channel})$. Assuming that the gate voltage exceeds the threshold voltage, the charge per unit area in the conducting channel at a point, $y_{channel}$, is represented by

$$Q_l(y_{channel}) = C_{ox}\left(V_{GS} - V(y_{channel}) - V_{th}\right) \tag{1}$$

where C_{ox} is the oxide capacitance per unit area, V_{GS} is the gate-source voltage, $V(y_{channel})$ is the voltage induced in the channel with respect to source, and V_{th} is the threshold voltage [Muller and Kamins, 1986; Hodges and Jackson, 1988]. The resistance, $dR_{channel}$ of this channel of length, $dy_{channel}$, is

$$dR_{channel} = \frac{dy_{channel}}{w_{channel}\mu_n Q_l(y_{channel})} \tag{2}$$

where $w_{channel}$ is the width of the channel perpendicular to the length, and μ_n is the average mobility of the electrons in the channel [Muller and Kamins, 1986; Hodges and Jackson, 1988]. The voltage drop along the channel of length $dy_{channel}$ is [Muller and Kamins, 1986; Hodges and Jackson, 1988]

$$dV_C = I_D dR_{channel} = -\frac{I_D dY}{w_{channel}\mu_n Q_n(y_{channel})} \tag{3}$$

Integrating along the path from the source to the drain for I_D as a function of applied voltage (for $V_{GS} \geq V_{th}$; and $V_{DS} \leq (V_{GS} - V_{th})$),

$$I_D = \frac{k_{tc}}{2}\left(2(V_{GS}-V_{th})-V_{DS}^2\right) \tag{4}$$

where V_{th} is the threshold voltage, and k_{tc} is the transconductance parameter $(k_{tc} = (\mu_n C_{ox} w_{channel})/L)$ [Muller and Kamins, 1986; Hodges and Jackson, 1988]. Equation 4 is the expression for drain current in the linear region of operation. In the saturation region, where $V_{GS} \geq V_{th}$; $V_{DS} \geq V_{GS} - V_{th}$, the drain current is represented as [Muller and Kamins, 1986; Hodges and Jackson, 1988]

$$I_D = \frac{k_{tc}}{2}(V_{GS}-V_{th})^2 \tag{5}$$

The total mobile electron charge is represented as [Muller and Kamins, 1986; Hodges and Jackson, 1988]

$$Q_n = -C_{ox}(V_G-V_{FB}-V_B-2|\phi_p|) + \sqrt{2\epsilon_s e N_a(2|\phi_p|+V_C-V_B)} \tag{6}$$

The threshold voltage is then represented by

$$V_{th} = V_{FB}+V_C+2|\phi_p| + \frac{1}{C_{ox}}\sqrt{2\epsilon_s e N_a(2|\phi_p|+V_C-V_B)} \tag{7}$$

where V_{FB} is the flatband voltage, V_C is the voltage in the MOS channel, ϕ_p is the potential in the p-type material (i.e., the difference between the Fermi level and the intrinsic concentration), C_{ox} is the oxide capacitance, N_a is the number of acceptors, e is the electronic charge, V_B is the base voltage, and ϵ_s is the silicon permittivity. The temperature dependence of threshold voltage is represented as

$$V_{TN}=P_o T+Q_o \tag{8}$$

$$V_{TP}=-P_o T-Q_o \tag{9}$$

where the constant P_o= -2.4 x 10^{-3} and Q_o = 1.72 (for the temperature range of 25-250°C) [Shoucair, 1986]. This equation applies to a typical MOSFET with V_t equal to 1 volt at 27°C. Typically, the threshold voltage of both NMOS and PMOS enhancement-mode transistors decreases in magnitude by 1.5 to 2 mV/°C with increasing temperature. Figure 5.1 (a) shows the threshold voltage variation versus temperature for n- and p-channel MOSFETs with bulk concentration 10^{15} cm^{-3}; 37(b) shows this for bulk concentration 3 x 10^{16} cm^{-3} [Wang, 1971]. However, the change in threshold voltage does not produce a very significant change in circuit performance, because even a 200 mV change in V_T does not cause a very large percentage change in $V_{GS} - V_T$ [Hodges and Jackson, 1988]. The threshold voltage of devices with heavily doped bulk is more sensitive to temperature variation than lightly doped substrates [Blicher, 1981].

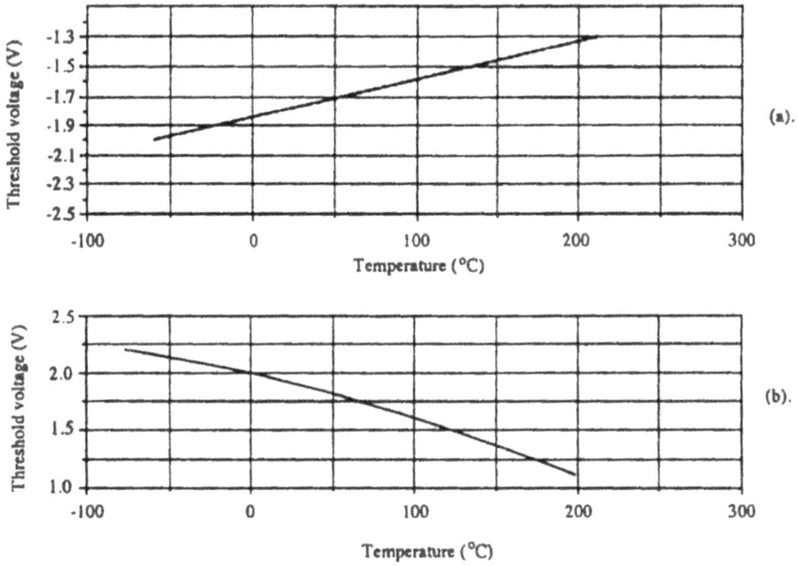

Figure 5.1 Threshold voltage variation of n- and p-channel MOSFETs (a) with bulk concentration 10^{15} cm^{-3} and (b) with bulk concentration 3 x 10^{16} cm^{-3} [Wang, 1971]. The change in threshold voltage produces a very significant change in circuit performance, because even a 200 mV change in V_T does not cause larger percentage change in V_{GS}-V_T [Hodges and Jackson, 1988].

1.2 Mobility

The mobility of carriers in a channel of an MOS transistor is an inverse function of absolute temperature. The temperature dependence is represented by

$$\mu_{effn} = \mu_{no} \left[\frac{T}{T_0} \right]^{-1.5} \tag{10}$$

$$\mu_{effp} = \mu_{po} \left[\frac{T}{T_o} \right]^{-1.5} \tag{11}$$

where μ_{no} and is μ_{po} the mobility at 27°C in both cases and T_o is 27°C [Shoucair, 1986; Tsividis and Antognetti, 1985; Muller and Kamins, 1986; Hodges and Jackson, 1988]. Thus, a 100°C rise in temperature may decrease the mobility by as much as 40% (Figure 5.2). The result is a proportional decrease in the drain current, for fixed applied voltages. The current consumption of the whole circuit may decrease considerably at high temperature. The maximum speed of operation thus decreases in proportion. The change in threshold voltage

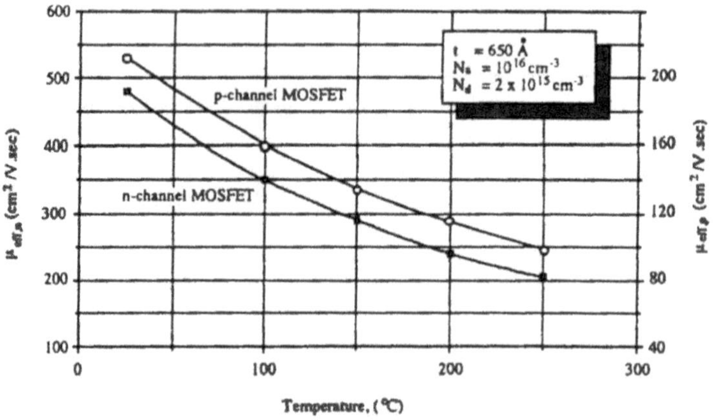

Figure 5.2 Effective channel mobility characteristics for n and p channel MOSFETs. Mobility in an inverse function of absolute temperature [Shoucair, 1986].

and mobility affect the drain current, the transconductance, and the drain-source on resistance (Figures 5.3 and 5. 4).

1.2.1 Drain Current

The drain current decreases with temperature (Figure 5.5). The temperature dependence of the drain current can be represented by

$$I_{ds} = \frac{A_{emp}}{1 + B_{emp}T} \tag{12}$$

where I_{ds} is the drain current, A_{emp} and B_{emp} are empirical constants, and T is the temperature (centigrade) [Chang, 1987]. The temperature degradation coefficient, B_{emp}, in Equation 12, is a function of gate voltage (V_{gs}) and drain voltage (V_{ds}). Typical values of the temperature degradation coefficient, B_{emp}, range from 1.03 x 10^{-3}/°C to 6 x 10^{3} /°C. This model fits experimental data for variations in drain current versus temperature for 10°C to 120°C. Theoretically, the temperature dependence of drain current versus temperature is represented by

$$I_{ds} = C_{fit}(T + 273)^{-D_{fit}} \tag{13}$$

where C_{fit} and D_{fit} are fitting parameters and T is the temperature (centigrade). The relationship between the experimental data represented by Equation 12 and theoretical calculations represented by Equation 13 is

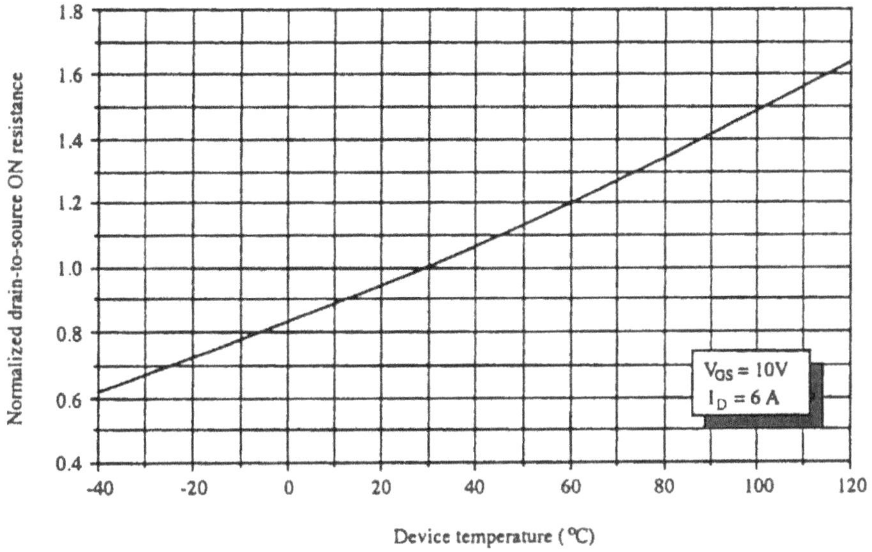

Figure 5.3 The variation in the drain-to-source ON resistance. The ON resistance increases by a factor of 3 over a temperature range of -40 °C to 120°C [Blicher, 1981].

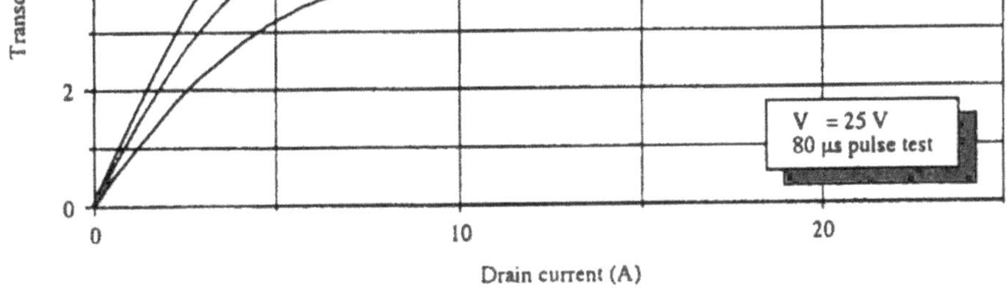

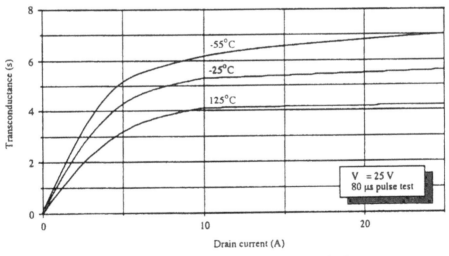

Figure 5.4 The variation in transconductance versus temperature in the temperature range of -55 to 120°C [Blicher, 1981].

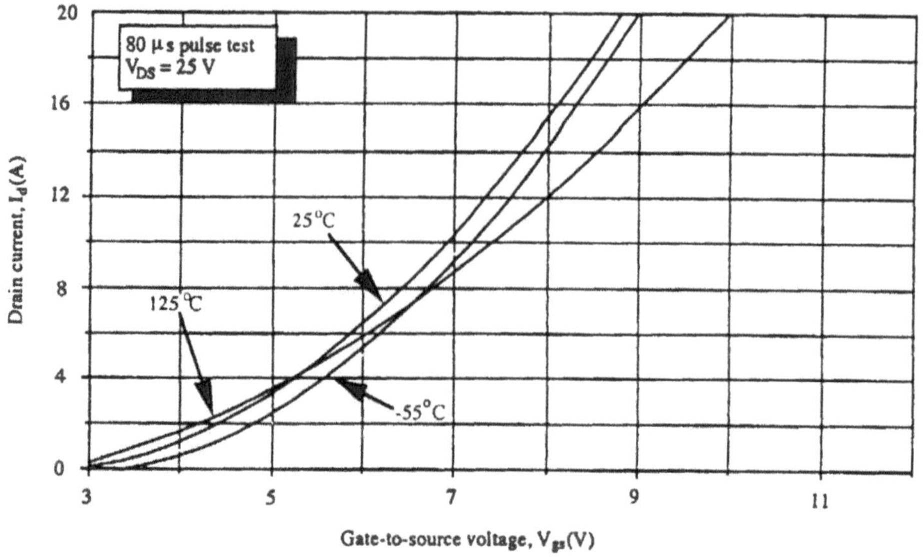

Figure 5.5 The variation in drain current in the equipment operating range of -55 to 125°C is not very significant [Blicher, 1981].

$$D_{fit} = 203B_{emp} + 0.25 \tag{14}$$

The correlation between the simplified first-order experimental equation and the theoretical equation is 0.998. The variation in drain current with temperature is not very significant in the temperature range of -55°C to 125°C (Figure 5.5) [Blicher, 1981].

1.2.2 Time Delay

In a MOS circuit, the delay time due to charging and discharging of the load capacitor through a turn-on has been represented by Chang as [1987]:

$$t_{dc} = C_l \int \frac{dV_{ds}}{I_{ds}(V_{ds}, T)} \tag{15}$$

where C_l is the load capacitance, V_{ds} is the drain voltage, I_{ds} is the drain current, and T is the temperature. The temperature dependence of time delay is represented by the linear temperature coefficient of delay time, K_{delay}, which is the time-change rate of delay time versus temperature

$$K_{delay} = \frac{dt}{t_o \, dT} \tag{16}$$

where t_o is the time delay at reference temperature [Chang, 1987]. Substituting into Equation 16 for I_{ds} from Equation 12 and for delay time, t_{dc}, from Equation 15,

$$K_{delay} = \frac{\int_{V_l}^{V_{dsat}} \frac{B_{emp}}{I_o} dV_{ds} + \int_{V_{dsat}}^{V_h} \frac{B_{emp}}{I_o} dV_{ds}}{\int_{V_l}^{V_{dsat}} \frac{dV_{ds}}{I_o} + \int_{V_{dsat}}^{V_h} \frac{dV_{ds}}{I_o}} \tag{17}$$

where I_o is the I_{ds} at 0°C, $V_l = FV_{cc}$, $V_h = (1-F)V_{cc}$ for $0 < F < 1$; V_l and V_h are the lower and upper bounds for the voltage swing used in delay-time calculation [Chang, 1987]. The total swing during device switching has been divided into linear and saturation regions. Equation 17 is simplified as

$$K_{delay} = \frac{B_d\left(\ln\left(\frac{V_{dsat}}{V_l}\right)V_{dsat} + V_l\right) + B_{sat}\left(V_{dsat} - V_l + \ln\left(\frac{V_{cc}-V_{dsat}}{V_{cc}-V_h}\right)(V_{cc}-V_{dsat})\right)}{\left(\ln\left(\frac{V_{dsat}}{V_l}\right) - \ln\left(\frac{V_{cc}-V_{dsat}}{V_{cc}-V_h}\right)\right)V_{dsat} + \ln\left(\frac{V_{cc}-V_{dsat}}{V_{cc}-V_h}\right)V_{cc}} \tag{18}$$

where K_{delay} is the linear temperature coefficient of delay time (the rate of delay time with respect to temperature), B_d is the temperature degradation coefficient at temperature 0°C, V_{dsat} is the drain voltage at saturation, V_l and V_h is the lower and higher bounds of the voltage swing during device switching [Chang, 1987]. The model can be extended to calculate the delay time for a CMOS ring oscillator, which consists of either seventeen or thirty-seven stages of directly coupled inverters. Each inverter consists of a pair of n-channel and p-channel MOSFETs. The total delay time, t_{fr}, of a CMOS inverter can be represented by

$$t_{fr} = t_f + t_r = t_{fo}(1 + K_r T) \tag{19}$$

where the subscripts represent the rise and fall times, respectively, and t_{fo} and t_{ro} are the corresponding delay times at 0°C [Chang 1987]. The total time-delay coefficient, K_{fr}, is

$$K_{fr} = \frac{K_f}{1 + \frac{t_{ro}}{t_{fo}}} + \frac{K_r}{1 + \frac{t_{fo}}{t_{ro}}} \tag{20}$$

Another model for gate delay versus temperature dependence is given by Shoucair [1987]. The gate delay is defined as the time between the input and output waveforms at the $V_{dd}/2$ points of the gate in the ring oscillator (V_{dd} is the drain supply voltage) at temperature T. The gate delay as a function of temperature is given by

$$t_d(T) \approx \frac{1.8C_L(T)V_{DD}}{\mu_N(T)C_{ox}\left(\frac{W}{L}\right)_N} \text{ X } \left(\frac{1}{\left(1 - \frac{V_{THN}(T)}{V_{DD}}\right)^2} + \frac{1}{\left(1 - \frac{|V_{THP}(T)|}{V_{DD}}\right)^2}\right) \tag{21}$$

$$t_D(T) = t_D(T_o)(1 + c(T)(T - T_o)) \tag{22}$$

where T_o is the reference temperature, $c(T)$ is the temperature rate of change of gate delay (also called the delineation factor, $1/K$ or 1/°C), V_{THN} is the threshold voltage for n-channel, V_{THP} is the threshold voltage for the p-channel, and $C_L(T)$ is the total load capacitance. The delay increases linearly with temperature and can be calculated using the delineation factor, $c(T)$, at any temperature ranging from 0.004/°C to 0.006/°C. At 250°C, the cell is half as fast as at 25°C. The delineation factor quantifying the temperature dependence is larger for cells driving loads dominated by temperature-dependent junction capacitances than for temperature-independent capacitive loads. The delineation factor is constant for a given cell over a temperature range of 25°C to 250°C [Shoucair, 1987]:

$$c(T) = \frac{1}{t_D(T_o)} \frac{dt_D(T)}{dT} \tag{23}$$

$$\frac{dt_D(T)}{dT} = t_D(T)\left(\frac{1.5}{T} + \frac{2}{V_{DD} - V_{TH}(T)} \frac{d_{TH}(T)}{dT} \right) \tag{24}$$

Shoucair estimated the temperature dependence of the rise and fall times, $t1$ and $t0$, on a sample size of seven inverters [1984]. Assuming that $|V_{TH}(27°C)| \sim 1.5$ V; $|dV_{TH}(T)/dT| \sim 3.5$ mV/°C, $V_{DD} = +5$V, the ratio of rise and fall times, $t1$ and $t0$, at 27°C and 250°C was estimated to be 1.67 ($R0 = t0(250\,°C)/t0(27\,°C)$; $R1 = t1(250\,°C)/t1(27\,°C)$; $(R1+R2)/2 = 1.67$). This is the factor by which the speed of the stage will decrease in a high-clock-rate application.

1.2.3 Leakage Currents

Leakage currents represent the limiting factor for high temperature functionality of MOSFETs. I_D versus V_{GS} curves represent a significant upward trend at temperatures above 250°C because drain leakage currents have become comparable to drain channel currents. The drain leakage current is typically four to five orders of magnitude smaller at 25°C than at 250°C. The drain or source junction leakage current is generally dominated by generation-recombination leakage over a temperature range of 25°C to $T_{transition}$ °C, and by diffusion leakage above $T_{transition}$ °C. Typically, in most CMOS processes, $T_{transition}$ °C lies in the range of 130°C to 150°C.

(I) Leakage Currents in Strong Inversion and Deep Depletion The leakage current components in an n-channel MOSFET in strong inversion comprise a drain and source well-bottom wall component, $I_B(T)$, which may be expressed as

$$I_B(T) = J_B(T).A_B \approx J_B(T).(w_{well}.d_{well}) \tag{25}$$

where $I_B(T)$ is the component of leakage current through the bottom wall of the well, w_{well} is the well width, and d_{well} is well length; a drain and source well-sidewall component, $I_{SW}(T)$, given as

$$I_{SW}(T) = J_{SW}(T).A_{SW} \approx J_{SW}(T).(w_{well}.x_j) \tag{26}$$

where w_{well} is the well width, and x_j is the well depth; a channel inversion-layer to body-junction leakage current component, $I_{CH}(T)$, given as

$$I_{CH}(T) = J_{CH}(T).A_{CH} \approx J_{CH}(T).(w_{well}.L_{eff}) \tag{27}$$

where L_{eff} is the effective channel length; and a substrate-to-well component, $I_{npwl}(T)$, which can be resolved into bottom and side wall components. For a device biased in deep depletion, no inversion layer is formed; thus the channel component of leakage current is eliminated. In a temperature range where the leakage currents are non-negligible, $T > 200°C$, the temperature dependence of drain-to-body leakage currents is represented as [Shoucair, 1984]

$$I_R(T) = eA_{db}\frac{D_p}{L_p}.\frac{n_i^2(T)}{N_d}.\alpha_{cons}A_{db}T^3 e^{\frac{-E_g(T)}{K_BT}} \tag{28}$$

where A_{db} is the area of the drain-to-body p-n junction, D_p is the diffusion constant (subscript p for p-channel or n for n-channel), L_p is the diffusion length (subscript p for p-channel or n for n-channel), $n_i(T)$ is the intrinsic carrier concentration, N_d is the substrate donor doping concentration (donor for p-channel or acceptor for n-channel), α_{cons} is the constant of proportionality, and $E_g(T)$ is the band-gap energy, $\approx 1.2 - 2.73 \times 10^{-4}\ T$.

(II) Subthreshold Leakage Currents Subthreshold leakage current flows in a leakage path through the transfer device channel while the cell (word line) gate voltage is down and the device is off. The magnitude of the leakage current depends on how well the transfer device channel can be turned off during standby, which in turn depends on the gate voltage and threshold voltage of the device. Even at the threshold voltage, the drain current is not truly zero and is of a 10^{-8} order of magnitude, multiplied by the width-to-length ratio of the device channel [Noble, 1984]. Subthreshold leakage current is highly temperature-sensitive. In the subthreshold bias region, $|V_{GS}| < |V_{TH}(T)|$, and $|V_{DS}| >> K_BT/q$, the channel of the MOSFET is weakly inverted, and the drain current results from carriers diffusing from source to drain. The temperature dependence of drain current is represented by

$$I_D(T) \approx \mu(T)\left(\frac{W}{L}\right)\frac{C_{ox}^2}{C_{ox}+C_d}(n_o(T)K_BT)^2.e^{\frac{e(V_{GS}-V_{TH}(T))}{n_o(T)K_BT}-1} \tag{29}$$

where C_d is the depletion capacitance of the space-charge region per unit area; and the temperature-dependent parameters are mobility $\mu(T)$, threshold voltage $V_{TH}(T)$, and subthreshold parameter $n_o(T)$ [Shoucair, 1989]. The value of $n_o(T)$ increases exponentially with temperature beyond temperatures in the neighborhood of 150°C to 200°C (Figure 5.6). Beyond this range, the diffusion leakage currents completely dominate the weak inversion drain-current characteristics. Thus, the temperature at which the subthreshold parameter starts to increase exponentially denotes the reasonable upper operating limit of the device, above which no turn-off can be achieved. To make sure that the device does not make a significant contribution to node leakage at high temperatures, the device off-current must be maintained at 10^{-13} A, or five decades below the current at the voltage threshold level [Noble, 1984].

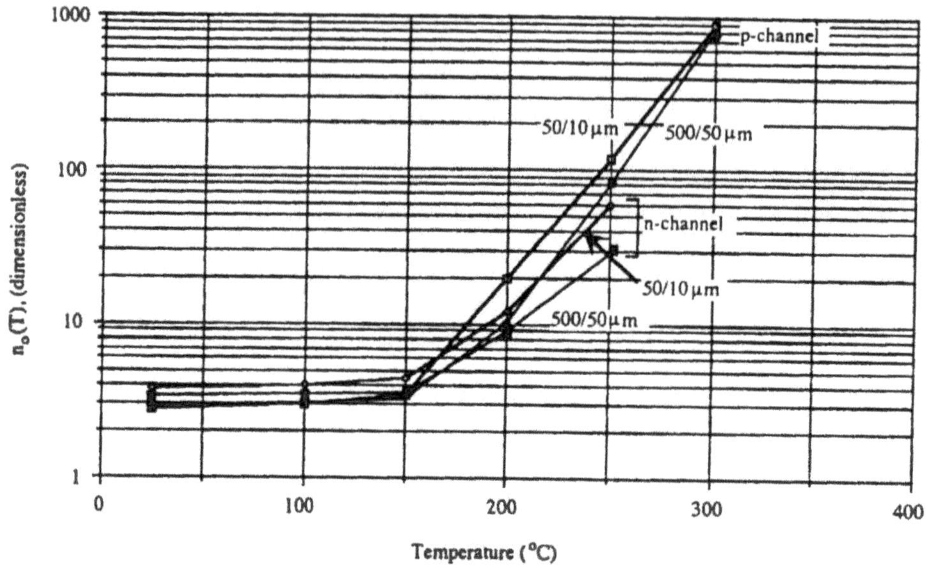

Figure 5.6 Subthreshold parameter versus temperature. At a temperature greater than 150-200 °C, diffusion leakage currents completely dominate the weak inversion drain characteristics, causing the subthreshold parameter $n_o(T)$ to exponentially increase with temperature. Depending on the device design, the temperature at which the subthreshold currents start to decrease exponentially, typically represents the practical upper operation junction temperature above which no reasonable device turn-off can be achieved [Shoucair, 1989].

1.2.4 Chip availability

The signal loss due to leakage current is proportional to time. To prevent the signal from falling below the magnitude needed for sensing, the cells in a dynamic random access memory (DRAM) need to be refreshed periodically. Chip availability is defined as the time that the chip is actually available for read and write operations, mathematically represented as

$$Availability = \frac{t_1 - 2n.t_{cy}}{t_1}.100\% \tag{30}$$

where

$$t_1 = \frac{V_1.C_s}{I_{LJ}} \tag{31}$$

where t_l is the refresh interval for a given cell, and t_{cy} is the amount of time spent actually performing the refresh operation [Noble, 1984]. If a device with a 30-nm cell-oxide thickness and a 200-mV loss of cell voltage due to leakage is maintained at temperatures below 85°C to 95°C, the chip availability is in the neighborhood of 98% to 99%. Once a chip designed for operation at 85°C to 95°C is operated at temperatures greater than 100°C, the chip leakage current becomes so high that most of the time is used just to maintain the data in the cells [Noble, 1984].

1.2.5 DC Transfer Characteristics

D.C. transfer characteristics maintain their integrity up to temperatures in the neighborhood of 270°C [Shoucair, 1984; Marazas, 1992; Zhu, 1992]. At 275°C, the transfer characteristic degrades into a flat line, due to the onset of pn-pn latchup phenomenon (Figure 5.7). The peak low-level output voltage, V_{OLP}, decreases and the peak high-level output voltage, V_{OHP} increases with temperature in the range of -50°C to 150°C. The noise margins thus increase with an increase in temperature (Figures 5.8 and 5.9).

The safe operating area of a typical MOSFET is shown in Figure 5.10. The d.c. or pulse drain currents are limited only by the power dissipation slope (I_D = max, allowable power dissipation/V_{DS}) and maximum allowable temperature. The slope in Figure 5.10 is for a device protected from second breakdown by either an external or a built-in clamp like the zenner diode, which breaks in advance of the MOS transistor drain breakdown.

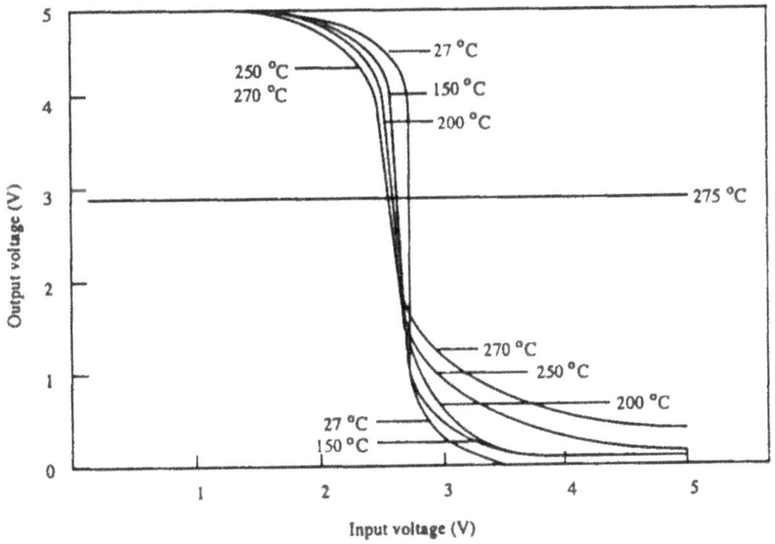

Figure 5.7 DC transfer characteristic of a typical CMOS inverter with temperature as a parameter. At temperature > 270°C, curves degenerate into a flat line due to onset of pnpn latchup phenomenon [Shoucair, 1984].

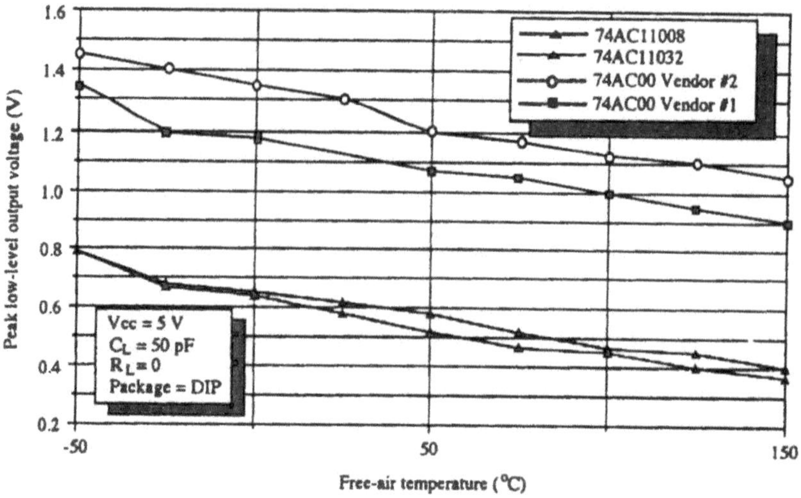

Figure 5.8 The variation in the peak low-level output voltage versus steady state temperature. The reduction in output voltage indicates an increase in the noise margins at higher temperatures [Texas Instruments, 1987].

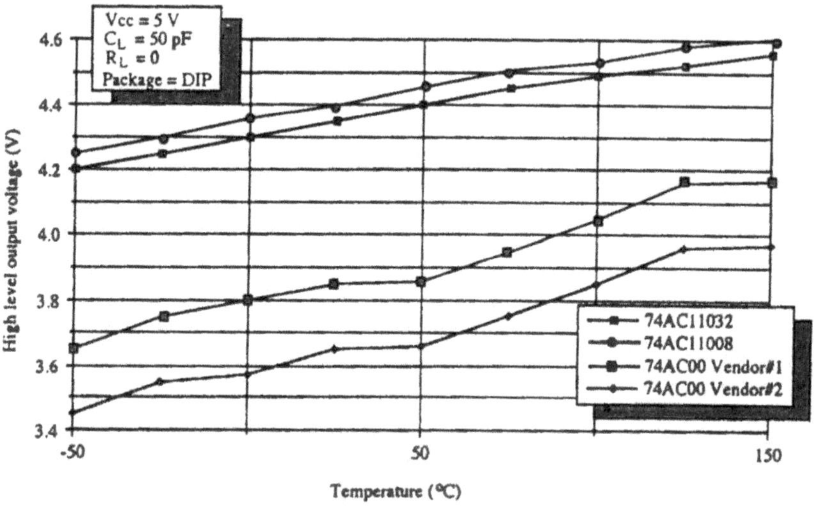

Figure 5.9 The variation in the peak high level output voltage versus steady state temperature. The increase in the output indicates an increase in noise margins at high temperatures [Texas Instruments, 1987].

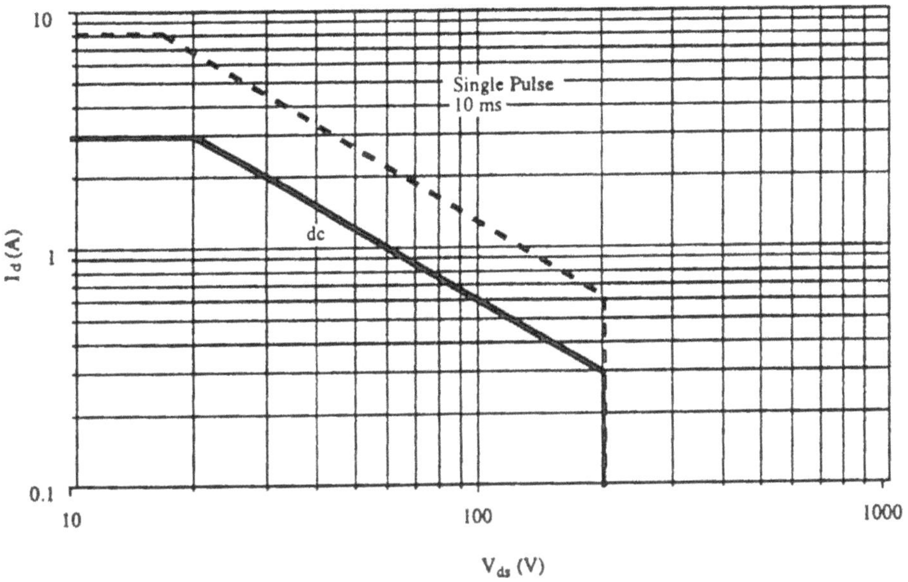

Figure 5.10 Typical safe operating area (SOA) of a power MOST, with device protected from second breakdown [Blicher, 1981].

Chapter 6

A PHYSICS-OF-FAILURE APPROACH TO IC BURN-IN

Screening is a process that detects defects in a sub-population of devices that may exhibit early failure characteristics unlike the main population. Such defects occur due to multiple variabilities detected either through non-stress screens or by stress screens, including burn-in.

This section examines the problems with existing burn-in methods and presents a physics-of-failure approach to burn-in.

1. BURN-IN PHILOSOPHY

Burn-in has been used as a screen that subjects devices to extended operation at high temperatures to precipitate early failures and eliminate the infant mortality region. Traditionally, burn-in has been based on the bathtub curve failure pattern. The bathtub curve is used to determine the screening magnitude and level. However, these failure patterns have become outdated and, as a result, their relevance has diminished.

The goal of burn-in is to prevent failures from occurring in the field. Burn-in is typically a requirement imposed by the customer to demonstrate higher product reliability; manufacturers have different burn-in procedures for the same class of components for military and commercial customers.

The typical burn-in procedure consists of placing parts in a thermal chamber for a specific amount of time under an electrical bias. During and/or after thermal environmental exposure, functional tests are conducted. Parts that fail the screen are discarded; parts that survive can be used.

2. PROBLEMS WITH PRESENT APPROACH TO BURN-IN

A review of the burn-in practices used by some leading IC manufacturers reveals that even though burn-in has been regarded as a method for eliminating marginal devices with defects from manufacturing aberrations, the specifics of burn-in vary (Table 1). Most companies have their own burn-in specifications for commercial products; MIL-STD-883 is used to satisfy burn-in requirements for military products. Other companies use only MIL-STD-883, but the selection of Method 1015 burn-in procedures for quality assurance also seems to be arbitrary. The emphasis is on empirical analysis, without any pointers to cost-effective application or subsequent manufacturing or assembly process modifications.

Burn-in at present is a generic procedure consisting of a combination of time, steady-state temperature, and electrical stress. Burn-in procedures are often conducted without any prior identification of the nature of the defects to be precipitated, the failure mechanisms active in the device, or their sensitivity to steady-state temperature stress or without any quantitative evidence of the improvement achieved by the process.

By looking at the data from various companies burn-in has shown that it is ineffective for precipitating many failures. Data collected from various procedures sources shows that the majority of failures precipitated by burn-in are not valid. Valid failures include such things as mechanical damage, broken bond and broken package pins. Non-valid failures include such things as handling damage, for example, electrostatic discharge. It has been shown that burn-in detects less than 0.5% of failures of which less than 0.002% were valid (my burn-in article). Therefore, the failures burn-in precipitates are unlikely to occur in the field which defeats the purpose of burn-in.

Burn-in may not precipitate many failures because it is performed under the widespread trust that the failure mechanisms are steady-state temperature dependent. However, it has generally not been determined whether steady-state temperature, temperature change, the rate of temperature change, or temperature gradients induce failures. An example that indicates that the failures are not steady-state dependent is presented. TriQuant Semiconductor found while testing their GaAs ICs that burn-in was ineffective in actuating any failure mechanisms [Allen 1991]. The reasons were traced to dive architecture and failure mechanisms that had no dependence on steady-state temperature.

Another reason could be because burn-in is conducted without prior knowledge of what failure mechanism is to be precipitated. The use of burn-in without attention to the dominant failure mechanisms and the nature of their temperature dependencies is a misapplication of reliability concepts. Such use of burn-in may cause failure avoidance efforts, without yielding anticipated overall results, or expensive system implementations whose costs and complexities exceed the anticipated benefits in reliability.

Since burn-in does not precipitate many failures, some believe that it is more effective if it is applied for longer duration's. However, Motorola concluded, after numerous tests, that after 160 hours, the effectiveness of burn-in decreases significantly with close to zero failures in the succeeding 1000 hours (Motorola 1990). Other problems result due to burn-in which include:

- palladium damage
- increase in leakage currents
- damage induced by additional handling

Burn-in has the potential to damage palladium lead finishes which can cause solderability problems. What happens is the finish on the leads disappears leaving a surface that cannot be soldered upon. It has also been shown that plastic parts degrade more severely than their ceramic counterparts after exposure to various radiation dosage levels (my article). The leakage current increases to the point where the performance is altered to an undesirable state. The reason for this is due to the presence of various materials present in the encapsulant which is not present in ceramic parts. During the burn-in process, parts are inserted and withdrawn from sockets, temperature chambers which makes them susceptible to additional handling damage. Handling damages that lead to failures include mechanical damage (e.g. bent leads), electrostatic discharge (ESD), and electrical overstress (EOS) failures. Many studies have been performed verifying the fact that burn-in is the source of EOS/ESD damage.

Table 1 Burn-in time and temperatures

Source	Min. Temp. T_a (°C)	Time (hours)	Test Condition	Comments
MIL-STD-883 Method 1015	100, 105, 110, 115, 120	Class B: 352, 300, 260, 220, 190	Hybrids only	Either of the combinations of the cited temperature and time is used for burn-in of hybrids.
	125	Class S: 240 Class B: 160	A - E	Any of the specified combinations of temperature and time can be used for burn-in according to Method 1015 of MIL-STD-883. The various conditions of burn-in are defined by the electrical stress, steady state temperature (ranging from 100°C to 250°C), and time period (12 to 352 hours). Conditions include: ● Test condition A: steady state temperature, reverse bias ● Test condition B: steady state temperature, forward bias ● Test condition C: steady state temperature, power and reverse bias ● Test condition D: parallel excitation ● Test condition E: ring excitation ● Test condition F: temperature accelerated test [MIL-STD-883C, 1983: last revision incorporated 1990]
	130	Class S: 208 Class B: 138	A - E	
	135	Class S: 180 Class B: 120	A - E	
	140	Class S: 160 Class B: 105	A - E	
	145	Class S: 140 Class B: 92	A - E	
	150	Class S: 120 Class B: 80	A - E	
	175	Class B: 48	A - E	
	200	Class B: 28	A - E	

Source	Min. Temp. T_a (°C)	Time (hours)	Test Condition	Comments
MIL-STD-883 Method 1015 (contd.)	225	Class B: 16	A - E	
	250	Class B: 12	A - E	
INTEL Corporation				
Intel Spec.	125°C	Memory products: 48 hours		Dynamic burn-in [Intel 1989, Intel 1990]
MIL-STD-883 Method 1015	125°C	Military products: 160 hours	Method 1015, Condition C or D	
Advanced Micro Devices Inc.				
MIL-STD-883 Method 1015	125°C min	Military products: 240 hours	Method 1015, Condition C or D	[Advanced Micro Devices 1990]

Source	Min. Temp. T_a (°C)	Time (hours)	Test Condition	Comments
LSI Logic Corporation MIL-STD-883 Method 1005	125°C	48 hours	Static or DC burn-in; dynamic burn-in	The results of production burn-in are measured as a percentage fallout rate or PDA (Percent Defective Allowable). The PDA calculation is simply the reject rate, the number of failures divided by the total number of devices in the lot, and the result compared against target PDA [LSI Logic 1990].

Source	Min. Temp. T_a (°C)	Time (hours)	Test Condition	Comments
Texas Instruments Inc.				
MIL-STD-883 Method 1015	125°C min.	MOS memory and LSI		PDA = 5%
MIL-STD-883 Method 1015		JAN S, monitored line, SEQ: 240 hours		[Texas instruments 1988]
Power burn-in MIL-STD-750 Method 1039	25°C	Optocoupler screening: JAN, JANTX, JANTXV, 4N22, 4N23, 4N24JAN, 4N22A, 4N23A, 4N24A: 168 hours	V_{cc} = 20 Vdc V_{ce} = 10±5 Vdc PT=275±25 mW I_F = 40 mA	

Source	Min. Temp. T_a (°C)	Time (hours)	Test Condition	Comments
Motorola Inc. monitored burn-in	125°C 85°C	6 hours 10 minutes	electrical and parametric measurement	Monitored burn-in is a technique in which devices are operated at elevated temperature and voltage for an extended period, followed by short duration at a lower temperature and voltage during which parametric and functional tests are performed [Eachus, MOTOROLA 1984]. The test sequence consists of two stages which include: • Temperature is increased from ambient to 125°C in 15 minutes and kept at that temperature for 6 hours • Temperature is lowered from 125°C to 85°C and allowed to stabilize for 10 minutes before electrical and parametric measurements are performed [MOTOROLA 1991]
Integrated Device Technology Inc. MIL-STD-883 Method 1015 IDT spec. MIL-STD-883 Method 1015	125°C min. 125°C min. 125°C min.	Military products: 160 hours Commercial products: 16 hours Military hermetic modules: 44±4 hours	Method 1015, Condition D Method 1015, Condition D	[Integrated Device Technology 1989]

Source	Min. Temp. T_a (°C)	Time (hours)	Test Condition	Comments
Cypress Semiconductor Inc.				Either of the conditions is used
Cypress Semiconductor spec.	150°C	Level 2 plastic and hermetic parts: 12 hours		[Cypress Semiconductor 1990]
MIL-STD-883 Method 1015	125°C min., or 150°C	JAN, SMD/Military grade products: 160 hours 80 hours	Method 1015, Condition C or D	
	125°C	Military grade modules: 48 hours		

Components inserted into and extracted from sockets, temperature chambers, and pre-and post- test procedures can suffer additional handling damage in the form of bent leads and electrostatic discharge. An example of burn-in-actuated failures was observed by Swanson and Licari [1986], who examined the effects of burn-in on hybrid package hermeticity. They found that low leak rates before screening in many hybrid packages increased by over an order of magnitude after burn-in. Packages that maintained a leak rate of 10^{-9} atm-cc/sec after burn-in had a moisture content of 3,000 to 5,000 ppm after screening, compared with less than 1,000 ppm before. The increase in leak rates was attributed to the failure of glass-to-metal lead seals and lid seals, which were found to open during burn-in and close again. The increase in moisture was further attributed to outgassing of die and substrate attach epoxies inside the package.

Historically, ionic contamination has been the dominant mechanism precipitated by burn-in [Allen 1991]. Sodium, potassium, or ions in the oxide of silicon MOS devices under bias and temperature lead to junction leakage and threshold voltage shifts that cause failure. Cool-down under bias and retest within 96 hrs. of burn-in produces a relaxation of bias-induced charge separation. GaAs MESFET-based ICs have no oxide between the gate metallization and the surface of the channel -- the interface is a Schotkky diode. Similarly, the MESFET device is unipolar, majority-carrier conducting, and reliant on the semi-insulating bulk material to achieve device isolation. There are no junctions and no leakage to consider.

In an effort to improve reliability, microelectronic manufacturers have often subjected devices to increasingly longer periods of burn-in. However, Motorola noted that most of the failures precipitated by burn-in occur in the first 160 device hours, with few or no failures over the next 1,000 hours. This is controlled by the fact that the long-term projected failure rate, based on the number of failures over 1,000 hours, is of the same order of magnitude as the actual measured failure rate over 1,000 hours [Motorola 1990].

3. A PHYSICS-OF-FAILURE APPROACH TO BURN-IN

3.1 Understanding Steady-state Temperature Effects

The use of burn-in without attention to the dominant failure mechanisms and the nature of their temperature dependencies is a misapplication of reliability concepts. Such use of burn-in may cause failure avoidance efforts, without yielding anticipated overall results, or expensive system implementations whose costs and complexities exceed the anticipated benefits in reliability. While burn-in can be used for certain types of failures that can be accelerated by steady-state temperature effects, more insight into the failure mechanism can yield better solutions in terms of design and processes.

Microelectronic design and corrective action are often misdirected because of the confusion between quality and performance. Quality is a measure of the ability of a device to fulfill its intended function. Device performance, defined by electrical parameters such as threshold voltage, leakage currents, and propagation delay, is dependent on steady-state temperature. While burn-in may not affect the reliability of the device, it may uncover the fact that at high temperature, the device does not meet performance requirements. This may serve as an indicator for a design change, or for the unsuitability of the technology for high-temperature operation. Burn-in, where performance is checked after the temperature is lowered, will not uncover this type of problem.

3.1.1 Setting up the Burn-in Profile

A physics-of-failure approach to burn-in considers the potential material defects, design inconsistencies, and manufacturing variabilities for each process that could cause defects in the product. The burn-in methodology is an iterative process consisting of the following major steps:

- *Identify failure mechanisms, failure sites and failure stresses.*
  Development of a burn-in program encompasses identifying potential failure mechanisms, failure sites, and failure stresses, active in a device technology. The burn-in process must be tailored to specific failure mechanism(s) at specific potential failure site(s) in order to be effective. The failure mechanism(s) and failure sites(s) depend on the materials and product processing technologies. Burn-in conditions are therefore specific to the manufacturing technology and hardware. The manufacturing sequence should be studied and possible defects, introduced due to the processing variabilities at each manufacturing stage, should be identified. The dominant failure stresses accelerating the failure mechanisms can be identified based on knowledge of the damage mechanics. The burn-in stress sequence will encompass those stresses that serve as dominant accelerators of the failure mechanisms.

- *Identify the combination of stresses to activate the identified failure mechanisms cost-effectively.*
  Typically, there may be a number of failure mechanisms dominant in a device technology; each may have a dominant dependence on a different stress. Thus, in order to activate all the failure mechanisms, the dominant stresses need to be applied simultaneously and cost-effectively. To quantitatively determine the magnitude of stresses necessary to activate the failure mechanisms and arrive at the desired cost- effective combination of stresses, models must be developed for each failure mechanism, as a function of stresses, device geometry, material, and magnitude of defects and design inconsistencies. The quantitative models, however, can aid only in relating the magnitude of a particular type of stress to manufacturing flaws and design inconsistencies, based on physics-of-failure concepts. The real case is more complex, involving interactions of stresses causing failure earlier than predicted by superposition of different stresses acting separately. There may be more than one failure mechanisms in a device technology. Each of the mechanisms will have its own dependence on steady state temperature, temperature cycle, temperature gradient, and time dependent temperature change. An ideal case will be when an optimized combination of the relevant stresses is used to activate the failure mechanisms in a cost effective manner. The desired combination of stresses is a function of the physics of the failure mechanism, and response of package material and configuration. The approach to arriving at the desired stress level consists of subjecting the components to discrete stress levels of steady state temperature, temperature cycle, temperature gradient, time dependent temperature change, and voltage. The selection of the temperature stress level should be based on the knowledge of designed-for temperature of the device, the temperature of the device during operational life and the thresholds for various failure mechanisms. Conducting step stress and HAST tests for various magnitudes of each type of stress applied separately will give failure results in terms of number of cycles to failure, failure mechanisms activated, and failure sites. From the test results the stress levels required for activation of failure mechanisms will be identified.

- *Conduct burn-in and evaluate effectiveness*
  Burn-in should be assessed based on root cause failure analysis of the failed components, revealing the failure mechanisms, failure modes, and failure sites. Inappropriate burn-in stresses will either damage good components by activating mechanisms not otherwise noticed in operational life, or allow defective parts to go through. To make sure the stress level and duration is right, the amount of damage (or the life consumed) for products without defects ("good" products) must be evaluated. If necessary, the stress levels modified if necessary.

Product reliability (due to the design improvements) must be used as the index for subsequent burn-in decisions. Physics of failure approach is used to determine the effective acceleration of device dominant failure mechanisms and is given by: $A_{eff} = (A_T A_V A_x ...)$. Acceleration factors for dominant failure mechanisms are used to determine burn-in time (t_{bi}) and temperature (T_{bi}). The effective burn-in time is given by: $t_{eff} = (A_T A_V A_x ...) t_{bi}$. [Jensen 1982]

- *Decisions regarding in-process monitors and burn-in modifications*

The above steps should be repeated until all products have the required expected life, with an optimized return on investment. The burn-in process should be augmented (supplemented or complemented) with in-line process controls to attain the desired quality and reliability. The physics of failure models along with the burn-in results will determine the optimal manufacturing stress levels and process parameters for minimal defect levels. The in-line process controls will ensure that the process parameters are maintained at their optimal values to minimize the occurrence of defects.

Economic analysis should indicate whether the burn-in should be continued or modified. A cost-effective burn-in program addresses all the relevant failure mechanisms by employing a minimum set of devices. Burn-in is recommended for all products which are in the development stage and do not have a mature manufacturing process. Burn-in at this stage not only improves the quality and reliability of the products but also assists in determining product and process (manufacturing, assembly, testing) corrective actions. Products with a standardized design and a relatively mature process need burn-in only if the field failure returns indicate infant mortality. A cost analysis and return on investment is conducted to calculate the economics of the burn-in program. Analysis of cost and return of investments based on the customer satisfaction and the hidden factory costs (the costs associated with the factory-inputs which do not add value to the product, like product inspection, testing, rework, etc.) determine the profits to an organization. Burn-in economics are critical in convincing management about the benefits that accrue from burn-in and provide a benchmark for making improvements in the next burn-in "cycle".

Chapter 7

GUIDELINES FOR TEMPERATURE-TOLERANT DESIGN AND USE OF MICROELECTRONIC DEVICES

Manufacturers of microelectronic devices often specify supply voltage limits and threshold values for power dissipation, junction temperature, frequency, and output current. Given these rated values, the equipment designer often elects to lower these specifications. This practice is known as "derating" and it has been perceived to provide a reliability safety factor. However, the validity of the derating factors and the utility of the practice of derating in terms of improving the reliability of microelectronics is not well founded. This chapter presents an overview of the practice of derating and then provides a science-based method for establishing effective design and operating profiles for microelectronic devices.

1. PROBLEMS WITH THE PRESENT APPROACH TO DEVICE DERATING

Derating is a technique through which either the stresses acting on a part are reduced, or the strength of the part is increased, in correspondence with allocated or rated strength-stress factors. When the equipment designer decides to select an alternative component with higher rated junction temperature or make a design change that maintains the temperature, consistently below the rated level, the component is said to have been derated for thermal stress. Thermal derating is one of the most common derating methodologies. The reliability of electronic systems is believed to be very sensitive to component derating and exactly how the derating is applied. Guidelines for derating of devices have emphasized lowering the steady state operating temperature [Brummet, 1982; Eskin, 1984; Naval Air Systems Command AS-4613, 1976; Westinghouse, 1986]. Some examples are presented in Table 1.

Table 1 shows that the thermal derating criteria vary among government agencies and companies. For example, Westinghouse has a set of derating guidelines for electronic components, in which ECL, TTL, and CMOS digital components, linear devices, and hybrids are equally derated for thermal stress. Derating for thermal stress involves reducing the maximum operating temperature to around 0.6 $T_{m,j}$ (maximum junction temperature). The Air Force Rome Laboratories has a set of derating guidelines in which microcircuits are grouped as: complex, digital, hybrid, linear, and memory. The exact value of the maximum allowable junction temperature depends on the derating level. There are three derating levels, depending on the criticality of the component. In all cases, derating for thermal stress involves lowering the maximum junction temperature. There are some shortcomings with this approach of derating. These are discussed in this chapter. An approach based on physics-of-failure is also discussed.

1.1 Dependency on Other Thermal Parameters

Implicit in the derating methodologies of Table 1 is the assumption that steady-state temperature is a dominant accelerator of microelectronic device failure mechanisms. Assigning a generic value of lower temperature to a technology (i.e. TTL, CMOS) or a packaging class (i.e. hybrids), based on the assumption that all devices operating at that lower temperature will be more reliable, is arbitrary.

Table 1 Guidelines for thermal derating

Microcircuit manufacturer/ acquisition agency	Component type		Derating criteria (max. allowable junction temperature)
Westinghouse Electric [Westinghouse, 1986]	digital: TTL/ECL		$0.6\ T_{J(rated)}$
	digital: CMOS		$0.6\ T_{J(rated)}$
	linear, amplifiers		$0.6\ T_{J(rated)}$
	linear, regulators		$0.6\ T_{J(rated)}$
Rome Laboratories [RADC-TR-82-177 RADC-TR-84-254]	complex (LSI, VLSI, VHSIC)	derating level 1	85°C
		derating level 2	100°C
		derating level 3	125°C
	digital	derating level 1	85°C
		derating level 2	100°C
		derating level 3	110°C
	hybrid	derating level 1	85°C
		derating level 2	100°C
		derating level 3	125°C
	linear	derating level 1	80°C
		derating level 2	95°C
		derating level 3	105°C
	memory	derating level 1	85°C
		derating level 2	100°C
		derating level 3	125°C
Naval Air Systems Command [AS-4613, 1976]	digital: TTL/ECL	class C	$T_{max(rated)}$ - 20°C
		class B	$T_{max(rated)}$ - 25°C
		class A	$T_{max(rated)}$ - 30°C
	digital: CMOS	class C	$T_{max(rated)}$ - 20°C
		class B	$T_{max(rated)}$ - 20°C
		class A	$T_{max(rated)}$ - 30°C
	linear, amplifiers	class C	$T_{max(rated)}$ - 20°C
		class B	$T_{max(rated)}$ - 25°C
		class A	$T_{max(rated)}$ - 30°C

Thermal derating guidelines, such as those given in Table 1 must account for the dominant failure mechanisms or their temperature dependencies. In particular, temperature cycle effects, which are failure accelerator at mating interfaces, must be accounted for in device derating criteria. The maximum number of temperature cycles that the device can endure is a function of fatigue failure mechanisms, such as wire-interconnect fatigue, die fracture, or die and substrate attach fatigue and solder joint fatigue. These mechanisms may or may not be dominant in the device architecture, depending on the stresses strains or plastic work generated at the mating interfaces, which are a function of interface geometry, and on material characteristics, including CTE mismatches.

Localized temperature gradients can exist in the chip metallization, chip, substrate, and package case, due to variations in the conductivities of material produced by defects in the form of voids or cracks. The locations of maximum temperature gradients in chip metallization are sites for mass transfer mechanisms, including electromigration. Current thermal derating guidelines do not provide any limits on these types of thermal stresses.

1.2 Interaction of Thermal and non-Thermal Stresses

Thermal and non-thermal stresses are not independent of each other in precipitating failure as implicitly assumed in most derating guidelines. For example, electromigration is accelerated by temperature and current density. Therefore, device reliability, for mechanisms with a dominant dependence on more than one operating stress, (temperature and non-temperature) complicated by dependence on magnitudes of manufacturing defects, needs to be maximized by more than just lowering temperature, as existing derating criteria do.

The interaction of various temperature and non-temperature stresses modifies the dominant dependence of the failure mechanisms on one or more of the stresses. Temperature transients generated by the duty cycle (ON/(ON+OFF)) modify the dependence of the metallization corrosion on steady-state temperature. At low duty cycle values, metallization corrosion has a dominant dependence on steady state-temperature. However, at higher values (≈ in neighborhood of 1.0), metallization corrosion has a dominant dependence on duty-cycle and a mild dependence on steady-state temperature.

1.3 Low Temperature Device Degradation

Thermal-stress derating guidelines rule out the option of reliable system designs at higher temperatures. In addition, they give the designer a false sense of security about achieving increased reliability at lower temperatures. Lower temperatures may not necessarily increase reliability, since some of the failure mechanisms are inversely dependent on temperature; for example, device technologies with hot carriers as the dominant failure mechanism may have lower reliability at lower temperatures.

Apart from the hot carrier induced failure mechanisms, there are other situations where very low temperature can adversely effect performance. Huang [1993] found larger degradation of current gain at lower temperature while stress testing bipolar transistors. Some other phenomena like large increase in leakage current in Poly-SOI MOSFET [Bhattacharya, 1994], logic swing loss in BiCMOS circuits [Rofail, 1993], kink and hysteresis [Kasley, 1993] occur at cryogenic temperatures.

1.4 Variations in Device Types

Many of the existing derating guidelines had been developed in mid seventies and remained largely unchanged since then. The breakdown of the devices into different categories for thermal derating follows those historical patterns. For example, Westinghouse and Naval Air Systems Command guidelines [see Table 1] derate in a similar manner for all CMOS devices. There are many variations in the CMOS technology with respect to their temperature tolerance. For example, twin tub CMOS technology is much more tolerant to latch up breakdown than other types of CMOS devices [Uyemura, 1988].

2. AN ALTERNATIVE APPROACH FOR THERMALLY TOLERANT DESIGN

The approach for thermally tolerant design guidelines is different from the current derating methodology in both its goals and its methods of arriving at the values of the guidelines. Instead of derating the values provided by device manufacturers, this approach directly determines the safe operating envelope. The values such obtained are not for a device class or type, but are for a particular device being used in a specific architecture. The goals of obtaining this appropriate thermal boundary are two-fold. The first goal is to achieve a desired mission life for the device (as opposed to improving MTBF). The second goal is to maintain critical electrical parameters of the device under consideration. Also, this methodology does not limit itself to considering only the effects of junction temperature, but also includes the effects of other (and often more critical) thermal parameters such as temperature cycling patterns, spatial and temporal temperature gradients. The last, and possibly the most important difference of this approach from the current derating method is the process through which the limits are established. The failure and performance degradation mechanisms for a particular device are evaluated and the limiting values of the allowable stresses which effect those mechanisms are determined.

The thermally tolerant design problem for dominant stress accelerators for a particular device architecture requires controlling the stress with maximum sensitivity with respect to life in order to achieve optimization. Some stresses can be easily controlled, while others may be hard to determine because they are a function of device architecture. Therefore, the practical situation may disclose that the most effective parameter is not the easiest to derate. The approach presented here allows the evaluation of the relative sensitivity of operating life to various stresses. Once the dominant stresses have been identified, the limiting values of the stresses which strongly effect life and are also easiest to control are determined (Figure 7.1).

The limiting values of steady-state temperature, temperature cycle magnitude, temperature gradient, and time-dependent temperature change, including non-temperature operating stresses, are determined for a desired device mission life. Calculated limiting levels of temperature and non-temperature operating stresses for devices using approach are specific to a design architecture and cannot be generalized to a device technology. The thermally tolerant design problem can thus be stated as follows:

$$
\begin{array}{l}
\textit{constraints:} \\
\quad \textit{desired mission life (hours)} \\
\quad \textit{device architecture (material, geometry)} \\
\quad \textit{performance parameters } (P_i;\ i = 1 \text{ to } m)) \\
\quad \textit{worst case manufacturing defect magnitudes} \\
\textit{problem:} \\
\quad \textit{find limiting } \left(T,\ \Delta T,\ \nabla T,\ \dfrac{\partial T}{\partial t} \right) \textit{ and non-temperature stresses}
\end{array}
\qquad (1)
$$

where P_i ($i = 1$ to m) are the critical device performance parameters. Determining limiting stress values involves addressing both performance and reliability requirements. Determining limiting stresses for performance requires evaluation of stress influence on critical device parameters. Determining limiting stress values for reliability involves evaluation of stress influence on device life under dominant failure mechanisms.

Device parameters considered in thermally tolerant design for performance are a function of device technology, which, for BJT devices, include device thermal voltage, current gain, and the invertor voltage transfer characteristic and for MOSFET devices include, threshold voltage, the invertor voltage transfer characteristic, and the invertor propagation delay:

$$Performance = \begin{cases} f(V_{thermal}, \beta_d, IVTC, \ldots) & \text{for BJT devices} \\ f(V_{th}, IVTC, IPD, \ldots) & \text{for MOSFET devices} \end{cases} \tag{2}$$

where $V_{thermal}$ is the thermal voltage, β_d is the device current gain, V_{th} is the threshold voltage, *IVTC* is the invertor voltage transfer characteristic, and IPD is the invertor propagation delay. Threshold values for temperature and non-temperature stresses that cause the critical parameters of the device technology to exceed acceptable ranges are calculated based on closed-form models relating device architecture to critical parameters.

$$S_p(i,j) = \begin{pmatrix} T_{p(1)} & \Delta T_{p(1)} & \nabla T_{p(1)} & \frac{\partial T_{p(1)}}{\partial t} & J_{p(1)} & E_{p(1)} & \cdots & S(1,n) \\ T_{p(2)} & \Delta T_{p(2)} & \nabla T_{p(2)} & \frac{\partial T_{p(2)}}{\partial t} & J_{p(2)} & E_{p(2)} & \cdots & \cdots \\ T_{p(3)} & \Delta T_{p(3)} & \cdots & & & & & \cdots \\ \cdots & & & & & & & \cdots \\ \cdots & & & & & & & \\ S(m,1) & \cdots & \cdots & \cdots & \cdots & \cdots & \cdots & S(m,n) \end{pmatrix} \tag{3}$$

where S_p is a $(m \times n)$ matrix of threshold stress values for device performance. The m rows correspond to m (i.e., $P(i)$, for $i = 1$ to m) critical parameters for the device technology (such as threshold voltage, the invertor voltage transfer characteristic, and the invertor propagation delay, for MOSFET devices). The n columns correspond to the n temperature and non-temperature stresses that affect the critical parameters. The acceptable range of stress values for device performance is calculated as follows:

$$\begin{aligned} &For\ i = 1\ to\ m;\ j = k \\ &If\ \left.\frac{\partial P_i}{\partial S_p(i,j)}\right|_{\substack{for\ i = 1\ to\ m \\ j=k}} > 0\ ;\ calculate\ \min(S_p(i,j))\Big|_{\substack{for\ i = 1\ to\ m \\ j=k}}\ for\ P_i < P_{max} \end{aligned} \tag{4}$$

$$\begin{aligned} &For\ i = 1\ to\ m;\ j = k \\ &If\ \left.\frac{\partial P_i}{\partial S_p(i,j)}\right|_{\substack{for\ i = 1\ to\ m \\ j=k}} < 0\ ;\ calculate\ \max(S_p(i,j))\Big|_{\substack{for\ i = 1\ to\ m \\ j=k}}\ for\ P_i > P_{min} \end{aligned} \tag{5}$$

where k represents a particular stress value.

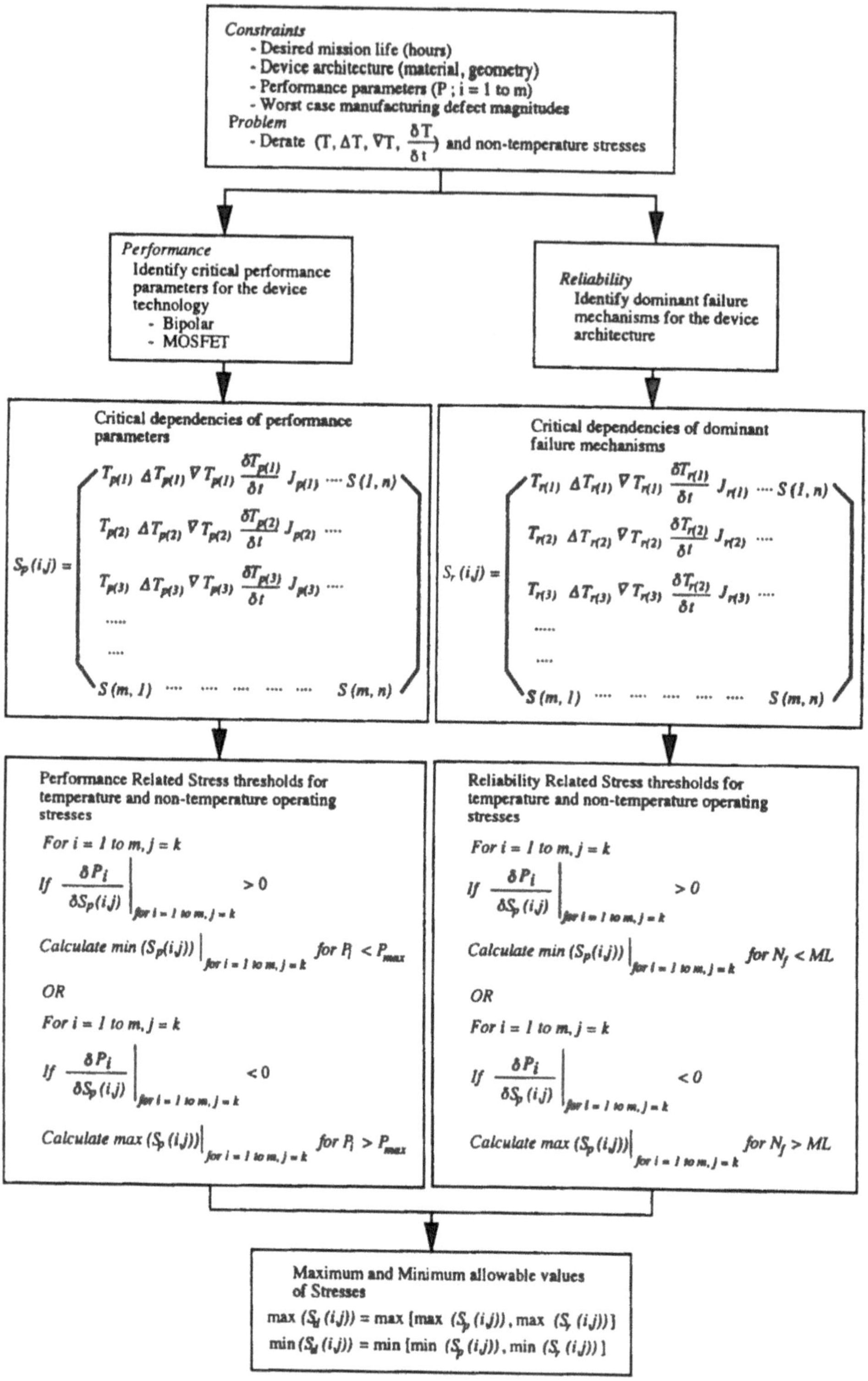

Figure 7.1 Derating methodology for temperature tolerant design.

The device life, determined by dominant failure mechanisms actuated by temperature and non-temperatures stresses, is used to determine limiting stress values for reliability. The influence of stresses on device life is quantified by closed-form models. The threshold stress values that result in a device life less than the desired mission life are calculated by

$$S_r(i,j) = \begin{pmatrix} T_{r(1)} & \Delta T_{r(1)} & \nabla T_{r(1)} & \frac{\partial T_{r(1)}}{\partial t} & J_{r(1)} & E_{r(1)} & \dots & S(1,n) \\ T_{r(2)} & \Delta T_{r(2)} & \nabla T_{r(2)} & \frac{\partial T_{r(2)}}{\partial t} & J_{r(2)} & E_{r(2)} & \dots & \dots \\ T_{r(3)} & \Delta T_{r(3)} & \dots & & & & & \dots \\ \dots & & & & & & & \dots \\ \dots & & & & & & & \\ S(m,1) & \dots & \dots & \dots & \dots & \dots & \dots & S(m,n) \end{pmatrix} \tag{6}$$

where S_r is a matrix of threshold stress values for device reliability (m x n). The m rows correspond to m (i.e., $P(i)$, for $i = 1$ to m) dominant failure mechanisms. The n columns correspond to the n temperature and non-temperature stresses that affect device life limitations resulting from the different failure mechanisms. The acceptable range of stress values for device reliability is calculated as follows:

$$For\ i = 1\ to\ m;\ j = k$$

$$If\ \left.\frac{\partial N_f}{\partial S_r(i,j)}\right|_{\substack{for\ i = 1\ to\ m \\ j=k}} > 0\ ;\ calculate\ \min(S_r(i,j))\Big|_{\substack{for\ i = to\ m \\ j=k}}\ for\ N_f > ML \tag{7}$$

where k represents a particular stress value, i represents a particular failure mechanism, N_f is the predicted time to failure due to a failure mechanism, and ML is the required device mission life. The maximum and minimum allowable values of stresses are then determined as:

$$For\ i = 1\ to\ m;\ j = k$$

$$If\ \left.\frac{\partial N_f}{\partial S_r(i,j)}\right|_{\substack{for\ i = 1\ to\ m \\ j=k}} < 0\ ;\ calculate\ \max(S_r(i,j))\Big|_{\substack{for\ i = to\ m \\ j=k}}\ for\ N_f > ML \tag{8}$$

$$\begin{aligned} \max(S_d(i,j)) &= \max[\max(S_p(i,j)),\ \max(S_r(i,j))] \\ \min(S_d(i,j)) &= \min[\min(S_p(i,j)),\ \min(S_r(i,j))] \end{aligned} \tag{9}$$

where $\max(S_r(i,\ j))$ is the maximum allowable stress value derived from reliability considerations, $\max(S_p(i,j))$ is the maximum allowable stress value derived from performance considerations, and $\max(S_d(i,j))$ is the maximum allowable stress value for the device. Min represents the minimum values of the individual stresses. Maximum values are specified for stresses directly proportional to detrimental effects on package reliability and performance. Minimum values are specified for stresses with an inverse influence on package performance or reliability. In the event that the limits such obtained are impossible to achieve or maintain in the operating condition, different device types, geometry or architecture need to be considered.

Physics-of-failure concepts have been used to relate allowable operating stresses to design strengths through quantitative models for failure mechanisms. Failure models have been used to assess the impact of stress levels on the effective reliability of the component for a given load. The quantitative correlations outlined between stress levels and reliability will enable designers and users to tailor the margin of safety more effectively to the level of criticality of the component, leading to better and more cost-effective utilization of the functional capacity of the component.

3. STRESS LIMITS FOR FAILURE MECHANISMS IN DIE METALLIZATION

Chemical reactions leading to loss of or property variation of die metallization are dependent on thermal effects. The mechanisms are primarily mass transfer type and those have complex dependency on temperature and non-temperature operating parameters. The following subsections describe some failure mechanism models for major failure mechanisms in die metallization.

3.1 Corrosion of Die Metallization

Metallization corrosion has a dominant dependence on two temperature stresses - steady-state temperature and time-dependent temperature change. Temperature transients generated by the duty cycle (ON/(ON+OFF)) modify the dependence of the metallization corrosion on steady- state temperature. At low values of the duty cycle, metallization corrosion has a dominant dependence on steady-state temperature; however, at higher values (≈ in neighborhood of 1.0), metallization corrosion has a dominant dependence on the duty cycle and a mild dependence on steady-state temperature. Time to failure due to metallization corrosion has been modeled as the sum of induction time and corrosion time [Pecht, 1990]. Calculate T_{max} and $(\partial T/\partial t)_{max}$ such that :

$$T_{operation} < T_{max}$$

$$\frac{\partial T}{\partial t} < \left(\frac{\partial T}{\partial t}\right)_{max} \tag{10}$$

$$TF_{corr} > mission\ life \tag{11}$$

where

$$K_3 = \frac{1}{(1 - DC)^{10DC - 1}} \tag{12}$$

$$K_4 = \frac{(RH_{ref})^{n_{corr}} \exp(E_{a-corr}/K_B T_{ref})}{(RH)^{n_{corr}} \exp(E_{a-corr}/K_B T)} \tag{13}$$

$$TF_{corr} = \tau_{INDUCTION} + \tau_{CORROSION}$$

$$\tau_{INDUCTION} = -\frac{4L^2}{\pi^2 D}\ln\left(1 - \frac{P_{in}}{P_{out}}\right) \qquad \textit{For nonhermetic package}$$

$$= \textit{From figure} \qquad \textit{For hermetic package} \tag{14}$$

$$\tau_{CORROSION} = \frac{K_1 K_2 K_3}{K_4}\,\frac{w_{cond}^2 h_{cond} n_{chem} d_{cond} F}{4M_{met} V_{met}}\,\frac{\rho_{elec}}{Z_{elec}}$$

where RH_{ref} is a reference relative humidity (%), E_{a-corr} is the activation energy for corrosion (eV), K_B is the Boltzmann constant (eV/K), n_{chem} is a material constant, T_{ref} is a reference temperature (K), DC is the duty cycle, M_{met} is the atomic weight of a metal conductor, with width w_{cond}, thickness h_{cond}, and chemical valence n_{chem}, ρ_{elec}/Z_{elec} is the sheet resistance of the electrolyte, V_{met} is the voltage applied, K_1 is the physical and chemical properties of the metallization material (from Table 2), K_2 is the coating integrity index (from Table 3), K_3 is the mission profile correction factor, and K_4 is the environmental stress correction factor.

Limiting stress values can be calculated from the curve of constant life versus temperature and duty cycle, obtained from Equation 12. (Figure 7.2 shows curves of constant life versus temperature and duty cycle.) It is evident that dependence of metallization corrosion on steady-state temperature and duty cycle is non-linear, and that reducing dominant stress magnitudes beyond a certain limit may not produce any observable benefit in terms of added reliability, because the time to failure is far beyond the wear-out life of the device. The paradigm of associating higher reliability with lower temperature is also misleading, since the same mission life of twenty-nine years can be obtained for any temperature from 40°C to 160°C, depending on the duty cycle (Figure 7.2).

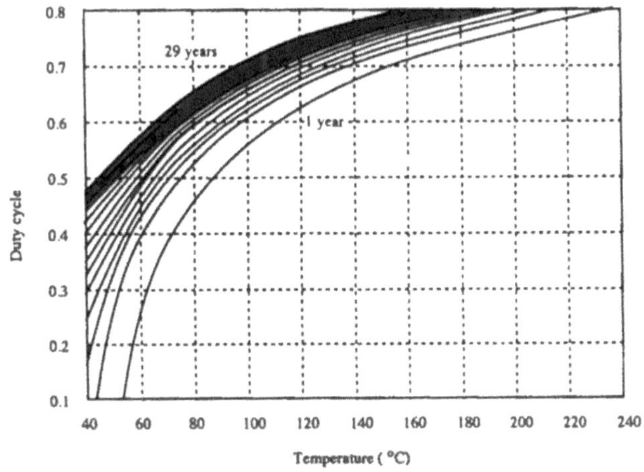

Figure 7.2 Limiting stress curves of life under corrosion versus temperature and duty cycle. The curves identify the various combinations of temperature and duty cycle which will result in desired life. The paradigm of higher reliability associated with lower temperature is misleading, since the same mission life of 29 years can be obtained for any temperature from 40 to 160 ^{0}C depending on the duty cycle.

Table 2 Physical and chemical properties index, k_p, of metallization materials [Pecht and Ko, 1991]

Material	Physical property	Atomic weight	Density	Chemical valence
Aluminum	0.1	27	2.7	3
Copper	0.7	64	8.9	2
Gold	1.0	197	19.32	3

Table 3 Coating integrity index [Pecht, 1991]

Coating Type	Coating Integrity Index (K_2)
No coating	1
Partially bonded	10 - 50
Completely bonded	100

3.2 Electromigration

Electromigration, a mass-transfer mechanism with a dominant dependence on current density, steady-state temperature, and temperature gradient, is a grain-boundary diffusion mechanism. The time to failure due to temperature stress for T < 150°C is much greater than the wear-out life. Typically, electromigration failures at T < 150°C occur at sites of maximum temperature gradient or structural non-uniformity. Steady-state temperature is a dominant stress accelerator for electromigration at T > 150°C for typical metallization geometries. The maximum temperature and current density combination that can be used for a device with a given metallization can be calculated from one of the following models:

- Black's Model [Black, 1982]
- Shatzkes and Lloyd Model [Shatzkes and Lloyd, 1986]
- Venables and Lye Model [Venables and Lye, 1972]

Black's Model [Black, 1982]. Calculate T_{max} (max. temperature) and j_{max} (current density), where

$$T_{operation} < T_{max} \qquad j_{operation} < j_{max} \tag{15}$$

such that

$$MTF > mission\ life \tag{16}$$

where

$$MTF = \frac{w_{met}\, h_{met}}{j_{met}^{n_{e,Black}}\, A_{g,Black}\, e^{-\frac{E_a}{K_B T}}} \tag{17}$$

Table 4 Black's constants for various metallization materials

Metallization Material	$A_{g,\ black}$	E_a (eV)
Al-2%Si	0.3119 x 10^{-14}	0.558
Al-4%Cu-4%Si	7.292 x 10^{-15}	0.703

and *MTF* is the mean time to failure (hours); w_{met} is the metallization width (cm); h_{met} is the metallization thickness (cm); $A_{g,\ black}$ is a parameter depending on sample geometry, physical characteristics of the film and substrate, and protective coating; j_{met} is the current density (A/cm^2); $n_{e,\ black}$ is an experimentally determined exponent; E_a is the activation energy (eV); and T is the steady-state temperature (K).

The value of the current exponent derived in studies varies from 1.5 to 7.0.

$n_{e,\ Black} = 1$ to 3	Chabra and Ainslie [1967]
$n_{e,\ Black} = 1.5$	Attardo [1972]
$n_{e,\ Black} = 1.7$	Danso and Tullos [1981]
$n_{e,\ Black} = 2$	Black [1983]
$n_{e,\ Black} = 6$ to 7	Blair et al. [1970]

The pre-exponential, current exponent, and activation energy for some common metallization materials reported in literature are presented in Tables 4 and 5. Figure 7.3 shows the derating curves for electromigration stress; these identify various combinations of current density and temperature that will result in desired life.

Shatzkes and Lloyd Model [Shatzkes and Lloyd, 1986]. Calculate T_{max} (maximum temperature) and j_{max} (maximum current density), where

$$T_{operation} < T_{max} \qquad (18)$$
$$j_{operation} < j_{max}$$

such that

$$TF_{elec} > mission\ life \qquad (19)$$

$$TF_{elec} = \left(\frac{2C_{vc,f}}{D_{0,met}}\right)\left(\frac{K_B}{Z^*_{met}\ e\ \rho_{met}}\right)^2 T^2\ j_{met}^{-2}\ e^{\frac{E_a}{K_B T}} \qquad (20)$$

and where $C_{vc,f}$ is the critical value of vacancy concentration at which failure occurs; $D_{o,\ met}$ is the pre-exponential factor for grain-boundary self-diffusivity; K_B is the Boltzmann constant; Z^*_{met} is the effective charge; e is the electronic charge; ρ_{met} is the resistivity; T is the steady-state temperature; j_{met} is the current density; and E_a is the activation energy.

Equation 20 differs from Black's equation in that it has a T^2 pre-exponential term, but it agrees with Black's data equally well.

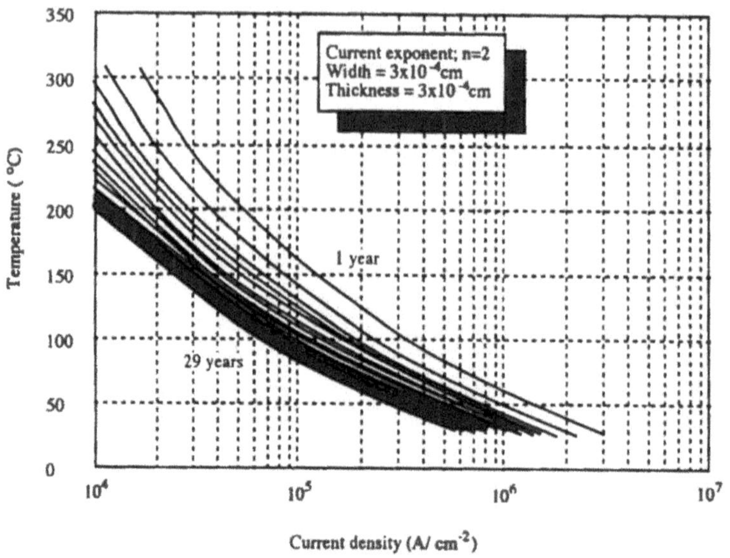

Figure 7.3 Limiting stress curves for electromigration stresses. The curves identify various combinations of current density and temperature which will result in desired life [Black, 1982]

Venables and Lye Model [Venables, 1972]. Given the allowable temperature rise of ΔT_R in the metallization stripe, calculate T_{max} (maximum temperature) and j_{max} (current density), where

$$\begin{aligned} T_{operation} &< T_{max} \\ j_{operation} &< j_{max} \end{aligned} \tag{21}$$

such that

$$TF_{elec} > mission\ life \tag{22}$$

where

$$TF_{elec} = \frac{1}{2C_{p,vl}n_{gb,met}}\left(\frac{\tau_o K_B T_o}{j_{o,met}\rho_{o,met}D_{o,met}e^{-\frac{E_a}{K_B T}}}\right)\int_{x_o}^{x_1}\frac{e^{\frac{-E_a x}{K_B T_o}}}{x^2(1-x+x\alpha T_o)}dx \tag{23}$$

and

$$\tau_0 = \Delta T_R\left(1+\alpha(T_o-300)\right)\left(\frac{1}{T_0}\right)\left(\frac{j_{0,met}}{1x10^6}\right)^2 \tag{24}$$

and

$$\begin{aligned} x &= \frac{\tau_o}{(1-p_{p,met})^2+\tau_o(1-\alpha_{r,met}T_o)} \\ &= 1-\left(\frac{T_o}{T}\right) \end{aligned} \tag{25}$$

$$x_o = \frac{\tau_o}{1-\tau_o(\alpha_{r,met}T_o-1)} \qquad @t=0$$

$$x_1 = 1-\left(\frac{T_o}{T_{m,met}}\right) \qquad @t=T_F \tag{26}$$

where ΔT_R is the temperature increase in the metallization stripe: $\alpha_{r,met}$ is the thermal coefficient of resistivity for metallization (/°C); T_o is the ambient temperature (K); $T_{m,met}$ is the melting temperature of the metallization stripe (K); E_a is the activation energy for grain-boundary diffusion (≈ 0.558 eV); K_B is Boltzmann's constant (8.617 x 10^{-5} eV/K or 1.38 x 10^{-23} J/K); $p_{p,met}$ is the porosity of the metallization stripe (0.1 < $p_{p,met}$ < 1.0); j_{met} is the current density (A/cm^2); $C_{p,vl}$ is the constant for the metallization material; $n_{gh,met}$ is the density of grain-boundary nodes (/cm^3); $\rho_{0,met}$ is the resistivity of metallization material at ambient temperature (Ω cm); and $D_{0,met}$ is the grain-boundary diffusivity of the metallization material (cm^2/s).

The current exponent for Black's equation (Equation 17) changes with the current density, j_{met}, in the metallization stripe. A constant value of the current exponent, $n_{e,black}$, is valid only over a small range of current densities [Venables, 1972]. The Venables and Lye model accounts for the relationship of time to failure and baseline temperature, which is often simplistically represented by an Arrhenius relationship to give an activation energy. The model also addresses the dependence of current exponent on current density.

3.3 Hillock Formation

Hillocks in die metallization can form as a result of electromigration or extended periods under temperature cycling conditions (thermal aging). The phenomenon of simultaneous voiding and hillock formation occurs at temperatures in the neighborhood of 140 to 200°C [Thomas, 1983]. Hillock formation due to extended periods under temperature cycling conditions is believed to be due to a self-diffusion process that occurs in the presence of strains within the metallization [LaCombe and Christou, 1982]. Coating the metal with silicon nitride prevents failures from occurring and voids and hillocks from forming for at least 500 hours at 360°C. Hillocks form at random in aluminum films heated to temperatures around 400°C during fabrication.

Table 5 Temperature acceleration and current exponents for various metallization materials

Material	Metallization dimensions (μm) w = width l = length t = thickness	Average grain size (μm)	Test conditions		Activation energy E_a, (eV)	Current density exponent (n)	Reference
			Current density (A/cm^2)	Steady-state temperature (C)			
Single layer metallizations							
Aluminum (Al)	w = 10 l = 400 t = 1.2		1.3 - 3	75-350 C	0.41		[Satake, 1973]
	w = 15.4 t = 0.6	8	0.55 - 2.02	109-260°C	0.84	2	[Black, 1969]
	t = 1.0	-	-	-	0.56	2	[Black, 1982]

	w = 1.3 l = 1524 t = 0.8	0.1	0.75	175-275°C	0.34		[Wu, 1983]
	w = 2.3	0.1	-		0.47		[Wu, 1983]
	w = 3.3	0.1	-		0.58		[Wu, 1983]
	t = 0.7	0.7-1.8	2	125-300°C	0.24-0.57		[Reimer, 1984]
	w = 4.5 l = 2000 t = 0.75	-	2	125-300	0.43		[Towner, 1983]
	w = 10 l = 800 t = 1	-	1	160-250°C	0.57		[Schreiber, 1981]
	w = 37 t = 1.2	8	0.46 - 0.99	109-260°C	1.2	2	[Black, 1969]
	w = 15-25 l = 800 t = 0.5	0.3	2	105-180°C	0.55	3	[Van Gurp, 1971]
	w = 7-20 l = 500-580 t = 1	0.8	0.5-2	130-200°C	0.7	2.5-4	[Saito, 1974]
	w = 10	10	0.2-3	110-245°C	0.55	2	[Sim, 1979]
	l = 25000-32000	3.5	0.2-3	110-245°C	0.6	2	[Sim, 1979]
	t = 1	2	0.2-3	110-245°C	0.4	2	[Sim, 1979]
	w = 6 l = 380 t = 0.8	0.7-1.4	1	150-215°C	0.43		[Ghate, 1981]
	w = 7.5 l = 275 t = 0.5	-	0.13-1	82-192°C	0.35-0.85		[Lloyd, 1987]
	-	-	1.3-2.9	30-141°C	0.56	2.5	[Wild, 1988]
	w = 12.5 l = 1390 t = 0.7	1.2	0.5-2.88	109-260°C	0.48	2	[Black, 1969]
Al-(0.79-3%)Si	w = 10 l = 762 t = 0.7-0.73	1.4	0.66-2	110-210°C	0.54-0.55	2	[Black, 1978]
Al-1%Si	w = 5 l = 400 t = 1.1	-	0.5-2.5	162-215°C	0.58	1.53	[Schafft, 1985]
	w = 3 l = 1000 t = 1	-	3.3	200°C	-	-	[Maiz, 1989]
	w = 3.7-4.1 l = 400 t = 1.2	-	0.8-3	277°C	1.67-2.56	0.5	[Suehle, 1989]
	w = 1.4 l = 160 t = 1.1	-	3.2	205-261°C	0.96	1	[Fantini, 1989]
Al-1.5%Si	-	3.5	1	145-210°C	0.71		[Tatsuzawa, 1985]
Al-2%Si	w = 1.2-2.2 l = 100-1600 t = 0.8	-	10-60	240-360°C	0.33	7.5	[Liew, 1989]
	w = 10 l = 1000 t = 1	0.3	1.3-3.2	150-250°C	0.4 - 0.56	1.8 (j < 2)	[Nagasawa, 1980]
	w = 10 l = 1000 t = 1	0.8	1.3-3.2	150-250°C	0.4	2.2	[Nagasawa, 1980]

Al-0.3%Cu	w = 7-8 l = 250 t = 0.5-1	-	4	121-222°C	0.65		[d'Heurle, 1972]
	w = 12.7 l = 1270 t = 1	2-3	-	-	0.63-0.7	-	[Rodbell, 1983]
Al-0.5%Cu	w = 1-4 l = 25000 t = 0.8-1.1	1.2-4.5	1.6-2	195-250°C	0.5	2	[Sim, 1979]
	w = 2-5 l = 1000 t = 1	1.8	2.1	209-264°C	0.73	-	[Levi, 1985]
	w = 2-5 l = 1000 t = 1	2.9	1.62	232-304°C	0.84	-	[Levi, 1985]
	w = 2-5 l = 1000 t = 1	1.8	1.82	205-283°C	0.68	-	[Levi, 1985]
	lw = 7.5 l = 275 t = 0.5	-	0.13-1	173-238°C	0.53-0.92	-	[Lloyd , 1987]
Al-2%Cu	w = 6 l = 380 t = 0.8	1-2.3	1	150-215°C	0.7	-	[Ghate, 1981]
Al-4%Cu	w = 7-8 l = 250 t = 0.5-1	-	4	105-175°C	0.83	-	[d'Heurle, 1972]
	w = 15 l = 1250 t = 1	-	0.8	220-275°C	0.76	-	[Learn, 1975]
Al-(5.1-5.4)%Cu	w = 10 l = 250-3380 t = 1.5	-	1.3-1.5	142-157°C	0.48	-	[Kakar, 1973]
Al-1%Si-2%Cu	w = 6 l = 380 t = 0.8	0.25	1	150-215°C	0.5	-	[Ghate, 1981]
Al-2%Si-4%Cu	t = 1.0	-	-	-	0.7	2	[Black, 1982]
Al-Si-0.5%Cu	w = 3.3-7.7 t = 0.62-0.67	1.8-3.6	0.5-1	200-240°C	0.76	-	[Bukkett, 1984]
Al-Si-2%Cu	w = 2.2-7.7 t = 0.28-0.71	1.5-3.1	0.5-1	180-240°C	0.86	-	[Bukkett, 1984]
Al-1%Si-(0.35-1.1)%Ti	w = 4 l = 2000	-	2	125-300 C	0.5-0.7	-	[Towner, 1986]
Double-layer metallizations							
Al-1%Si/TiW	w = 5.5 l = 2550 t = 0.6/0.22	-	0.5-2.2	100-200°C	0.49	2.5	[Onduresk, 1988; Hoang, 1988]
Al-0.3%Cu/TiW	w = 12.7 l = 1270 t = 0.8/0.2	2-3	1.4	100-300°C	0.65	-	[Rodbell, 1983]
Al-4%Cu/TiW	w = 12.7 l = 1270 t = 0.8/0.2	5	1.4	100-300 C	0.75	-	[Rodbell, 1983]

3.4 Metallization Migration

DiGiacomo [1982] proposed a model to calculate the maximum allowable steady-state temperature to avoid metallization migration, where:

$$T_{operation} < T_{max} \tag{27}$$

and

$$j_{tip} > j_{critical}$$

$$\frac{\partial Q_{tip}}{\partial t} > \frac{\partial Q_{critical}}{\partial t} \tag{28}$$

such that

$$TF_f > mission\ life \tag{29}$$

where

$$TF = \frac{\left(\frac{d_{dndt}}{M_{at}C_{i,bulk}}\right)\left(\frac{2r_{dndt}}{Z_{met}FD_{o,met}}\right)l_{d,e}}{\beta_{an,ox}(V_{app} - V_{th})\frac{1}{K_BT}e^{-\frac{E_a}{K_BT}}erfc\frac{1}{2\sigma_{sd}}\ln\left(\frac{K_BT\ln(RH)r_{av,pore}}{2\gamma_{st}V_{mol,vl}}\right)^2} \tag{30}$$

where

$$r = \frac{2\gamma_{st}V_{mol,vl}}{K_BT\ \ln\left(\frac{P_{sat,men}}{P_{sat,flat}}\right)} \tag{31}$$

$$erfc\left(\frac{1}{2\sigma_{sd}}\right) = 1 - \frac{2}{\sqrt{\pi}}\int_0^{\left(\frac{1}{2\sigma_{sd}}\right)} e^{-u^2}\,du \tag{32}$$

and where M_{at} is the atomic weight; $C_{i,bulk}$ is the ionic concentration; Z_{met} is the ionic charge or the valence of the metal ions; $D_{o,met}$ is the diffusivity coefficient; $l_{d,c}$ is the distance between electrodes; r_{dndt} is the dendrite radius; $r_{av,pore}$ is the average pore radius; γ_{st} is the surface tension; $V_{mol,vl}$ is the molar volume; $p_{sat,men}$ is the saturated vapor pressure above the meniscus; $p_{sat,flat}$ is the saturated vapor pressure above the flat surface RH approaching $p_{sat,men}/p_{sat,flat}$ is the relative humidity at which condensation occurs; σ_{sd}= ln $[r_{50} / r_{16}]$ sigma; $\beta_{an,ox}$ is the fraction of metal surface at the anode that is susceptible to metal oxidation and promotes metal migration; V_{app}= potential difference applied; V_{th}= critical value of potential difference for dendritic growth; j_{tip} = current density at the whisker's tip; $j_{critical}$ = critical current density at the whisker's tip; Q_{tip} = $(j_{tip} / Z_{met}F)$ = mass transport rate; $Q_{critical}$ = critical mass transport rate; and *erfc* is the error function as defined in Equation 32.

3.5 Constraint Cavitation of Conductor Metallization

Okabayashi Model [Okabayashi 1991]: The lifetime due to SDDV becomes minimum at a certain temperature T_m. The increase in lifetime when T increases or decreases from T_m (temperature where the minima occurs) results from a decrease in stress in the metallization for $T \geq T_m$, or a decrease in diffusivity for $T \leq T_m$, respectively. The allowable operation temperatures are calculated from

$$
\begin{aligned}
T_{operation} &< T_{max} \qquad for\ T < T_{psv,dp} \\
T_{operation} &> T_{max} \qquad for\ T > T_{psv,dp}
\end{aligned}
\tag{33}
$$

such that

$$TF_{sddv} > mission\ life \tag{34}$$

where

$$TF_{sddv} = \frac{K_B G_{met} L_{rlx,vd} w_{met} T e^{E_a/K_B T}}{2A_{sddv,ok} E D_{0,sd}(n_{ok}-1)} \left[\left(\frac{\sigma_{met}(0)}{G_{met}} - \frac{E w_{met} \tan\psi_{vd,hf}}{G_{met} L_{rlx,vd}} \right)^{1-n_{ok}} - \left(\frac{\sigma_{met}(0)}{G_{met}} \right)^{1-n_{ok}} \right] \tag{35}$$

$for\ n_{ok} \neq 1$

$$TF_{sddv} = \frac{K_B G_{met} L_{rlx,vd} w_{met} T e^{E_a/K_B T}}{2A_{sddv,ok} E D_{0,sd}} \left[\ln\left(\frac{\sigma_{met}(0)}{G_{met}} \right) - \ln\left(\frac{\sigma_{met}(0)}{G_{met}} - \frac{E w_{met} \tan\psi_{vd,hf}}{G_{met} L_{rlx,vd}} \right) \right] \tag{36}$$

$for\ n_{ok} = 1$

$$\sigma_{met}(t) = \sigma_{0,met}(T_{psv,dp} - T) \left(\frac{w_{0,met}}{w_{met}} \right)^{m_{ok}} \left(\frac{h_{0,met}}{h_{met}} \right)^{p_{ok}} \tag{37}$$

Where K_B is the Boltzmann's constant, G_{met} is the shear modulus of metallization, $(L_{rlx,vd}/2)$ is the length over which stress due to void formation is relaxed, ω_{met} is the width of the metallization, T is the steady state temperature, $A_{sddv,ok}$ is a factor independent of stress, temperature, and time, E is Young's modulus for the metallization, $D_{o,sd}$ is the diffusivity constant, E_a is the activation energy for diffusion, n_{ok} is the stress dependency (Okabayashi) exponent, $\sigma_{met}(0)$ is the stress at time =0, $\sigma_{met}(t)$ is the stress at time t, $\Psi_{vd,hf}$ is the half angle of the void, $T_{psv,dp}$ is the passivation deposition temperature, $\sigma_{0,met}$ is the metallization stress due to a temperature change of 1°C, $\omega_{o,met}$ is the normalization thickness of metallization, $h_{0,met}$ is the normalization thickness of metallization, h_{met} is the thickness of metallization, p_{ok} is the Okabayashi metallization thickness exponent (=0.5), and m_{ok} is the Okabayashi metallization width exponent (=0.5).

Kato and Niwa Model [Kato, 1990, Niwa, 1990]. This model accounts for plastic deformation and diffusional relaxation. While plastic deformation occurs almost instantaneously after the passivation is laid on the metallization, diffusion is practically inoperative at low temperatures. (Low temperature has been defined as the temperature at which the time for diffusional relaxation is greater than 10^3 seconds) [Kato, 1990, Niwa, 1990]. Kato et al. found that stresses did not become hydrostatic after relaxation due to plastic deformation. Time to failure due to stress-driven diffusive voiding is considered to be the time during which the area fraction, A, increases from A_{min} to A_{max} (Area fraction = $r_{0,vd}^2/l_{int,vd}^2$, $r_{0,vd}$ is the void radius, and $2l_{int,vd}$ is the void separation). The variation in time to failure versus aspect ratio and temperature shows that SDDV changes its temperature dependence from steady-state to inverse temperature dependence at a temperature threshold which is a function of passivation temperature (Figure 7.4). The steady-state temperature limits to avoid SDDV can be calculated from

$$\begin{aligned} &T_{operation} < T_{\max} && \mathit{for}\ T < T_{psv,dp} \\ &T_{operation} > T_{\max} && \mathit{for}\ T > T_{psv,dp} \end{aligned} \tag{38}$$

such that

$$TF_{sddv} > \mathit{misssion\ life} \tag{39}$$

where

$$\mathit{for} \quad \sigma_{met} \geq \frac{2\Gamma_{sf,ur}}{r_{o,vd}}$$

$$TF_{sddv} = \frac{K_B T l^{3}{}_{int,vd}}{20 D_{bo,met} w_{b,met} \sigma_{met} V_{at,vl}} \tag{40}$$

$$\begin{aligned} &l_{int,vd} = \frac{a_2}{r_{met}} && \mathit{for}\ r_{met} \geq 1 \\ &l_{int,vd} = a_2 && \mathit{for}\ 0 \leq r_{met} \leq 1 \end{aligned}$$

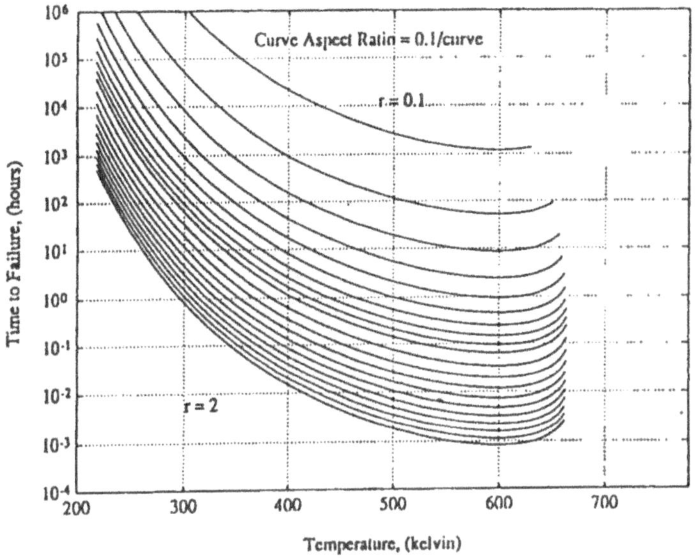

Figure 7.4 The variation in time to failure versus aspect ratio and temperature shows that SDDV changes its temperature dependence from steady state to inverse temperature dependence, at a temperature threshold which is a function of the passivation temperature.

$$\text{for} \quad \sigma_{met} < \frac{2\Gamma_{sf,ar}}{r_{o,vd}} \tag{41}$$

for no void growth

where

for equiaxed grain structure,

$$\sigma_{met} = \sigma^R$$

for bamboo grain structure, (42)

$$\sigma_{met} = \sigma_{33} \qquad t_{relax} \le 1{,}000 \text{ seconds}$$

$$\sigma_{met} = \sigma^R \qquad t_{relax} \ge 1{,}000 \text{ seconds}$$

the completely relaxed stress state is represented by [Kato, 1990, Niwa, 1990]

$$\sigma_{11}^R = \sigma_{22}^R = \sigma_{33}^R = \sigma^R$$

$$= \frac{-6G_{met}r_{met}(1+\nu_{met})\epsilon^T}{(2r_{met}^2 + r_{met} - 2) - 4r_{met}\nu_{met} - 2(1 - r_{met}^2)\nu_{met}^2} \tag{43}$$

$$\epsilon_{11} = \epsilon_{22} = \epsilon_{33} = \epsilon^T = \Delta\alpha(T - T_{psv,dp}) \tag{44}$$

For aspect ratios, r_{met}, such that $r_{met} > 0$ or $r_{met} < 1$, the strains after relaxation by plastic deformation are [Kato, 1990, Niwa, 1990]

$$\epsilon_{11}^{P} = \frac{-((C_6+C_2C_7)\epsilon^T+C_1(C_3+C_2C_5))}{(C_2(2C_3+C_2C_5)+C_4)} \tag{45}$$

and

$$\epsilon_{22}^{P} = C_1 + C_2\,\epsilon_{11}^{P} \tag{46}$$

where

$$C_1 = \frac{(1+r_{met})^2}{4r_{met}^2(1-\nu_{met})+r_{met}(5-4\nu_{met})+2} \times \left(\frac{2(1+\nu_{met})\gamma_{th}}{1+r_{met}} + \frac{(1-\nu_{met})S_{y-met}}{G_{met}}\right) \tag{47}$$

$$C_2 = \frac{2(1+r_{met})^2(1-\nu_{met})+r_{met}}{4r_{met}^2(1-\nu_{met})+r_{met}(5-4\nu_{met})+2} \tag{48}$$

$$C_3 = \frac{2r_{met}^2(1-\nu_{met})+r_{met}(5-4\nu_{met})+2(1-\nu_{met})}{(1+r_{met})^2} \tag{49}$$

$$C_4 = \frac{2r_{met}^2+r_{met}+(5-4\nu_{met})+4(1-\nu_{met})}{(1+r_{met})^2} \tag{50}$$

$$C_5 = \frac{4r_{met}^2(1-\nu_{met})+r_{met}(5-4\nu_{met})+2}{(1+r_{met})^2} \tag{51}$$

$$C_6 = -\frac{2r_{met}(1+\nu_{met})}{1+r_{met}} \tag{52}$$

$$C_7 = -\frac{2(1+\nu_{met})}{1+r_{met}} \tag{53}$$

$$\sigma_{11}^{R} = -\frac{G_{met}}{1-\nu_{met}}\left(\frac{2(1+\nu_{met})}{(1+r_{met})}\epsilon^{T} + \frac{r_{met}(1-2\nu_{met})+2(1-\nu_{met})}{(1+r_{met})^{2}}\epsilon_{11}^{R} + \frac{r_{met}(1-2\nu_{met})-2\nu_{met}}{(1+r_{met})^{2}}\epsilon_{22}^{R}\right) \tag{54}$$

$$\sigma_{22}^{R} = -\frac{G_{met}}{(1-\nu_{met})}\left(\frac{2r_{met}(1+\nu_{met})}{(1+r_{met})}\epsilon^{T} + \frac{r_{met}(1-2\nu_{met})-2\nu_{met}r_{met}^{2}}{(1+r_{met})^{2}}\epsilon_{11}^{R} + \frac{r_{met}(1-2\nu_{met})-2r_{met}^{2}(1-\nu_{met})}{(1+r_{met})^{2}}\epsilon_{22}^{R}\right) \tag{55}$$

$$\sigma_{33}^{R} = -\frac{G_{met}}{(1-\nu_{met})}\left(2(1+\nu_{met})\epsilon^{T} + \frac{-2r_{met}-2(1-\nu_{met})}{(1+r_{met})} \times \epsilon_{11}^{R} + \frac{-2-2r_{met}(1-\nu_{met})}{(1+r_{met})}\epsilon_{22}^{R}\right) \tag{56}$$

For $r_{met} \approx 1$

$$\epsilon_{11}^{P} = \epsilon_{22}^{P} = -\frac{1}{2}\epsilon_{33}^{P} = -\frac{(1+\nu_{met})\epsilon^{T} + \frac{(1-\nu_{met})\sigma_{y}}{G_{met}}}{(4\nu_{met}-5)} \tag{57}$$

$$\sigma_{11}^{P} = \sigma_{22}^{P} = -\frac{G_{met}}{(1-\nu_{met})}((1+\nu_{met})\epsilon_{T}(1-2\nu_{met})\epsilon_{11}^{P}) \tag{58}$$

$$\sigma_{33}^{P} = \frac{-2G_{met}}{(1-\nu_{met})}((1+\nu_{met})\epsilon^{T}+(-2+\nu_{met})\epsilon_{11}^{P}) \tag{59}$$

For $r_{met} \approx 0$

$$\begin{aligned} \epsilon_{11}^{P} &= \epsilon_{33}^{P} = -\frac{1}{2}\epsilon_{22}^{P} \\ &= -\epsilon^{T} - \frac{(1-\nu_{met})\,\sigma_y}{2G_{met}(1+\nu_{met})} \end{aligned} \tag{60}$$

$$\begin{aligned} \sigma_{11}^{P} &= \sigma_{33}^{P} = \sigma_y \\ \sigma_{22}^{P} &= 0 \end{aligned} \tag{61}$$

Where σ^R is the hydrostatic stress after diffusional relaxation, σ_{ii} is the stress along the width (1), thickness (2), and length (3), TF_{sldv} is the time to failure in seconds, K_B is the Boltzmann's constant, T is the steady-state temperature, $\Gamma_{sf,ar}$ is the surface energy per unit area, $D_{bd,met}$ is the grain boundary diffusivity (m^2/s), $\omega_{b,met}$ is the grain boundary thickness (m)$_{met}$, σ is the metallization stress, $V_{at,vl}$ is the atomic volume of conductor atoms (m^3), r_{met} is the aspect ratio of metallization line (thickness/width), $2a_1$ is the metallization width, $2a_2$ is the metallization thickness, G_{met} is the shear modulus of the metallization line, ν_{met} is the Poisson's ratio of metallization line, $\Delta\alpha$ is the difference in coefficient of thermal expansion between the metallization line and passivation, $T_{psv,dp}$ is the passivation deposition temperature. ϵ_{ii} are the strains in metallization line along its width (1), thickness (2), and length (3) respectively. ϵ^T is the thermal strain resulting from mismatch between passivation and metallization, σ_{ii} is the stress in metallization line along its width, and σ_y is the yield strength of metallization.

4. STRESS LIMITS FOR FAILURE MECHANISMS IN DEVICE OXIDE

4.1 Slow Trapping

Slow trapping, a phenomenon of MOS devices only, has a dominant dependence on the trap density of oxide. The trapping and detrapping kinetics of electrons cause threshold voltage shifts. The magnitude of threshold voltage shift is a function of oxide permittivity, oxide thickness, density, and location of oxide charge. Electrons, or holes trapped in oxide traps, are excited by the application of high steady-state temperature. The highest steady-state temperature for a maximum allowable flatband voltage shift can be calculated from

$$T_{operation} < T_{max} \tag{62}$$

such that

$$\Delta V_{FB} < \Delta V_{FB,max} \tag{63}$$

where

$$\begin{aligned} \Delta V_G^{+} &= \Delta V_{FB} = -\frac{x_m Q_{ot}}{\epsilon_o \epsilon_{ox}} && \textit{for positive gate} \\ \Delta V_G^{-} &= \Delta V_{FB} = -\frac{t_{ox}}{\epsilon_o \epsilon_{ox}}\left(1 - \frac{x_m}{t_{ox}}\right) Q_{ot} && \textit{for negative gate} \end{aligned} \tag{64}$$

$$Q_{ot}(t) = \left(Q_{ot(t=0)}\right) e^{-t\,C_{th,em}} \tag{65}$$

and

$$C_{th,em} = \left(\frac{2\sqrt{3}(2\pi)^{3/2} m_\eta^* K_B^2}{h^3}\right) A_{c,r} T^2 e^{-E_a/K_B T} \tag{66}$$

Where T is the steady state temperature (K), ΔV_{FB} is the flatband voltage shift, ϵ_0 is the free space permittivity (8.85 x 10^{-14} F cm^{1}), ϵ_{ox} is the relative permittivity of oxide (dielectric constant ≈3.9 (typical value), t_{ox} is the oxide thickness (cm), x is the distance away form the metal SiO_2 interface, x_m is the centroid of charge contained in oxide (between $x=0$ and $x=t_{ox}$), $C_{th,em}$ is the thermal emission coefficient (sec^{-1}), t is the time (sec), $E_a = E_{n,con} - E_{n,trp}$ (eV), where $E_{n,con}$ is the conduction band energy level, $E_{n,trp}$ is the trap energy level, K_B is the Boltzmann's constant (8.617 x 10^{-5} eV/K or 1.38 x 10^{-23} J/K), h is the Plank's constant (6.62 x 10^{-34} J s), and m^*_η is the effective mass of the electron's in SiO_2, Q_{ot} is the density of the oxide trapped charge over the thickness of the oxide, expressed per unit area of Si-SiO_2 interface (C cm^2) and $Q_{ot(t=0)}$ is the value of Q_{ot} at time =0.

The prime distribution moment, $(x_m\ Q_{ot})$ increases or decreases with the increase or decrease in x_m or Q_{ot}. If the trap distribution in the oxide is uniform parallel to the interfaces, the C-V curve shifts without distortion.

4.1.1 Gate oxide breakdown

Electrostatic discharge. Electrostatic discharge in MOS devices can produce gate-to-drain shorts, gate-to-source shorts, or gate-to-substrate shorts, depending on the nature of imperfection. The most likely sites for ESD damage in defect-free oxides are source or drain sites depending on the polarity of transient currents and the biasing of the device. Gate-to-substrate shorts are more prevalent in devices with pre-existing oxide defects, in the form of geometrical or dopant irregularities. For bipolar devices, typical ESD junction damage occurs in the device bulk under reverse-biased conditions resulting in degraded p-n junction characteristics. Failed p-n junctions appear as cracked glass across the junction on the surface of the chip. The device operating temperature affects the ESD damage threshold. The higher the operating temperature, the lower the damage threshold. The maximum allowable temperature for the device can be calculated from the Wunsch-Bell model [Wunsch, 1968]

$$T_{operation} < T_{max} \qquad for\ V_{ESD} > V_{protection} \tag{67}$$

where T_{max} is calculated from

$$T_i = T_{m,chp} - \frac{P_{gt}\sqrt{t}}{A_{gt}\sqrt{\pi k_{tc,chp} d_{chp} C_{p,chp}}} \tag{68}$$

and where P_{gt}/A_{gt} is the power per unit junction area, calculated from the protection structure breakdown voltage, $k_{tc,chp}$ is the thermal conductivity of the semiconductor, d_{chp} is the density of the semiconductor, $C_{p,chp}$ is the specific heat of the semiconductor, $T_{m,chp}$ is the melting point of the semiconductor, T_i is the operating ambient temperature, and t is the ESD pulse time (seconds).

The threshold voltage and power density, P_{gt}/A_{gt}, for the circuit can be calculated from the Speakman model [Speakman, 1974]:

$$A_{gt} = D_{bs} \times l_{emtr} \tag{69}$$

$$\tau_{RC} = (R_{by} + R_{dv} + R_{ct})C_{by} \tag{70}$$

$$I_p = \frac{V_{by} - V_{dv}}{R_{by} + R_{dv} + R_{ct}} \tag{71}$$

$$\begin{aligned} P_{av,rpn} &= \frac{1}{5\tau_{RC}} \int_0^{5\tau_{RC}} V_{dv} I_p \, e^{-t/\tau_{RC}} dt \\ &+ \frac{1}{5\tau_{RC}} \int_0^{5\tau_{RC}} R_{by} I_p^2 \, e^{-2t/\tau_{RC}} dt \\ &= \frac{V_{dv} I_p}{5}\left(1 - e^{-2}\right) + \frac{R_{by} I_p^2}{10}\left(1 - e^{-10}\right) \\ &\approx \frac{V_{dv} I_p}{5} + \frac{R_{by} I_p^2}{10} \end{aligned} \tag{72}$$

$$R_{dv} = R_{sh,bs} \times \frac{t_{be}}{l_{emtr}} \tag{73}$$

Where D_{bs} is the depth of the base region, l_{emtr} is the length of the emitter region, τ_{RC} is the time constant of the RC circuit, R_i is the resistance components (i = by,ct,dv for body contact and device resistances respectively), t is the time in seconds, C_{by} is the body capacitance, t_{be} is the base-emitter separation, V_{dv} is the device voltage, V_{by} is the discharge voltage on the human body, I_p is the Peak current of the discharge waveform, $R_{sh,bs}$ is the base-sheet resistance, $V_{protection}$ is the voltage capacity of the device protection circuit (see Table 6).

Time-dependent dielectric breakdown. Time dependent dielectric breakdown (TDDB) is the formation of low-resistance dielectric paths through localized defects in the MOS oxide. TDDB has a very weak dependence on temperature and a dominant dependence on the electric field across the oxide. The field acceleration itself is a function of steady-state temperature. The field acceleration is inversely dependent on steady-state temperature, and has been found to reduce from 6 decades/MV/cm at 25°C to 2 decades/MV/cm at 150°C. The maximum allowable temperature and electric field across the oxide can be calculated from:

- Fowler-Nordheim Tunneling-based Models [Lee, 1988; Moazzami, 1989; Moazzami, 1990]
- Thermodynamic Model [McPherson, 1985]
- Empirical Models [Anolick, 1981; Crook, 1979; Berman, 1981]

Table 6 Comparison of Typical ESD Protection Thresholds

Protective Circuit Device	Device Technology	Protection Voltage Threshold (Volt)	Reference
double implant field isolation device in well	CMOS	8000 volts	[Nischizawa, 1975]
thick oxide-diffused resistor-field plate	DRAM junction depth : 0.4 μm breakdown voltage : 20 V field oxide : 1 μm	5000 volts	[Duvvury, 1983]
	EPROM junction depth : 0.8 μm breakdown voltage : 26 V field oxide : 1.4 μm	6000 volts	[Duvvury, 1983]
diode-diffused resistor-field plate	EPROM junction depth : 0.8 μm breakdown voltage : 26 V field oxide : 1.4 μm	4500 volts	[Duvvury, 1983]
thick oxide polysilicon resistor-field plate	EPROM junction depth : 0.8 μm breakdown voltage : 26 V field oxide : 1.4 μm	3000 volts	[Duvvury, 1983]
diffused resistor-gated diodes and spark gap	CMOS/SOS type A CMOS/SOS type B	800 volts	[Palumbo, 1986]
diode-resistor circuit	CMOS/SOS type C	1800 to 2000 volts	[Palumbo, 1986]
thick oxide	CMOS, NMOS	> 6000 volts	[Rountree, 1988]
diode-resistor-diode	CMOS gate array	4000 volts	[Hull, 1988]

Fowler-Nordheim Tunneling-based Models for TDDB. The maximum allowable temperature and electric field can be evaluated by calculating predicted life versus steady-state temperature, worst-case manufacturing defect magnitude (t_{r-ox}), and electric field (Figures 7.5 a and b). Due to the non-linear dependence of life under TDDB on temperature, defect magnitude, or electric field, reducing the stress below a particular value may not result in a noticeable benefit in terms of increased life, because the time to failure is much greater than the wear-out life of the device. Furthermore, because of the multiplicity of dependencies of TDDB, reducing the temperature to a pre-specified value for a device technology, as is done now, may not result in the desired reliability. The dependence of time to failure under TDDB is given by

$$\begin{aligned} &T_{operation} < T_{max} && \textit{weak acceleration} \\ &V_G < V_{ox(max)} && \textit{strong acceleration} \end{aligned} \tag{74}$$

such that

$$t_f > \textit{mission life} \tag{75}$$

where

$$t_{bd}(T) = \tau_{ox,0}\, e^{\frac{G_{ox,bd} X_{eff,ox}}{V_{ox}}\left(1 + \frac{\delta_{ox,bd}}{K_B}\left(\frac{1}{T} - \frac{1}{300}\right)\right) - \frac{E_B}{K_B}\left(\frac{1}{T} - \frac{1}{300}\right)} \tag{76}$$

where

$$\begin{aligned} &\delta_{ox,bd} = 0.0167\ eV \quad && for \quad 25^{o}\ C < T < 125^{o}\ C \\ &E_B = 0.28\ eV && \\ &\delta_{ox,bd} = 0.024\ eV \quad && for \quad T > 150^{o}\ C \\ &E_B = 0.28\ eV && \end{aligned} \tag{77}$$

$$\delta_{ox,bd} = \frac{K_B}{G^0_{ox,bd}} \frac{dG_{ox,bd}(T)}{d(1/T)} \tag{78}$$

and where T is the steady-state temperature (K), $G^0_{ox,bd}$ is the slope of the $\ln(t_f)$ versus $1/E_{-ox}$ plot ($\approx$ 350 x 10^6 volt/cm), X_{ox} is the oxide thickness (cm), $X_{eff,ox}$ is the effective oxide thickness at the weakest spot in the oxide (cm), V_G is the gate voltage across the oxide (Volt), V_{ox} is the voltage across the oxide (Volt), $\tau_{ox,o}$ is the room-temperature value of the pre-exponential, $n_{prexp}(T)$ ($\approx 1\times10^{-11}$ sec), E_B is the activation energy of the pre-exponential (eV), and K_B is Boltzmann's constant (8.617 x 10^{-5} eV/K).

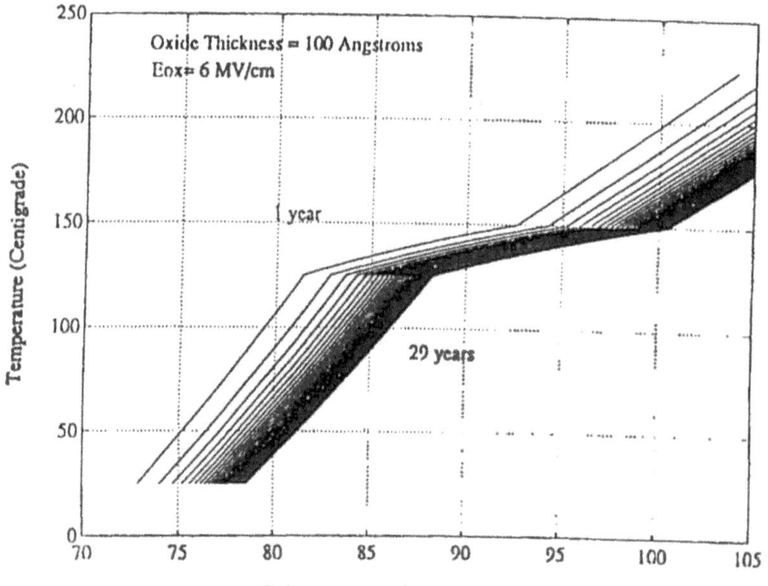

Figure 7.5a Due to the non-linear dependence of life under TDDB on temperature, worst case manufacturing defect magnitude, or electric field, reducing the stress below a particular value may not result in noticeable benefit in terms of increased life because the time to failure is much beyond wear-out life of the device. (a) Derating curve for TDDB versus steady state temperature and effective oxide thickness. The curves identify the various combinations of temperature T, and effective oxide thickness x_{eff}, which will result in desired life. [Lee, 1988; Moazzami, 1989; Moazzami, 1990].

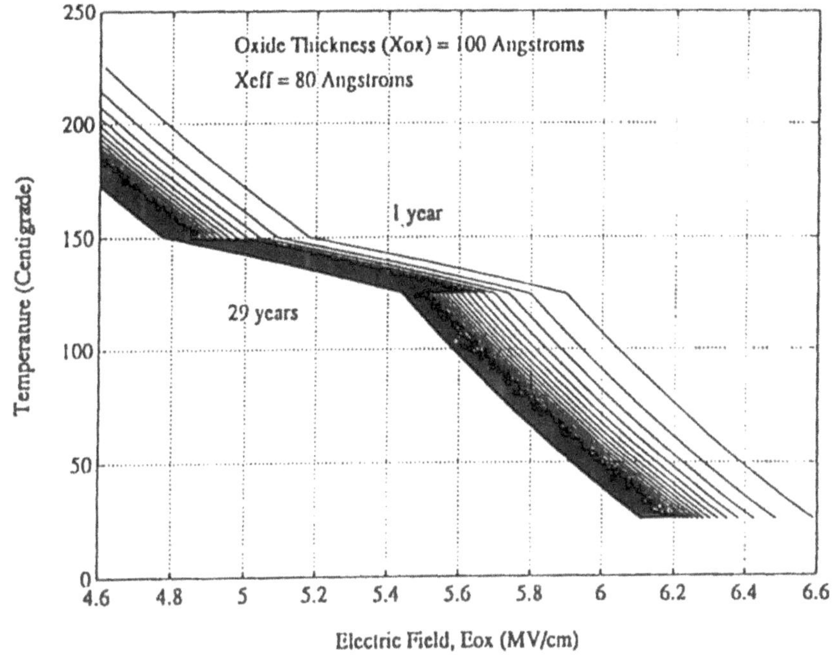

Figure 7.5b Derating curve for TDDB versus steady state temperature and electric field. The curves identify the various combinations of temperature T, and electric field E_{ox}, which will result in desired life [Lee, 1988; Moazzami, 1989; Moazzami, 1990].

Thermodynamic models [McPherson, 1985]. The thermodynamic model is based on the assumption that when the dielectric breaks down, it undergoes an irreversible phase transition transforming the material from an insulating phase to a conducting phase. The driving force for this transformation is the difference between the free energies of the conducting phase and the insulating phase. The maximum temperature and electric field for desired mission life can thus be calculated as follows:

$$T_{operation} < T_{max} \qquad \textit{weak acceleration}$$
$$V_G < V_{ox(\max)} \qquad \textit{strong acceleration} \tag{79}$$

such that

$$t_f > \textit{mission life} \tag{80}$$

where

$$t_{bd,f\%} = A_{ox,cf}\, e^{\left(\frac{\Delta H_{ox}}{K_B T}\right)} e^{(\gamma_{ef,acc}(T) S_{ox})} \tag{81}$$

The field acceleration parameter, $\gamma_{ef,acc}$, is the steady-state temperature-dependent parameter given by

$$\gamma_{ef,acc} = B + \frac{C}{T} \tag{82}$$

where T is the steady-state temperature, K_B is the Boltzmann constant (8.617×10^{-5} eV/K or 1.38×10^{-23} J/K), ΔH_{ox} is the change in enthalpy required to activate the poly filament growth at breakdown, B and C are constants, $S_{ox}=E_{bd,ox}-E_{ox}$, $E_{bd,ox}$ is the breakdown strength of the dielectric, and E_{ox} is the stressing in the dielectric.

Empirical models [Anolick, 1981, Crook, 1979, Berman, 1981]. The maximum temperature or the electric field across the oxide can be calculated from

$$\begin{aligned} &T_{operation} < T_{max} && \textit{weak acceleration} \\ &V_G < V_{ox(max)} && \textit{strong acceleration} \end{aligned} \tag{83}$$

such that

$$t_f > \textit{mission life} \tag{84}$$

where

$$t_{bd,f\%} = A_{ox,cf}\, e^{\frac{\Delta H_{ox}}{K_B T}}\, e^{\gamma_{ef,acc}(V_{bd,ox}(f) - V_{app,ox})} \tag{85}$$

where $t_{bd,f\%}$ is the time to failure for $f\%$ percent of the population, $\gamma_{ef,acc}$ is the voltage form factor (determined by life testing), ΔH_{ox} is the activation energy, K_B is Boltzmann constant, T steady-state temperature, $V_{bd,ox}(f)$ is the breakdown voltage for $f\%$ percent of population, $V_{app,ox}$ is the applied voltage, and $A_{ox,cf}$ is a constant. (Some typical values are given in Table 7)

5. STRESS LIMITS FOR FAILURE MECHANISMS IN THE DEVICE

5.1 Ionic contamination

Ionic contamination occurs predominantly in MOS devices and results in reversible degradation, in the form of threshold voltage shift and gain reduction due to the presence of mobile ions within the oxide or at the device-oxide interface. P-channel devices are less sensitive to ionic contamination than n-channel devices. The mobility of ions is steady-state temperature dependent. The threshold voltage shift increases with an increase in the steady-state temperature. However, a high-temperature storage bake in the neighborhood of 150°C to 250°C restores device characteristics. The maximum allowable steady-state temperature for the device is a function of the concentration of ionic contaminants in the oxide:

$$T_{operation} < T_{max} \tag{86}$$

such that

$$\Delta V_{th} < \Delta V_{th(max)} \tag{87}$$

where

Table 7 Empirical models for TDDB

Reference	Oxide Thickness	Experimental Conditions	Observations	Model Predictions
Anolick, Nelson, 1979	700 Å	E_{ox}= 1.3 MV/cm $E_{bd,ox} - \sigma_{\sim diel}$= 7 MV/cm	$(\Delta H_{ox})_{50\%} = 2.1$ eV $\gamma_{ef,ox} = B + \frac{C}{T}$	$(\Delta H_{ox})_{50\%} = 1.8$ eV $\gamma_{ef,ox} = B + \frac{C}{T}$
Crook, 1979	1100 Å	E_{ox} = 3.5 MV/cm $E_{bd,ox} - \sigma_{\sim diel}$= 3 MV/cm	$(\Delta H_{ox})_{50\%} = 0.3$ eV $\gamma_{ef,ox} = 7$ @ 25°C	$(\Delta H_{ox})_{50\%} = 0.34$ eV $\gamma_{ef,ox} = 6$ @ 25°C
Berman, 1981	≥400 Å	linear ramp	$(\Delta H_{ox})_{50\%} = 0.29\ (E_{bd,ox} - E_{ox})$ $\gamma_{ef,ox} = -5.4 + \frac{0.29}{T}$	$(\Delta H_{ox})_{50\%} = 0.29\ (E_{bd,ox} - E_{ox})$ $\gamma_{ef,ox} = -5.4 + \frac{0.29}{T}$
Hokari, 1982	100 Å	E_{ox} = 5 -7 MV/cm $E_{bd,ox} - \sigma_{\sim diel}$= 5 MV/cm	$(\Delta H_{ox})_{50\%} = 1.0$ eV @ 6 MV/cm $\gamma_{ef,ox} = 1.7$ @ 250°C	$(\Delta H_{ox})_{50\%} = 1.0$ eV @ 6 MV/cm $\gamma_{ef,ox} = 1.5$ @ 250°C
McPherson, 1985	100 Å	E_{ox} = 6 - 8 MV/cm $E_{bd,ox} - \sigma_{\sim diel}$= 3-5 MV/cm	$(\Delta H_{ox})_{50\%} = 0.3 - 1.0$ eV $\gamma_{ef,ox} = B + \frac{C}{T}$	$(\Delta H_{ox})_{50\%} = 0.3$ eV $\gamma_{ef,ox} = B + \frac{C}{T}$

$$\Delta V_{th} \propto E_{f,ch}^{1/2}\, t_{bs}^{1/2}\, e^{-E_a/K_B T} \tag{88}$$

and where ΔV_{th} is the change in threshold voltage (volt), $E_{f,ch}$ is the electric field across the oxide (volt/cm), t_{bs} is the time under bias (seconds), E_a is the activation energy (eV), and K_B is the Boltzmann constant (8.617 x 10^{-5} eV/K or 1.38 x 10^{-23} J/K), N-channel devices are often covered with phosphosilicate glass films to stabilize the threshold voltage against changes resulting from ionic contamination.

Surface-charge spreading. This failure mechanism, occurring mostly in MOS devices, involves the lateral spreading of charge from the biased metal conductors along the oxide layer or through moisture on the device surface. The failure mechanism is manufacturing-defect-activated, due to the presence of ionic contaminants on the die surface. Determining stress limits for this failure mechanism involves controlling the amount of contaminant on the die.

6. STRESS LIMITS FOR FAILURE MECHANISMS IN THE DEVICE OXIDE INTERFACE

6.1 Hot Electrons

The mechanism of hot electrons in MOS devices is inversely dependent on steady-state temperature. The minimum allowable steady-state temperature to avoid hot electrons can be calculated using the Lucky Electron Model [Ning, 1977, Garrigues, 1981]

$$T_{operation} < T_{min} \tag{89}$$

such that

$$P_{pr,eme} = P_{pr,ref}\, e^{-d_{mpl,e}/(\lambda_{mfp,e} \tanh(E_a/2K_B T))} \tag{90}$$

where

$$d_{mpl,e} = \sqrt{\frac{2\epsilon_{sm}\epsilon_o}{e N_{a,dp}}}\left(\sqrt{\Psi_s} - \sqrt{\Psi_s - \frac{\phi_{BS}}{e}}\right) \tag{91}$$

and where $P_{pr,eme}$ is the emission probability of an electron, $P_{pr,ref}$ is a constant (≈ 2.9 @ 300 K; 4.3 @ 77 K), $d_{mpl,e}$ is the minimum path length of an electron required to attain the critical energy, ϕ_{BS}, T is the steady-state temperature (K), E_a is the activation for mean free-path length (0.063 eV), $\lambda_{mfp,e}$ is a constant (≈ 108 Å), λ is the electron mean free path between lattice interactions, K_B is Boltzmann's constant (8.617 x 10^{-5} eV/K or 1.38 x 10^{-23} J/K), ϵ_{sm} is the dielectric constant of the semiconductor (F cm^{-1}), ϵ_o is the free-space permittivity (8.85 x 10^{-14} F cm^{-1}), $N_{a,dp}$ is the concentration of acceptor doping atoms (cm^{-3}), e is the magnitude of electronic charge (1.6 x 10^{-19} Coulomb), Ψ_s is the surface potential (volt), and ϕ_{BS} is the Si-SiO_2 barrier height for electrons, taking into account the lowering due to the Schottky effect.

Chapter 8

GUIDELINES FOR TEMPERATURE-TOLERANT DESIGN AND USE OF ELECTRONIC PACKAGES

Determining the allowable thermal limits for electronic systems involve finding such limits at each level of packaging. The focus of this chapter is determining the thermal stress limits on electronic packages for safe operation which ensures a desired useful mission life.

1. STRESS LIMITS FOR FAILURE MECHANISMS IN THE DIE AND DIE/SUBSTRATE ATTACH

1.1 Die Fracture

Westergaard-Bolger-Paris equation-based formulation

The maximum allowable temperature cycle should be less than the calculated temperature cycle magnitude:

$$\Delta T_{operation} < \Delta T_{\max} \tag{1}$$

such that

$$N_f > mission\ life \tag{2}$$

where

$$\Delta\sigma_{mdl} = 10^{-6} k \left| \alpha_{sub} - \alpha_{die} \right| \Delta T \sqrt{\frac{E_{sub} E_{att} L_{die}}{X_{da}}} \tag{3}$$

$$N_{f(--die)} = \frac{2}{(n_{paris,die} - 2) A_{paris,die} (\Delta\sigma_{mdl})^{n_{paris,die}} \pi^{\frac{n_{paris,die}}{2}}} \left(\frac{1}{a_i^{\frac{(n_{paris,die}-2)}{2}}} - \frac{1}{a_f^{\frac{(n_{paris,die}-2)}{2}}} \right) \tag{4}$$

and where $A_{paris,die}$ is the die-material coefficient (or Paris' coefficient), $n_{paris,die}$ is the die-material exponent, α_{sub} is the CTE of the substrate (/°C); α_{die} is the CTE of the die (/°C), $\Delta\sigma_{app}$ is the mode I applied-stress amplitude (psi), a_i is the initial crack size (inch), and a_f is the final crack length at failure, which may be taken to be equal to the critical crack size (inch).

Suhir-Paris equation-based formulation: The maximum allowable temperature cycle should be less than the calculated temperature cycle magnitude:

$$\Delta T_{operation} < \Delta T_{\max} \tag{5}$$

such that

$$K_{f,h\text{-}die} < K_{c,h\text{-}die} \tag{6}$$

$$K_{f,v\text{-}die} < K_{c,v\text{-}die} \tag{7}$$

and

$$\begin{aligned} \sigma_{mdl} &< \sigma_{Rupture\text{-}die} \qquad @\ die\ middle \\ \sigma_{1} &< \sigma_{Rupture\text{-}die} \qquad @\ die\ edge \end{aligned} \tag{8}$$

and

$$\min\left(N_{f(v\text{-}die)}, N_{f(h\text{-}die)}\right) > N_{required} \tag{9}$$

where

$$N_{f(v\text{-}die)} = \frac{1}{(1-\frac{n_{paris,die}}{2})\, A_{paris,die}^{n_{paris,die}}\, \sigma^{n_{paris,die}}\, \pi^{\frac{n_{paris,die}}{2}}} \left[a_{f,die}^{1-\frac{n_{paris,die}}{2}} - a_{i,die}^{1-\frac{n_{paris,die}}{2}} \right] \tag{10}$$

$$N_{f(h\text{-}die)} = \frac{1}{\left(1-\frac{n_{paris,die}}{2}\right) A_{paris,die}^{n_{paris,die}}\, p^{n_{paris,die}}\, \pi^{\frac{n_{paris,die}}{2}}} \left[a_{f\text{-}die}^{1-\frac{n_{paris,die}}{2}} - a_{i\text{-}die}^{1-\frac{n_{paris,die}}{2}} \right] \tag{11}$$

and

$$K_{f,vdie} = \sigma_{da} \sqrt{\pi\, a_{v\text{-}die}}\, F_{g} \tag{12}$$

$$K_{f,hdie} = P_{da} \sqrt{\pi\, a_{h\text{-}die}}\, F_{g} \tag{13}$$

$$F_g = \frac{2.85\,[0.953 - 2.369\,(\frac{a}{t_d}) + 2.74\tan(\frac{a}{t_d})]}{3 + \frac{a^2}{c^2}} \tag{14}$$

$$\sigma_{da} = \frac{\frac{1}{t_{die}} + 3\,(t_{die} + t_{sub})\,\frac{E_{die}\,t_{die}}{12\,(1-\nu_{die}^2)\,D_f}}{\frac{1-\nu_{die}}{E_{die}\,t_{die}} + \frac{1-\nu_{sub}}{E_{sub}\,t_{sub}} + \frac{(t_{die}+t_{sub})^2}{4\,D_f}} \left[1 - \frac{1}{\cosh(A_{da}L_{die})}\right](\alpha_{sub} - \alpha_{die})\,\Delta T \tag{15}$$

$$\tau_{da} = \left[\frac{1-\nu_{sub}}{E_{sub}\,t_{sub}} + \frac{1-\nu_{die}}{E_{die}\,t_{die}} + \frac{(t_{die}+t_{sub})^2}{4\,D_f}\right]^{-\frac{1}{2}} \left[\frac{t_{die}}{3\,G_{die}} + \frac{2t_{att}}{3\,G_{att}} + \frac{t_{sub}}{3\,G_{sub}}\right]^{-\frac{1}{2}} \tanh(A_{da}L_{die})\;(\alpha_{sub} - \alpha_{die})\,\Delta T \tag{16}$$

$$p_{da} = \left[\frac{t_{sub}\,t_{die}^3\,E_{die}}{12(1-\nu_{die}^2)} - \frac{t_{die}\,t_{sub}^3\,E_{sub}}{12(1-\nu_{sub}^2)}\right]\left[\frac{t_{die}}{3G_{die}} + \frac{2\;t_{att}}{3G_{att}} + \frac{t_{sub}}{3G_{sub}}\right]^{-1} \frac{1}{2D_f}(\alpha_{sub} - \alpha_{die})\Delta T \tag{17}$$

$$A = \sqrt{\frac{\frac{1-\nu_{sub}}{E_{sub}\,t_{sub}} + \frac{1-\nu_{die}}{E_{die}\,t_{die}} + \frac{(t_{die}+t_{sub})^2}{4\,D_f}}{\frac{t_{die}}{3\,G_{die}} + \frac{2t_{att}}{3\,G_{att}} + \frac{t_{sub}}{3\,G_{sub}}}} \tag{18}$$

$$D_f = \frac{E_{die}\,t_{die}^3}{12\,(1-\nu_{die}^2)} + \frac{E_{att}\,t_{att}^3}{12\,(1-\nu_{att}^2)} + \frac{E_{sub}\,t_{sub}^3}{12\,(1-\nu_{sub}^2)} \tag{19}$$

$$\sigma_1 = \frac{p_{pa}}{2} + \sqrt{\frac{p_{pa}^2}{4} + \tau_{da}^2} \tag{20}$$

and where E, G, ν, and α are the modulus of elasticity, the shear modulus, Poisson's ratio, and the coefficients of thermal expansion (CTE), respectively; t is the thickness. The subscripts *die*, *att*, and *sub* denote the die, attachment, and substrate, respectively. ΔT is the temperature change; L is the half-diagonal length of the die; σ is the tensile stress in the middle of the die; p_{pa} is the peeling stress at the die-attachment interface, τ_{da} is the shear stress at the die-attachment interface, σ_l is the principal stress at the die edge, $\sigma_{Rupture}$ is the modulus of rupture of the die; K_v and K_h are the stress-intensity factors for vertical cracks on the top surface of the die and for horizontal cracks at the edge of the die; K_C is the fracture toughness of the die; $a_{v\text{-}die}$ is the depth of the vertical crack on the top surface of the die; and $a_{h\text{-}die}$ is the length of the horizontal surface crack at the edge of the die.

1.2 Die Thermal Breakdown

The calculated junction temperature should be less than the allowable junction temperature of the device:

$$T_{junc} < T_{junc,allow} \tag{21}$$

where $T_{junc,allow}$ is the allowable junction temperature. T_{junc} is the effective junction temperature calculated from

$$T_{junc} = T_{amb,max} + Q_d \sum_{i=1}^{3} R_{th,i} \tag{22}$$

where T_{max} is the mean value of the maximum ambient temperature, and Q_d is the power dissipated by the devices. The effective thermal resistance of each layer, θ_i (die, attachment, and substrate), is determined from

$$R_{th,i} = \frac{1}{2k_i(l_i - w_i)} \left[\ln \frac{l_i(w_i + 2t_i)}{w_i(l_i + 2t_i)} \right] \tag{23}$$

where l_i, w_i, t_i are the lengths, widths, and thicknesses of each layer, and k_i is the thermal conductivity of each layer.

1.3 Die and Substrate Adhesion Fatigue

Suhir-Paris-(Coffin-Manson) equation-based formulation
For brittle-attach materials:

The maximum allowable temperature cycle should be less than the maximum calculated value, such that:

$$\Delta T_{operation} < \Delta T_{\max} \tag{24}$$

such that,

$$K_{hatt} < K_{C,att} \tag{25}$$

where

$$K_{hatt} = P_{da} \sqrt{\pi\, a_{hatt}}\, F_g \tag{26}$$

$$N_{f,hatt} = \frac{1}{\left(1 - \frac{n_{paris,att}}{2}\right) A_{paris,att}^{n_{paris,att}} P_{da}^{n_{paris,att}} \pi^{\frac{n_{paris,att}}{2}}} \left[a_{f-att}^{1-\frac{n_{paris,att}}{2}} - a_{i-att}^{1-\frac{n_{paris,att}}{2}} \right] \tag{27}$$

and where $a_{i\text{-}att}$ is the initial attach-crack size, $a_{f\text{-}att}$ is the final attach-crack depth, $A_{paris,att}$ and $n_{paris,att}$ are fatigue properties of the brittle attachment material; P_{da} is the cyclic peeling stress determined in Equation 16, F_g is the geometric correction factor given by Equation 14; $K_{C,att}$ is the fracture toughness of the attachment, and $K_{h,att}$ is the stress intensity factor for horizontal crack in the edge of the brittle attachment material.

For ductile-attach materials: The maximum allowable temperature cycle should be less than the maximum calculated value, such that:

$$\Delta T_{operation} < \Delta T_{\max} \tag{28}$$

such that

$$\begin{aligned} \sigma_{1,att} &< \sigma_{\mu t,att} \\ \tau_{m,att} &< \tau_{\mu t,att} \end{aligned} \tag{29}$$

where

$$\begin{aligned} \sigma_{1,att} &= \frac{3p_{da}}{2} + \sqrt{\left(\frac{3p_{da}}{2}\right)^2 + (4\tau_{da})^2} \\ \tau_{m,att} &= \sqrt{(3p_{da})^2 + 3(4\tau_{da})^2} \end{aligned} \tag{30}$$

$$N_{f,att} = C_{cf,att1} (\sigma_{1,att})^{m_{cf,att1}}$$
$$N_{f,att} = C_{cf,att2} (\tau_{m,att})^{m_{cf,att2}} \tag{31}$$

and where p_{sub} and τ_{sub} are the peeling and shear stresses given by Equations 16 and 17; $\sigma_{ult,att}$ and $\tau_{ult,att}$ are the tensile strength and the shear strength of the attachment material, $C_{cf,att1}$ and $m_{cf,att1}$ are the tensile fatigue constants, $C_{cf,att2}$ and $m_{cf,att}$ are the shear fatigue constants of the attachment material, $\sigma_{1,att1}$ is the local cyclic principal stress; and $\tau_{m,att}$ is the local cyclic von Mises' stress.

2. STRESS LIMITS FOR FAILURE MECHANISMS IN FIRST-LEVEL INTERCONNECTS

2.1 Wirebonded interconnections

Wire fatigue. The mechanism of wire fatigue involves the flexure of the wire about the reduced wire cross-section at the heel during temperature cycling. The mechanism has a dominant dependence on the magnitude of the temperature cycle and is independent of steady-state temperature. The maximum temperature cycle ($\Delta T_{operation}$) that the wire can be subjected to should be less than the calculated temperature cycle magnitude (Δt_{max}) [Pecht, 1989],

$$\Delta T_{operation} < \Delta T_{\max} \tag{32}$$

such that

$$N_f > mission\ life \tag{33}$$

where N_f can be calculated from either of the following models:
Pecht et al. Model [Pecht et al. 1989]

$$N_{f,wb} = A_{cf,wb}(\Delta\epsilon_f)^{n_{cf,wb}} \tag{34}$$

and

$$\Delta\epsilon_f = \frac{r_{wr}}{\rho_{o,wb}}\left[\frac{Cos^{-1}((Cos\lambda_{o,wb})(1 - (\alpha_{wr} - \alpha_{sub})\Delta T))}{\lambda_{o,wb}} - 1\right] \tag{35}$$

and where r_{wr} is the wire radius; $\lambda_{o,wb}$ is the angle of the wire with the substrate, α_{wr} is the coefficient of thermal expansion of the wire, α_{sub} is the coefficient of thermal expansion of the substrate, ΔT is the temperature cycle encountered by the structure, $\rho_{o,wb}$ is the initial radius of curvature of the wire, $N_{f,wb}$ is the mission life and $A_{cf,wb}$ and $n_{cf,wb}$ are material constants.

Hu et al. Model [Hu, Pecht, Dasgupta, 1991]

$$N_{f,wb} = C_{cf,wr}(\Delta\sigma_{wr})^{-m_{cf,wr}} \tag{36}$$

where

$$\Delta\sigma_{wr} = 6\,E_{wr}\frac{r_{wr}}{D_{wr}}\left(\frac{L_{wr}}{D_{wr}} - 1\right)^{\frac{1}{2}}\left(2\alpha_{sub} + \frac{\alpha_{sub} - \alpha_{wr}}{1 - \frac{D_{wr}}{L_{wr}}}\right)\Delta T \tag{37}$$

and where $2L_{wr}$ is the wire length, $2D_{wr}$ is the wire span, E_{wr} is the elastic modulus of the wire, α_{wr} and α_{sub} are the coefficients of thermal expansion of the wire and the substrate materials, respectively; ΔT is the temperature change encountered during operation, $\Delta\sigma_{wr}$ is the bending stress (already calculated), and $C_{cf,wr}$ and $m_{tf,wr}$ are fatigue properties determined by tensile fatigue tests of the wire material.

Wirebond fatigue: Bond pad shear fatigue: The maximum temperature cycle ($\Delta T_{operation}$) that the wire-wirebond assembly can be subjected to without causing shear fatigue failure of the bondpad should be less than the calculated temperature cycle magnitude (ΔT_{max}) [Hu, Pecht, Dasgupta, 1991]:

$$\Delta T_{operation} < \Delta T_{max} \tag{38}$$

$$\Delta T_{max} = \frac{\left(\frac{N_R}{C_{cf,bp}}\right)^{-\frac{1}{m^{f}_{bp}}}}{\left(\frac{G_{bp}}{t_{bp}\,Z_{wb}}\right)\left((\alpha_{wr} - \alpha_{sub}) - \frac{(\alpha_{sub} - \alpha_{bp})}{\left(1 + \frac{\frac{E_{sub}\,A_{cr,sub}}{E_{bp}\,A_{cr,bp}}}{(1 - \nu_{sub})}\right)}\right)} \tag{39}$$

and

$$Z^2_{wb} = \frac{G_{bp}}{t_{bp}}\left(\frac{r_{wr}}{E_{wr}A_{cr,wr}} + \frac{(1 - \nu_{sub})w_{bp}}{E_{sub}A_{cr,sub}}\right) \tag{40}$$

and where N_R is the mission profile requirement for the number of cycles to failure, G_{bp} is the shear modulus of the bond-pad material, t_{bp} is the bond-pad thickness, Z is given by Equation 40; $\Delta\ T$ is the temperature cycle magnitude; α_{sub}, α_{p}, α_{wr}, are the coefficients of thermal expansion for the substrate, pad, and wire, respectively; E_{bp}, and E_{sub} are the modulus of elasticity of the pad and the substrate, respectively, ν_{sub} is Poisson's ratio for the substrate materials, $A_{cr,bp}$ is the cross-sectional area of the pad, $A_{cr,sub}$ is the effective cross-sectional area of the substrate, equal to $h_{sub}(w_{bp}+w_{sub})/2$, and w_{bp} is the width of the bond pad.

Wire shear fatigue: The maximum temperature cycle ($\Delta T_{operation}$) that the wire-wirebond assembly can be subjected to without causing shear fatigue of the wire should be less than the calculated temperature cycle magnitude ($z\Delta T_{max}$) [Hu, Pecht, Dasgupta, 1991]:

$$\Delta T_{operation} < \Delta T_{max} \tag{41}$$

where

$$\Delta T_{max} = \frac{\left(\frac{N_R}{C_{cf,sub}}\right)^{-\frac{1}{m_{cf,sub}}}}{\left(\frac{r_{wr}^2}{4Z_{wb}^2 A_{cr,wr}^{\ 2}}\left(\frac{Cosh(z_{wb}\ x_{wr})}{Cosh(z_{wb}\ l_w)} - 1\right)^2 + \frac{Sinh^2(Z_{wb}\ x_{wr})}{Cosh^2(Z_{wb}\ l_w)}\right)^{\frac{1}{2}} C_{intl,wr}} \tag{42}$$

$$Z_{wb}^2 = \frac{G_{bp}}{t_{bp}}\left(\frac{r_{wr}}{E_{wr}A_{cr,sub}} + \frac{(1 - \nu_{sub})w_{bp}}{E_{sub}A_{cr,sub}}\right) \tag{43}$$

$$C_{intl,wr} = \left(\frac{G_{bp}}{t_{bp}\ Z_{wb}}\right)\left((\alpha_{wr} - \alpha_{sub}) - \frac{(\alpha_{sub} - \alpha_{bp})}{\left(1 + \frac{\frac{E_{sub}\ A_{cr,sub}}{E_{bp}\ A_{cr,bp}}}{(1 - \nu_{sub})}\right)}\right) \tag{44}$$

and where N_R is the mission profile requirement for the number of cycles to failure, r_{wr} is the wire radius, Z_{wb} is given by Equation 43, $A_{cr,wr}$ is the effective cross-sectional area of the wire at the bond = (0.6 wire diameter) x (1.5 wire diameter), x_{wr} is the position along the length of the bond (x=0 is the center of the bond), l_w = half the length of the bond (total bonded length = $2l_w$); $C_{intl,wr}$ is given by Equation 44, G_{bp} is the shear modulus of the bond pad, t_{bp} is the bond-pad thickness; w_{bp} is the width of the bond pad, E_{wr} is the elastic modulus of the wire, E_{bp} is the elastic modulus of the bond pad, E_{sub} is the elastic modulus of chip, A_{sub} is the effective cross-sectional area of the substrate, equal to $h_{sub}(w_{bp}+w_{sub})/2$, A_{bp} is the cross-sectional area of the pad, and α_{sub}, α_{bp}, α_w are the coefficients of thermal expansion for the substrate, pad, and wire, respectively.

Chip shear fatigue leading to cratering: The maximum temperature cycle ($\Delta T_{operation}$) that the wire-wirebond assembly can be subjected to without causing shear fatigue of the chip should be less than the calculated temperature cycle magnitude (ΔT_{max}) [Hu, Pecht, Dasgupta, 1991]

$$\Delta T_{operation} < \Delta T_{max} \tag{45}$$

where

$$\Delta T_{max} = \frac{\left(\frac{N_R}{C_{cf,bp}}\right)^{-\frac{1}{m_{cf,bp}}}}{\left(\left(\frac{w_{bp}C_{intl,wr}}{2Z_{wb}A_{cr,sub}}\left(1 - \frac{Cosh(Z_{wb}x_{sub})}{Cosh(Z_{wb}l_{sub})}\right) + \frac{(\alpha_{sub} - \alpha_{bp})}{\frac{(1-\nu_{sub})}{(E_{sub}A_{cr,sub})} + \frac{1}{(E_{bp}A_{bp})}}\right)^2 + C_{intl,wr}^2\frac{Sinh^2(Z_{wb}x_{sub})}{Cosh^2(Z_{wb}l_{sub})}\right)^{1/2}} \tag{46}$$

$$Z_{wb}^2 = \frac{G_{bp}}{t_{bp}}\left(\frac{r_{wr}}{E_{wb}A_{cr,sub}} + \frac{(1 - \nu_{sub})w_{bp}}{E_{sub}A_{sub}}\right) \tag{47}$$

$$C_{intl,wr} = \left(\frac{G_{bp}}{t_{bp}\, Z_{wb}}\right)\left((\alpha_{wr} - \alpha_{sub}) - \frac{(\alpha_{sub} - \alpha_{bp})}{\left(1 + \dfrac{E_{sub}\, A_{cr,sub}}{\dfrac{E_{bp}\, A_{cr,bp}}{(1 - \nu_{sub})}}\right)}\right) \tag{48}$$

and where N_R is the mission profile requirement for the number of cycles to failure, r_{wr} is the wire radius, Z_{wb} is given by Equation 47, $A_{cr,wr}$ is the effective cross-sectional area of the wire at the bond = (0.6 wire diameter) x (1.5 wire diameter), x_{sub} is the position along the length of the bond (x=0 is the center of the bond), l_{sub} is half the length of the bond (total bonded length = $2l_{sub}$), $C_{intl,wr}$ is given by Equation 48, G_{bp} is the shear modulus of the bond pad, t_{bp} is the bond pad thickness, w_{bp} is the width of the bond pad, E_{wr} is the elastic modulus of the wire, E_{bp} is the elastic modulus of the bond pad, E_{sub} is the elastic modulus of the chip, $A_{cr,sub}$ is the effective cross-sectional area of the substrate, equal to $h_{sub}(w_{bp}+w_{sub})/2$, $A_{cr,bp}$ is the cross-sectional area of the pad, and α_{sub}, α_{bp}, α_{wr} are the coefficients of thermal expansion for the substrate, pad, and wire, respectively.

Intermetallic formation. The maximum operating temperature for a bimetallic wirebond system can be obtained by using the parabolic relationship [Kidson, 1961]:

$$t = \frac{h_{int}^2}{k_{cr,int}^2} \tag{49}$$

The maximum allowable operating temperature is less than the calculated operating temperature:

$$T_{operation} < T_{max} \tag{50}$$

where

$$T_{max} = -\frac{E_a}{R\ \ln\left(\dfrac{h_{int}^2}{C_x t}\right)} \tag{51}$$

$$k_{rc,int} = C_x\, e^{-\frac{E_a}{K_B T}}$$

and where h_{int} is the critical intermetallic layer thickness (from Tables 1 and 2); t is time to failure due to intermetallic formation, $k_{rc,\,int}$ is the rate constant, depending on the interdiffusion coefficients of the bonded materials (from Table 1).

The critical intermetallic layer thickness is calculated based on the intermetallic compounds that form the fastest for the bimetallic combination. The dominant compounds for some of the common material combinations are given in Table 1. The composition of the intermetallic compound when critical layer thickness is reached, and rate constants for various material combinations, are given in Table 2. The time to failure is calculated as the time for complete consumption of the bond pad in the intermetallic reaction. For example, in gold-aluminum bonds, the compound Au_5Al_2 forms the fastest and is the dominant product during intermetallic formation. Thus, for gold bond pads, the critical layer thickness is defined as the

Table 1 The dominant intermetallic compounds for common bimetal combinations

Bimetal combination	Intermetallic compounds formed	Dominant intermetallic compounds
gold-aluminum (Au-Al)	Au_5Al_2, Au_2Al, $AuAl_2$, AuAl, Au_4Al [Philosky, 1970, Philosky, 1971]	Au_5Al_2, Au_2Al
copper-aluminum (Cu-Al)	$CuAl_2$, CuAl, $CuAl_2$, Cu_9Al_4 [Olsen and James, 1984, Pitt and Needes, 1981, Gershinskii, 1977, Campisano, 1978, Funamizu and Watanabe, 1971]	$CuAl_2$
gold-silver (Au-Ag)	significant silver diffusion into the wire bulk along the wire length. Rapid silver surface diffusion resulted in depletion at bond periphery [James, 1977]	
copper-gold (Cu-Au)	Cu_3Au, CuAu, $CuAu_3$ [Hall, 1975, Tu and Berry, 1972]	Cu_3Au

Table 2 Rate constants and activation energies for common material combinations

Material combination	E_a (cal mol^{-1})	Critical layer thickness (t_i = thickness of ith element)	Rate constant (k) ($cm^2\ sec^{-1}$); R =1.98719 cal $mol^{-1}\ K^{-1}$
gold-aluminum (Au-Al)	15,900	gold bond pad [Philosky, 1970, Philosky, 1971]: $t_{Au} + (2/5)\ t_{Au}$ aluminum bond pad: $t_{Al} + (2.5)\ t_{Al}$	$k_{rc,int} = 5.2 \times 10^{-4}\ e^{-\frac{E_a}{RT}}\ \frac{cm^2}{sec}$
gold-silver (Au-Ag)	15,000	t_{Ag} [James, 1977]	$k_{rc,int} = 7.65 \times 10^{-4}\ e^{-\frac{E_a}{RT}}\ \frac{cm^2}{sec}$

time to reach an intermetallic compound thickness of t_{Au} + (2/5) t_{Au}, where t_{Au} is the thickness of the gold bond pad.

The time to failure versus temperature is then plotted, based on Equation 53. The maximum operating temperature of the device for a given mission life is calculated from the graph (Figures 8.1 a, b, and c). A similar procedure can be followed for each of the intermetallic combinations. Based on Philosky's observations of the layer thicknesses of intermetallic compounds at various temperatures (Table 3), a graph of intermetallic layer thickness versus square-root time to failure can be drawn (Figure 8.2) [1971]. A horizontal line on the graph at a thickness equal to the wire thickness at the bond pad gives the time-temperature product that will result in failure after time equal to the abscise. The allowable time-temperature for a desired mission life can thus be computed. The slope of each of the lines in Figure 8.2 gives rate constants at various temperatures. Intermetallic formation in aluminum-gold systems can be eliminated either by using a fritless gold composition that impedes intermetallic formation or by using diffusion barrier disks attached between the aluminum (Al) wire and thick film [Palmer and Gaynard, 1978].

2.2 Tape Automated Bonds

Thermally induced solder-joint fatigue. The temperature cycle magnitude to which the device is subjected should be less than the maximum allowable:

$$\Delta T < \Delta T_{max} \tag{52}$$

such that

$$N_f > mission\ life \tag{53}$$

where N_f is the [Engelmaier 1983, 1988] cycles to failure for outer-lead TAB solder joint.

$$N_f = \frac{1}{2}\left(\frac{1}{2\epsilon_f'}\frac{K(L_D \Delta\alpha \Delta T_e)^2}{200 A_{eff} h_{solder}}\right)^{\frac{1}{c}} \tag{54}$$

Table 3 Measured thickness of gold-aluminum (Au-Al) intermetallic compounds [Philosky, 1971]

Temperature (°C)	Time (sec)	$AuAl_2$ cm x10^{-3}	AuAl cm x10^{-3}	Au_2Al cm x10^{-3}	Au_5Al_2 cm x10^{-3}	Au_4Al cm x 10^{-3}	Total cm x10^{-3}
200	8700	0	0	0	0.68	0	0.69
200	27300	0	0	0	1.22	0	1.25
200	57600	0	0	0	1.57	0	1.63
200	349200	0.21	0	0.53	2.68	0.21	3.63
250	7200	0	0	0	1.09	0	1.13
250	75600	0.24	0	0.8	2.47	0.24	3.75
300	15900	0.11	0	0.72	2.38	0.17	3.38
300	58200			1.29	4.75	0.23	6.5
350	22200	0.27	0	0.1	1.65	0.11	2.13
350	24000	0.42	0	0.52	6.31		7.38
400	300	0.15	0	0.09	0.89	0	1.14
400	1500	0.32	0	0.29	2.32	0	2.94
400	6000	0.36	0	0.93	4.57	0	5.88
400	14400	0.33	0.12	1.48	6.46	0.11	8.5
400	36000	0.26	0.27	2.36	7.03	0.21	10.13
460	1200	0.26	0.27	1.24	2.36	0	4.13
460	6000	0.21	0.35	1.4	7.14	0.28	9.38

$$c = -0.442 - 6\times10^{-4}T_{sj} + 1.74\times10^{-2}\ln\left[1 + \left(\frac{360}{t_D}\right)\right] \tag{55}$$

and

$$T_{sj} = \frac{1}{4}(T_c + 2T_o + T_s) \tag{56}$$

and where T_S and T_c are the steady-state operating temperatures of the substrate and component, respectively, T_O is the temperature during off half-cycle, t_D is the half-cycle dwell time in minutes, K is the diagonal flexural stiffness of the unconstrained lead, determined by finite element analysis [Barker, 1991] or strain-energy methods, $\Delta\alpha$ is the difference in the coefficients of thermal expansion of the die and the substrate, ΔT_{max} is the maximum allowable equivalent temperature range, ϵ_f' is the fatigue ductility coefficient, A_{eff} is the effective solder-joint area, which is two-thirds of the vertical projection of the solder-wetted lead area, h_{solder} is the height of the solder joint, and t_D is the half-cycle dwell time in minutes.

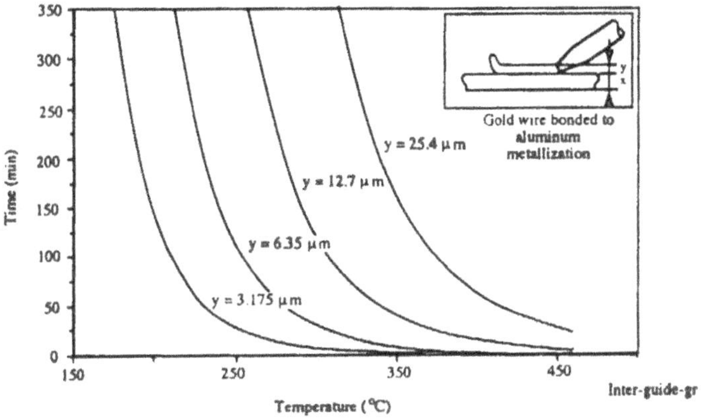

Figure 8.1a For gold wire bonded to aluminum metallization. Time for heel of the wire to transform to intermetallic versus temperature. The temperature at which intermetallic formation has a dominant dependence on steady-state temperature is a function of the bond geometry [Philosky, 1970, 1972].

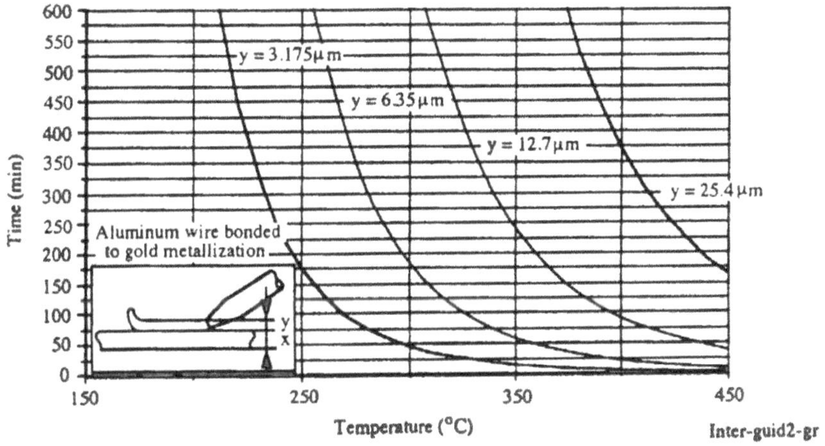

Figure 8.1b For aluminum wire bonded to gold metallization. Time for heel to transform to intermetallic versus temperature. The temperature at which intermetallic formation has a dominant dependence on steady state temperature is a function of the bond geometry.

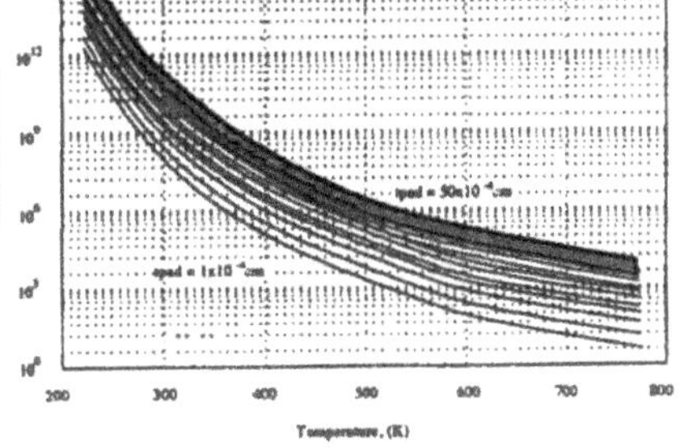

Figure 8.1.c Time for heel to transform to intermetallic versus temperature. The temperature at which intermetallic formation has dominant dependence on steady state temperature is a function of the bond geometry [Philosky, 1970, 1971]. Time to failure versus temperature and bond pad thickness. The time to failure at any temperature may be much greater than the mission life depending on the bond pad thickness.

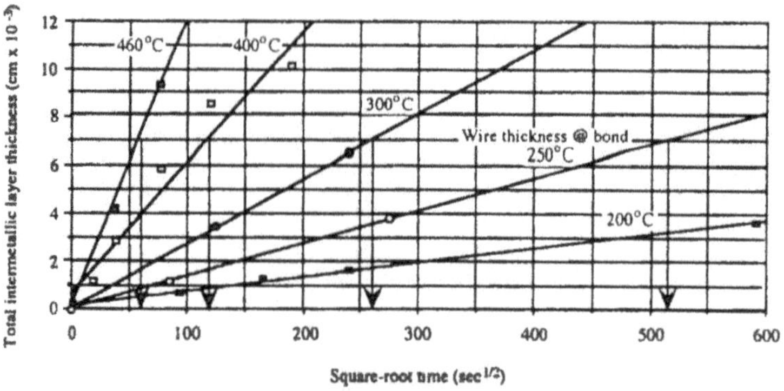

Figure 8.2 Various time-temperature products which will result in gold-aluminum intermetallic growth [Philosky, 1971]. A horizontal line on the graph at a thickness equal to the wire thickness at the bond pad gives the time-temperature product which will result in failure after time equal to the abscissa. The slope of each line gives the rate constants at various temperatures.

2.3 Flip-chip Bonds

The temperature cycle magnitude to which the device is subjected should be less than the maximum allowable:

$$\Delta T < \Delta T_{max} \tag{57}$$

such that

$$N_f > mission\ life \tag{58}$$

where cycles to point failure based on a modification of quasi-empirical, developed by Norris and Landzberg [Norris 1969], by Engelmaier [1983] is

$$N_f = \frac{1}{2}\left[\frac{\Delta\gamma}{2\epsilon_f}\right]^{\frac{1}{c}} \tag{59}$$

$$c = \alpha + \beta\, T_{SJ} + \gamma \ln\left(1 + \frac{360}{t_D}\right) \tag{60}$$

$$\begin{aligned}\Delta\gamma &= k_g \delta \\ &= k_g d[\alpha_c\{T_{jc} - T_{amb}\} - \alpha_{st}\{T_{jc} - \theta_{sol}(\frac{P_{chip}}{n}) - T_{amb}\}]\end{aligned} \tag{61}$$

$$k_g = \frac{1}{(\pi r^2_{critical})^{1/\beta} \int_0^h \frac{dy}{[\pi(R^2 - y^2)]^{1/\beta}}} \tag{62}$$

$$T_{SJ} = \frac{1}{4}[2T_{jc} - \theta_{sol}(\frac{P_{chip}}{n}) + 2T_{amb}] \tag{63}$$

$$\theta_{sol} = \frac{h}{k_{sol}\, A_{eff}} \tag{64}$$

and where ϵ_f is the fatigue ductility coefficient of the solder material, α, β, and γ are factors to be determined empirically, T_{SJ} is the mean cyclic solder-joint temperature (°C); t_D is half-cycle dwell time in minutes, k_g is the geometric parameter, δ is the total displacement, d is the distance of farthest solder joint from neutral axis, α_c and α_{st} are the coefficients of thermal expansion of the chip and the substrate, respectively, T_{jc} is the junction temperature, T_{amb} is the ambient temperature, θ_{sol} is the thermal resistance of the solder, P_{chip} is the chip power rating, n is the number of solder bumps on the chip, h is the height of the solder, $r_{critical}$ is the critical interface radius, k_{sol} is the thermal conductivity of the solder, and A_{eff} is the effective area of cross section of the solder.

3. STRESS LIMITS FOR FAILURE MECHANISMS IN THE PACKAGE CASE

3.1 Cracking in Plastic Packages

The maximum temperature during thermal shock to which the device can be subjected can be calculated from:

$$T < T_{max} \tag{65}$$

such that

$$\sigma_{max,mc} < \sigma_{crit,mc}(T_{max}) \tag{66}$$

where

$$\sigma_{max,mc} = 6K_f\left(\frac{a_{die}}{t_{mc}}\right)^2 P_c \tag{67}$$

$$P_c = RH_{sat}P_{sat,elev}(T_{max}) \tag{68}$$

and where RH_{sat} is the relative humidity of the saturation ambient prior to temperature shock, T_{max} is the peak temperature of the thermal shock, K_f is a dimensionless stress concentration factor that depends on the aspect ratio of the die pad (K=0.05 for a square pad), a_{die} is the length of the short side of the die pad, t_{mc} is the thickness of the molding compound under the pad, P_c is the vapor pressure in the cavity, and $\sigma_{crit,mc}$ is the value of critical stress.

3.2 Reversion or depolymerization of polymeric bonds

Reversion is a steady-state temperature-dependent phenomenon that actuates above the reversion temperature for the encapsulant or molding compound. Determining limiting stress for reversion involves limiting the maximum temperature for device operation to lower than the depolymerization temperature:

$$T < T_{depolymerization} \tag{69}$$

3.3 Whisker and dendritic growth

Cases coated with tin are prone to whiskers and dendrites. Whiskering can, however, be reduced by increasing the coating thickness during manufacture. Typically, hot-dipping is used to obtain a stress-free tin coating. If electroplating is used, tin is reflowed to remove residual stress, and 2 to 3% lead is added to the tin to retard whisker growth. This is a contamination-actuated mechanism and has no temperature dependence.

3.4 Modular case fatigue failure

The temperature cycle magnitude to which the device is subjected should be less than the maximum allowable:

$$\Delta T < \Delta T_{max} \tag{70}$$

such that

$$N_f > mission\ life \tag{71}$$

where

$$\Delta\tau_{max,case} = 2C_{cf,cs}(2N_{f,cs})^{m_{cf,cs}} \tag{72}$$

and

$$\Delta\tau_{max,case} = \left[\frac{1-\nu_{hdr}}{E_{hdr}h_{hdr}} + \frac{1-\nu_{wal}}{E_{wal}h_{wal}} + \frac{(h_{wal}+h_{hdr})^2}{4D_{f,cs}}\right]^{-\frac{1}{2}} \left[\frac{h_{wal}}{3G_{wal}} + \frac{2h_{att}}{3G_{att}} + \frac{h_{hdr}}{3G_{hdr}}\right]^{-\frac{1}{2}} \tanh(A_{cs}L_{jt})\Delta\alpha\Delta T \tag{73}$$

$$A_{cs} = \sqrt{\frac{\dfrac{1-\nu_{hdr}}{E_{hdr}t_{hdr}} + \dfrac{1-\nu_{wal}}{E_{wal}t_{wal}} + \dfrac{(t_{wal}+t_{hdr})^2}{4D_{f,cs}}}{\dfrac{t_{wal}}{3G_{wal}} + \dfrac{2t_{att}}{3G_{att}} + \dfrac{t_{hdr}}{3G_{hdr}}}} \tag{74}$$

The flexural rigidity of the structure, $D_{f,cs}$, is defined as

$$D_{f,cs} = \frac{E_{hdr}h_{hdr}^3}{12(1-\nu_{hdr}^2)} + \frac{E_{att}h_{att}^3}{12(1-\nu_{att}^2)} + \frac{E_{wal}h_{wal}^3}{12(1-\nu_{wal}^2)} \tag{75}$$

where ν is Poisson's ratio, E is the modulus of elasticity, h is the thickness, G is the shear modulus, L is the length of the joint, $\Delta\alpha$ is the difference in the CTEs of the case wall and header materials, and subscripts *hdr, wal,* and *att* denote the header, wall, and attach material, respectively.

4. STRESS LIMITS FOR FAILURE MECHANISMS IN LID SEALS

4.1 Thermal Fatigue of Lid Seal

Lid and case material are the same:

Most lid-seal failures are overstress failures fatigue is not a concern in most applications. The temperature cycle magnitude to which the device is subjected should be less than the maximum allowable:

$$\Delta T < \Delta T_{max} \tag{76}$$

such that

$$\sigma_1 < \sigma_{ult(seal)} \tag{77}$$

The maximum principal stress in the seal is calculated using Mohr's circle, where the shear stress at any seal cross-section at a distance, x, from the center cross-section is

$$\tau_{case}(x) = \frac{\Delta\alpha\Delta T}{\sqrt{C_{c,t}\,C_{c,jt}}}\left(\frac{\sinh\left(\sqrt{\frac{C_{c,t}}{C_{c,jt}}}x\right)}{\sinh\left(\sqrt{\frac{C_{c,t}}{C_{c,jt}}}l_{seal}\right)}\right) \tag{78}$$

and the normal stress at any cross-section at a distance x from the mid-section of the seal is

$$\sigma_{case}(x) = \frac{\Delta\alpha\Delta T}{C_{c,t}\,t_{seal}}\left[1-\frac{\cosh\left(\sqrt{\frac{C_{c,t}}{C_{c,jt}}}x\right)}{\cosh\left(\sqrt{\frac{C_{c,t}}{C_{c,jt}}}l_{seal}\right)}\right] \tag{79}$$

where $\Delta\alpha$ is the difference in the coefficients of thermal expansion of the lid and case material and the seal material, ΔT is the difference between the sealant melting point (stress-free temperature) and operating temperature, $C_{c,t}$ is the in-plane compliance of the joint (=[(1-ν_{case})/E_{case}(h_{cavity}+t_{lid})] + [(1-ν_{seal})/$E_{seal}t_{seal}$]), $C_{c,jt}$ is the interfacial compliance, given as the sum of the individual compliances for the seal and the case or lid material; that is, $C_{c,jt} = \kappa_{seal} + \kappa_{case}$, while $\kappa_{seal} = 2(1+\nu_{seal})t_{seal}/3E_{seal}$ and $\kappa_{case} = 2(1+\nu_{case})(h_{cavity}+t_{lid})/3E_{case}$. l_{seal} is half the seal length (or width).

Lid and case material are different: The temperature cycle magnitude to which the device is subjected should be less than the maximum allowable:

$$\Delta T < \Delta T_{max} \tag{80}$$

such that

$$\sigma_1 < \sigma_{ult(seal)} \tag{81}$$

where the principal stress in the seal is

$$p_{ls} = -\left[\frac{t_{case}t_{lid}^3E_{lid}}{12(1-\nu_{lid}^2)} - \frac{t_{lid}t_{case}^3E_{case}}{12(1-\nu_{case}^2)}\right]\frac{1}{2G_{ls}D_{f,ls}}\Delta\alpha\Delta T \tag{82}$$

$$\sigma_1 = \frac{p_{ls}}{2} \pm \sqrt{\frac{p_{ls}^2}{4} + \tau_{ls}^2} \tag{83}$$

$$\tau_{ls} = \left[\frac{1-\nu_{case}}{E_ct_c} + \frac{1-\nu_{lid}}{E_lt_l} + \frac{(t_{case}+t_{lid})^2}{4D_fls}\right]^{-1/2}(G_{ls})^{-1/2}\tanh(A_{ls}l_{seal})\ \Delta\alpha\Delta T \tag{84}$$

and where p_{ls} is the peeling stress, τ_{ls} is the shear stress in the seal, ν, E, l, and t denote Poisson's ratio, Young's modulus of elasticity, the length, and the thickness, respectively; subscripts l, c and s denote that the symbols refer respectively to the lid, case, and seal; $\Delta\alpha$ is computed as $(\alpha_c - \alpha_l)$, where α is the coefficient of thermal expansion, ΔT is the temperature excursion with respect to the stress-free state of the seal, or the sealing temperature; A_{ls}, $D_{f,ls}$, and G'_{ls} are defined as

$$A_{ls} = \sqrt{\frac{\frac{1-\nu_{case}}{E_{case}t_{case}} + \frac{1-\nu_{lid}}{E_{lid}t_{lid}} + \frac{(t_{case}+t_{lid})^2}{4D_{f,ls}}}{G_{ls}}} \tag{85}$$

$$D_{f,ls} = \frac{E_{case}t_{case}^3}{12(1-\nu_{case}^2)} + \frac{E_{lid}t_{lid}^3}{12(1-\nu_{lid}^2)} + \frac{E_{seal}t_{seal}^3}{12(1-\nu_{sasl}^2)} \tag{86}$$

$$G_{ls} = \frac{t_{case}}{3G_{case}} + \frac{2t_{seal}}{3G_{seal}} + \frac{t_{lid}}{3G_{lid}} \tag{87}$$

where G is the shear modulus of elasticity.

Chapter 9

CONCLUSIONS

The temperature dependencies of the failure mechanisms existing at various package elements have been investigated in terms of steady state temperature, temperature cycle, temperature gradient, and time dependent temperature change (Tables A1 and A 2, Appendix A). It has been found that temperature dependencies for the same mechanism are not the same at all operating temperatures. The mechanisms have been broadly grouped into the steady state temperature ranges of -55°C to 150°C, 150°C to 400°C, and T > 400°C (Tables A3 and A 4). The investigation demonstrates that there is no steady state temperature dependence for any of the failure mechanisms in the equipment operating range of -55°C to 125°C, but the steady state temperature dependence increases for temperatures above 150°C as more mechanisms assume a dominant steady state temperature dependence. An overview of the effects of temperature on microcircuits can be categorized as follows:

1. STEADY STATE TEMPERATURE EFFECTS

Changes in semiconductor characteristics due to:

- Steady state temperature dependence of semiconductor characteristics such as resistivity which can lead to failures in the form of thermal runaway (*e.g., electrical overstress*).
- Steady state temperature dependence on contaminants in the semiconductor that affect electronic functions or compatibility with the circuit parameters (*e.g., ionic contamination*).
- Manufacturing defects which will cause device failures after prolonged operation under voltage and steady state temperature (*e.g., TDDB*).

Increase in susceptibility to failure due to an environmental stress, at higher temperatures (T dependence) (*e.g. electrostatic discharge*).

- Increase in susceptibility to failure due to a failure mechanism at lowered temperature, i.e., the failure mechanisms are inversely dependent on steady state temperature (*e.g., stress driven diffusive voiding, and hot electrons*).
- Increase in susceptibility to failure under steady state temperature stress above a threshold value of temperature (*e.g., hillock formation, metallization migration, contact spiking, encapsulant reversion, electromigration*).

Temperature cycle magnitude effects.

- Temperature cycles result in cyclic fatigue failures due to mechanical stresses and dimensional changes caused by thermal mismatches between mating surfaces (*e.g., wirebond shear and flexure fatigue, die fracture, die adhesive fatigue*).

Temperature gradient effects.

- Sites of maximum temperature gradient provide most probable sites for failure due to mass transport mechanisms (*e.g., electromigration damage at high current occurs at sites of maximum temperature gradient*).

Time dependent temperature change effects.

- Time dependent temperature changes produce very large stress transients resulting in material failure (*e.g., encapsulant cracking*).
- Temperature change in time may activate or de-activate a failure mechanism (*e.g., duty cycle serves as an ON/OFF switch for the corrosion process by evaporation of the electrolyte at higher temperatures*).

The use of a simple Arrhenius expression to model microelectronic device reliability at all steady state temperatures is not correct because the temperature dependencies of the device are different at different steady state temperatures. It is also improper to assess the thermal acceleration of the device by stress tests at elevated temperatures and extrapolate the results to lower temperatures, because the failure mechanisms (Table A.4) are not uniformly active for all steady state temperatures.

The generalized association of lowered temperature with higher reliability may not be true for all device technologies. Two possible exceptions may exist. First, the mechanism may not be dependent on steady state temperature, and second, the mechanism may have an inverse dependence on steady state temperature (e.g., hot electrons). In either case, the steady state temperature will not be a driver for microelectronic reliability.

In cases where temperature can be directly related to parameter drift, premature aging, and even catastrophic failure, the temperature influence on reliability can be unarguable. Although examinations of case histories indicate that temperature's relationship to reliability may be more a case of exposing incompatibilities between operational requirements and design or manufacturing processes. In other words, the product as designed is not suitable for operation in the desired environment without changes. Failure mechanisms such as reversion or depolymerization, contact spiking, and metallization migration occur at high temperatures encountered during fabrication or assembly are well above normal operating temperatures. In such cases, certain questions must be addressed before taking action:

- Will lowering the maximum operating temperature by itself avoid or de-accelerate the experienced failures? If so, how much should it be lowered and how is the conclusion reached?
- Will lowering the magnitude of the temperature cycle or change avoid the experienced failures? If so, how much should it be lowered and how is the conclusion reached?

Thus, for technologies exhibiting an obvious dependence of reliability on temperature, lowering the maximum operating temperature may be one of the measures for enhanced reliability, however, a generalization of the concept for universal application is incorrect.

Burn-in has been routinely used as a screen, with higher reliability attributed to burned-in parts without questioning the objectives, necessity, or benefits. It has been used as a customer imposed requirement to supposedly ensure higher reliability. The emphasis has been on empirical analysis, without any analysis as to the real or "root" cause of failure in terms of improper manufacturing parameters or design inconsistencies. The stresses applied in the burn-in process have not been tailored to the dominant stress dependencies of the failure mechanisms in the device technology.

The goal of screening, and burn-in in particular, is to remove defective devices which would fail abnormally early in the field, and to implement corrective actions to avoid the occurrence of defects. Devices with mature design and manufacturing processes should experience few if any such failures. In many cases current material manufacturing and processing have reached such a state of maturity that few failures occur within the specified device lifetime.

The device derating criteria have been critically evaluated. A physics of failure based device derating criteria has been proposed to aid the designer to enhance device life by means other than lowering steady state. The derating approach allows the user to examine the critical stresses in multichip module packages and reduce their value to obtain a desired mission life. The acceptable values of stress are represented as graphs of life versus magnitudes of dominant operating stresses. Such graphs are called derating curves. Derating to achieve desired life involves examining the dominant failure mechanisms. By varying the operating point on the derating curve, different stress combinations which would result in desired life can be identified. The operating point can be the functionally most acceptable combination.

To evaluate the sensitivity of device life towards the temperature and non-temperature stresses, the user can plot the device life versus percentage change from a nominal stress value (Figures 9.1 and 9.2). This menu allows the user to identify stress derating thresholds below which lowering stress magnitudes will produce no additional benefit in terms of added life. These thresholds can be identified as values of stresses for which the projected time to failure is well beyond the specified mission life of the module.

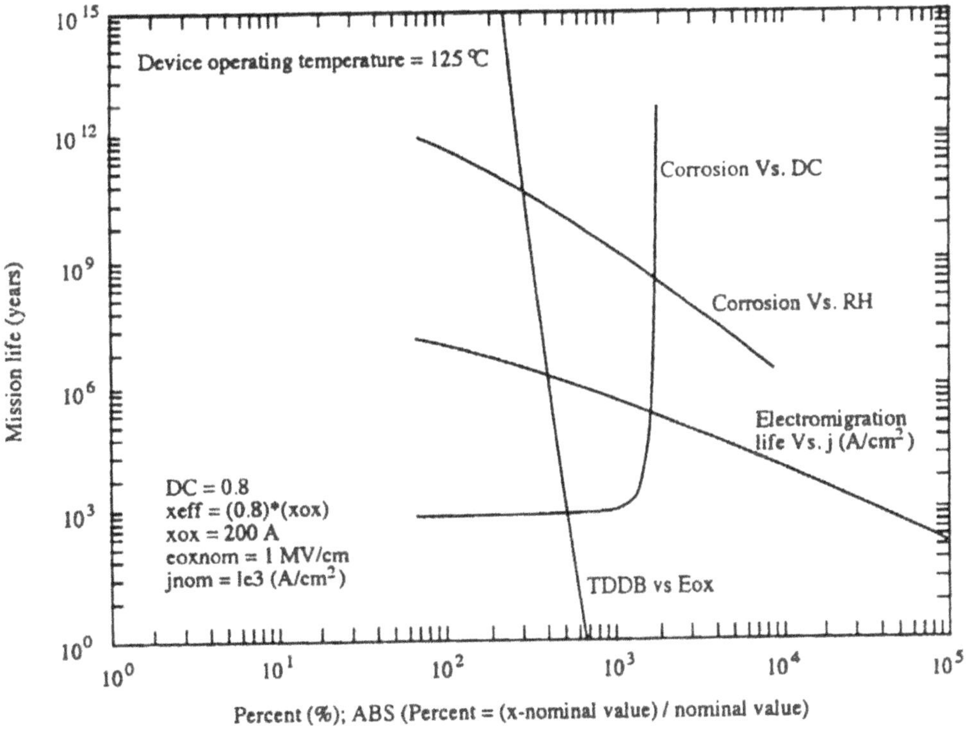

Figure 9.1 Derating curve for mission life versus the operational parameters for dominant failure mechanisms at constant operating temperature at 125 °C. This derating plot can be used to derate non-temperature operational stresses for cost-effective designs which are reliable at high temperature.

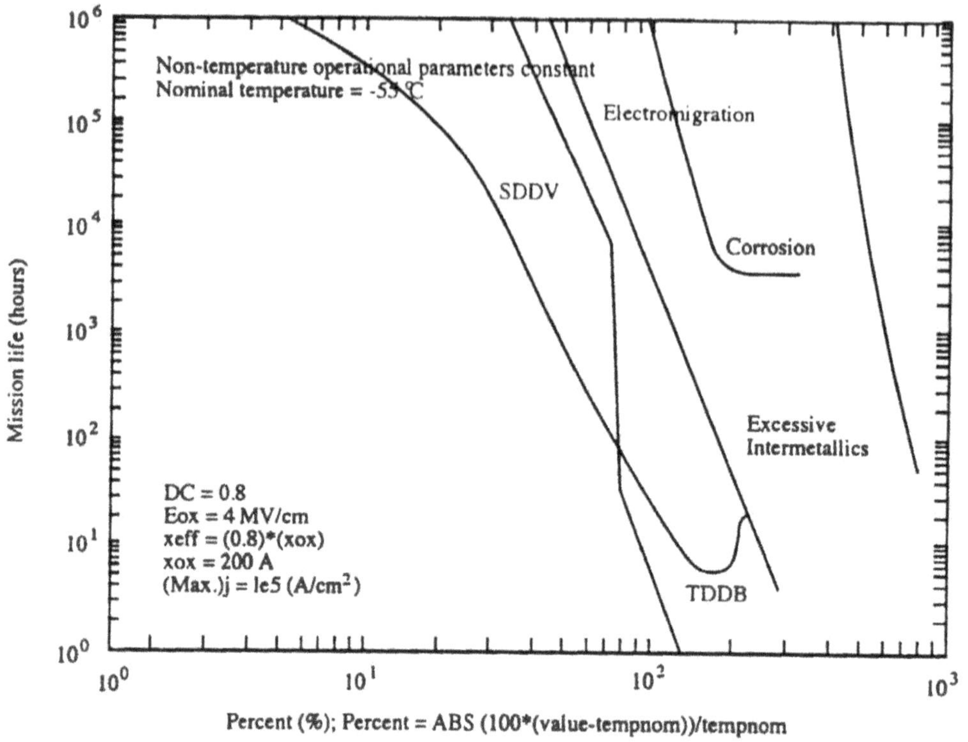

Figure 9.2 Derating curve for mission life versus steady state temperature for dominant failure mechanisms, at specified values of non-temperature operational stresses. This derating plot can be used to derate non-temperature operational stresses for cost-effective designs which are reliable at high temperature.

Appendix A

Table A 1 Steady state temperature dependence of device failure sites (normal operation is assumed at -55°C to 125°C)

Failure site	Failure mechanism	Dominant temperature dependence	Nature of steady state temperature dependence	References
wire	flexure fatigue	ΔT	Independent of steady-state temperature function under normal operation	[Gaffeny 1968], [Villela 1970], [Ravi 1972], [Phillips 1974], [Pecht 1989], [Harman 1974]
wirebond	shear fatigue	ΔT	independent of steady-state temperature	[Philosky 1973], [Pecht 1989], [Newsome 1976], [Philosky 1970, 1971], [Gerling 1984], [Khan 1986] [Pinnel 1972], [Feinstein 1979], [Pitt 1982]
	Kirkendall voiding	T	independent of steady-state temperature below 150°C; independent of steady-state temperature above lower temperatures ($T < 150°C$) in presence of halogenated compounds	[Newsome 1976], [Philosky 1970, 1971], [Gerling 1984], [Khan 1986] [Pinnel 1972], [Feinstein 1979], [Pitt 1982], [Khan 1986], [Villela 1971]
die	fracture	ΔT, ∇T	primarily dependent on temperature cycle	[Tan 1987], [Hawkins 1987]
die adhesive	fatigue	ΔT	independent of steady-state temperature under normal operation	[Chiang 1984], [Mahalingham 1984]
encapsulant	reversion	T	Independent of steady-state temperature below 300°C (glass transition for typical epoxy molding resin for plastic packages)	[Tummala 1989]
	cracking	ΔT	independent of steady state temperature (T) below the glass-transition temperature of the encapsulant; a ΔT, ∇T-driven mechanism.	[Nishimura 1987], [Fukuzawa 1985], [Kitano 1988]

package	stress corrosion	dT/dt	mildly steady-state temperature dependent under normal operation	[Tummala 1989]
	corrosion	dT/dt	only occurs above dew point temperature; mildly steady-state temperature dependent under normal operation	[Pecht 1990], [Commizalli 1980], [Inayoski 1979], [Sim 1979], [White 1969], [Schnable 1969]
	electromigration	∇ T, T	steady-state temperature dependent above 150°C	[Schnable 1988], [Onduresk 1988], [Oliver 1970], [Schwarzenberger 1988], [Chabra and Ainslie 1967], [Attardo 1972], [Danso 1981], [Black 1983], [Blair 1970], [Ghate 1981], [Partridge 1982], [LaCombe 1986], [Canali 1984], [Kinsborn 1978]
die metallization	hillock formation	T	Hillocks in die metallization can form as a result of electromigration or extended periods under temperature cycling conditions (thermal aging). Extended periods in the neighborhood of 400°C produce hillocks.	[LaCombe 1982], [Thomas 1983], [Amerasekera 1987]
	metallization migration	T	independent of steady-state temperature below 500°C	[Diagiacomo 1982], [Bart 1969], [Lane 1970]
	contact spiking	T	independent of steady-state temperature below 400°C	[Chang 1988], [Farahani 1987], [T.I. 1987], [DeChairo 1981], [Christou 1980, 1982], [Ballamy 1978],
	constraint cavitation	T	steady-state temperature dependent above 25°C	[Yost 1988, 1989]
die	electrical overstress	T	independent of steady-state temperature below 160°C (the temperature at which the coefficient of thermal resistance changes sign)	[Alexander 1978], [Canali 1981], [Smith 1978], [Runayan 1965], [Pancholy 1978]
device oxide	slow trapping	T	steady state temperature dependent above 175°C	[Nicollian 1974], [Woods 1980], [Gottesfeld 1984]

	electrostatic discharge	T	ESD voltage (resistance to ESD) reduces with temperature increase (from 25°C to 125°C); not a dominant mechanism in properly protected devices.	[Kuo 1983], [Hart 1980], [Moss 1982], [Amerasekera 1986, 1987], [Scherier 1978]
device oxide (contd.)	Time Dependent Dielectric Breakdown	T	steady-state temperature dependence is very weak; TDDB is a dominant function of voltage	[Anolick 1979], [Crook 1979], [Crook 1978], [Schnable 1988], [McPherson 1985], [Boyko 1989], [Swartz 1986], [Lee 1983]
device	Ionic Contamination	T^{-1}	steady-state temperature-dependent above 200°C	[Brambilla 1981], [Johnson 1976], [Hemmert 1980, 1981], [Bell 1980], [Wager 1984]
	Forward Second Breakdown	T	independent of steady-state temperature below 160°C	[Beatty 1976], [Chen 1983], [Hower 1970], [Hu 1982]
	Reverse Second Breakdown	T^{-1}	insignificant dependence of steady-state temperature; the breakdown voltage increases from 650V to 680V when the temperature increases from 25°C to 150°C	[Beatty 1976], [Chen 1983], [Hower 1970]
	Surface Charge Spreading	T	steady-state temperature dependent above 150°C	[Edwards 1982], [Blanks 1980], [Stojadinovic 1983], [Lycoudes 1980]
device substrate-oxide interface	Hot Electrons	T^{-1}	steady-state temperature dependence decreases above -55°C. Temperature-independent in range of 20°C to 100°C.	[Stojadinovic 1983], [Woods 1980] [Ning 1979], [Takeda 1983], [Matsumoto 1981], [Ko 1980], [Hu 1983], [Sze 1981], [Tam 1983], [Hsu 1984]

Table A2 T, ΔT, ∂T/∂t, ∂T/∂x dependence of failure mechanisms and failure models in the temperature range -55°C to 125°C

Failure mechanism	Nature of temperature dependence
bondpad-substrate shear fatigue	$N_{f,wb} = C_{cf,bp} (\Delta\tau_{max})^{-m_{cf,bp}}$ $\Delta\tau_{max} = C_{intl,wr}\, \Delta T$ $C_{intl,wr} = \left(\frac{G_{bp}}{t_{bp} Z_{wb}}\right)\left((\alpha_{wr} - \alpha_{sub}) - \frac{(\alpha_{sub} - \alpha_{bp})}{\left(1 + \frac{E_{sub} A_{cr,sub}}{\left(\frac{E_{bp} A_{cr,bp}}{1 - \nu_{sub}}\right)}\right)}\right)$ Δ T-dependent, T-independent
constraint cavitation	$\log_{10} D_{s,met} = 5.07 - \frac{1.09x10^4}{(273+T)} + \frac{2.89x10^6}{(273+T)^2} \quad \frac{3.24x10^8}{(273+T)^3}$ T^{-1} for temperatures above 25°C
corrosion	$TF_{corr} = \left(\frac{K_1 K_2 K_3}{K_4}\right)\left[\frac{w^2_{cond}\, h_{cond}\, n_{chem}\, d_{cond}\, P_{elec}}{4M_{met}\, V_{met}\, Z_{elec}}\right]$ $K_4 = \frac{(RH_{ref})^{n_{corr}} \exp(E_{a-corr}/K_B T_{ref})}{(RH)^{n_{corr}} \exp(E_{a-corr}/K_B T)}$ ∂T/∂t dependent

die and substrate adhesion fatigue	$$N_f = 0.5\left(\frac{\Delta\gamma_{p,att}}{\varepsilon_{cf}}\right)^{\frac{1}{c_{cf}}}$$ $$\gamma_{p,att} = \frac{L_{die}\lvert\alpha_{sub}-\alpha_{die}\rvert\Delta T}{X_{da}}$$ $$N_f = 0.5\left[\frac{L_s(\alpha_{case} - \alpha_{sub})\Delta T}{\varepsilon_{cf}\, x_{sa}}\right]^{\frac{1}{c_{cf}}}$$ Δ T-dependent, T-independent
die fracture	$$\Delta\sigma_{md1} = 10^{-6}\lvert\alpha_{sub}-\alpha_{die}\rvert\Delta T\sqrt{\frac{E_{sub}E_{att}L_{die}}{X_{da}}}$$ $$\frac{da}{dN} = A_{parris}\left(\Delta\sigma_{md1}\sqrt{\pi a}\right)^{n_{parris}}$$ $$N_f = \frac{2}{(n_{paris,die}-2)\,A_{paris,die}(\Delta\sigma_{md1})^{n_{paris,die}}\,\pi^{\frac{n_{paris,die}}{2}}}\left(\frac{1}{a_i^{\frac{(n_{paris,die}-2)}{2}}}-\frac{1}{a_f^{\frac{(n_{paris,die}-2)}{2}}}\right)$$ Δ T-dependent, T-independent

<table>
<tr>
<td>electromigration</td>
<td>

thin film metallizations

$$TF_{elec} = \frac{1}{2C_{p,vl}\, n_{gb,met}}\left(\frac{\tau_o k_B T_o}{j_{o,met}\rho_{o,met}D_{o,met}\, e^{-\frac{E_a}{k_B T}}}\right)\int_{x_o}^{x_1}\frac{e^{\frac{-E_a x}{k_B T_o}}}{x^2(1-x+x\alpha T_o)}dx$$

where

$$\tau_o = \frac{\Delta T_o}{T_o} = \frac{j_{o,met}\rho_{o,met}}{h_{met}T_o}$$

and

$$x = \frac{\tau_o}{1-p_{p,met}^2+\tau_o(1-\alpha_{r,met}T_o)}$$

$$= 1 - \left(\frac{T_o}{T}\right)$$

$$x_o = \frac{\tau_o}{1-\tau_o(\alpha_{r,met}T_o-1)} \qquad @t=0$$

$$x_1 = 1-\left(\frac{T_o}{T_{m,met}}\right) \qquad @t=T_F$$

multilayered metallizations

$$\frac{\partial R_{T,met}}{\partial T} = R_{im,met}\,\alpha_{r,met}\frac{L_{met}-L_{void,met}(t)}{L_{met}}+R_{ir,met}\alpha_{r,refct}\frac{L_{void,met}(t)}{L_{met}}$$

Dependent on structural non-uniformity and temperature gradient at temperatures lower than 150°C. Steady-state temperature-dependent for temperatures above 150°C, even though dependencies on structural non-uniformity and temperature gradient still exist.

</td>
</tr>
<tr>
<td>hillock formation</td>
<td>ΔT-dependent, steady-state temperature dependent in the neighborhood of 400°C</td>
</tr>
<tr>
<td>hot electrons</td>
<td>

$$\Delta V_{th} = C_{st,fac}\left[C_{mt,fac}\left(I_{sub}\right)^{\alpha_{r}}\right]^{\beta_{st}}$$

T^{-1}-dependent in the neighborhood of -55°C, and independent of steady-state temperature at higher temperatures (20°C to 100°C)

</td>
</tr>
</table>

wire-bondpad shear fatigue	$N_{f,wb} = C_{cf,bp}(\Delta\tau_{max,wr})^{-m_{cf,bp}}$ $N_{f,wb} = C_{cf,bp}(\Delta\tau_{max,wr-sub})^{-m_{cf,bp}}$ $\Delta\tau_{max,wr} = \left(\frac{r_{wr}^2}{4Z^2_{wb}\, A^2_{cr,wr}}\left(\frac{Cosh(z_{wb}\, x_w)}{Cosh(z_{wb}\, l_w)} - 1\right)^2 + \frac{Sinh^2(Z_{wb}\, x_w)}{Cosh^2(Z_{wb}\, l_w)}\right)^{\frac{1}{2}} C_{int1,wr}\Delta T$ $\Delta T_{max,wr-sub} = ((\frac{W_{bp}C_{int1,wr}}{2Z_{wb}A_{cr,sub}}(1 - \frac{Cosh(Z_{wb}X_{sub})}{Cosh(Z_{wb}l_{sub})}) + \frac{(\alpha_{sub}-\alpha_{bp})}{\frac{(1-\nu_{sub})}{(E_{sub}A_{sub})}+\frac{1}{(E_{bp}A_{bp})}})^2 +$ $C^2_{int1,wr}\frac{Sinh^2(Z_{wb}x_{sub})}{Cosh^2(Z_{wb}l_{sub})})^{\frac{1}{2}}\Delta T$ $Z^2_{wb} = \frac{G_{bp}}{t_{bp}}\left(\frac{r_{wr}}{E_{wr}A_{cr,wr}} + \frac{(1 - \nu_{sub})w_{bp}}{E_{sub}A_{cr,sub}}\right)$ Δ T-dependent, T-independent
wire fatigue	$N_{f,wb} = A_{cf,wb}(\Delta\epsilon_p)^{n_{cf,wb}}$ $\Delta\epsilon_f = \frac{r_{wr}}{\rho_{\alpha,wb}}\left[\frac{Cos^{-1}((Cos\lambda_{\alpha,wb})(1 - (\alpha_{wr} - \alpha_{sub})\Delta T))}{\lambda_{\alpha,wb}} - 1\right]$ Δ T-dependent, T-independent

Table A 3 A Dominant temperature dependency over steady-state temperatures from -55°C to 500°C

		Temperature	-55°C < T < 150°C	150°C < T < 400°C	T > 400°C
Package element	Mechanism		Nature of temperature dependence	Nature of temperature dependence	Nature of temperature dependence
wire	flexure fatigue		Δ T	Δ T	Δ T
wirebond	shear fatigue		Δ T	Δ T	Δ T
	Kirkendall voiding		temperature-independent	T-dependent for T > 150°C	T-dependent
die	die fracture		Δ T	Δ T	Δ T
	electrical overstress		Temperature-independent	T-dependent for T > 160°C	T-dependent
die adhesive	die adhesive fatigue		Δ T	Δ T	Δ T
encapsulant	encapsulant reversion (plastic package only)		Temperature-independent	T-dependent for T > 300°C	T-dependent
	encapsulant cracking (plastic package only)		Δ T, ∂ T/∂ t (≥ 25°C/sec.)	T-dependent (for T > 215°C), Δ T, ∂ T/∂ t, (≥ 25°C/sec.)	T-dependent Δ T ∂ T/∂ t
package	stress corrosion		∂ T/∂ t	∂ T/∂ t	∂ T/∂ t
die metallization	corrosion		mildly T-dependent, ∂ T/∂ t	∂ T/∂ t	∂ T/∂ t
	electromigration		structural non-uniformity dependent, ∇ T	T-dependent (for T > 150°C), ∇ T-dependent	T-dependent (for T > 150°C), ∇ T-dependent
	hillock formation		Δ T	Δ T	T-dependent (for T > 400°C)
	metallization migration		Temperature-independent	Temperature-independent	T-dependent (for T > 500°C)
	contact spiking		Temperature-independent	Temperature-independent	T-dependent (for T > 400°C)
	constraint cavitation		T-dependent	T^{-1}-dependent	T^{-1}-dependent
	slow trapping		Temperature-independent	T-dependent (for T > 175°C)	T-dependent
	electrostatic discharge		Temperature-independent (in presence of protection circuits)	Temperature-independent (in presence of protection circuits)	Temperature-independent (in presence of protection circuits)
	time-dependent dielectric breakdown		voltage-dependent, weak T dependence	voltage-dependent, weak T dependence	voltage-dependent, weak T dependence

Device	ionic contamination	T^{-1}-dependence (device not operational)	T^{-1}-dependence (device not operational)	T^{-1}-dependence (device not operational)
	forward second breakdown	Temperature-independent	T-dependent (for $T > 160°C$)	T-dependent
	reverse second breakdown	Mild T^{-1}-dependence	mild T^{-1}-dependence	mild T^{-1}-dependence
	surface-charge spreading	Temperature-independent	T-dependent (for $T > 150°C$)	T-dependent (for $T > 150°C$)
Device/oxide interface	hot electrons	T^{-1}-dependent	Temperature-independent	Temperature-independent

Table A 4 A Variation of the nature of temperature dependency of the device from -55°C to 500°C

Nature of stress dependence	$-55°C < T < 150°C$	$150°C < T < 400°C$	$T > 400°C$
	Number of Mechanisms	Number of Mechanisms	Number of Mechanisms
steady-state temperature (T)	1	8	11
temperature cycle (ΔT)	6	6	5
temperature gradient (∇T)	1	1	1
time dependent temperature change ($\partial T/\partial t$)	3	3	3
voltage (V)	1	1	1
inverse temperature dependent (T^{-1})	3	3	3
temperature independent (form of temperature dependence)	9	4	2

REFERENCES

Abbott, D.A. and Turner, J.A. Some Aspects of GaAs MESFET Reliability. *IEEE Transactions Microwave Theory and Technology*, MTT, 24, 317, 1976.

Adamson, A.W. *Physical Chemistry of Surfaces*, New York: Interscience Publishers, 1990.

Advanced Micro Devices Inc. *Memory Products 1989/1990 Databook*, Sunnyvale, CA, 2-187, 1990.

Agarwala, B.N. Electromigration Failure in Au. *Proceedings of 23rd International Reliability Physics Symposium*, 108-114, 1985.

Agarwala, B.N. Thermal Fatigue Damage in Pb-In Solder Interconnections. *23rd Annual Proceedings of International Reliability Physics Symposium*, 198-205, 1985.

Agarwala, B., Attardo, M.J., and Ingraham, A.P. The Dependence of Electromigration Induced Failure Time on the Length of Thin Film Conductors. *Journal of Applied Physics*, 41, 3954, 1970.

Ajiki, T., Sugumoto, M., Higuchi, H., and Kumada, S. A New Cyclic Biased T.H.B. Test for Power Dissipating IC's. *International Reliability Physics Symposium*, 118-126, 1979.

Alexander, D.R. An Electrical Overstress Failure Model for Bipolar Semiconductor Components. *IEEE Transactions on Components, Hybrids and Manufacturing Technology*, 1, 345-353, 1978.

Allen, D.A. Stability of Schottky Barrier at High Temperature for Use in GaAs MESFET Technology. *IEEE Proceedings Part I*, 133, 18-24, 1991.

Amerasekera, E.A. and Campbell, D.S. Electrostatic Pulse Breakdown in NMOS Devices. *Quality and Reliability Engineering International*, 2, 107-116, 1986. Amerasekera, E. A. and Campbell, D. S. Failure Mechanisms in Semiconductor Devices, John Wiley & Sons, 12-96, 1987.

Anolick, E.S. and Chen, L.Y. Applications of Step Stress to Time Dependent Breakdown. *Proceedings of International Reliability Physics Symposium*, 23-27, 1981.

Anolick, E.S. and Nelson, G.R. Low Field Time Dependent Dielectric Integrity. *Proceedings 17th Annual IEEE Reliability Physics Symposium*, 1, 1979.

Antognetti, Paolo and Giuseppe, M. *Semiconductor Device Modeling with Spice*. New York, McGraw-Hill Company, 1988.

Arrhenius, S.Z. *Physik.Chem.*, Vol. 4, 226, 1889.

Attardo, M.J. and Rosenberg, R. Electromigration Damage in Aluminum Film Conductors. *Journal of Applied Physics*, 41, 2381, 1972.

Baglee, D.A. Characteristics and Reliability of 100Å Oxides. *International Reliability Physics Symposium*, 152-155, 1984.

Baliga, B.J. *Modern Power Devices*, New York, John Wiley and Sons, 1987.

Ballamy, W.C. and Kimmerling, L.C. Premature Failure in Pt-GaAs IMPATTs-Recombination Assisted Diffusion as a Failure Mechanism. *IEEE Transactions Electron Devices*, 25, 746-752, 1978.

Balland, B. and Barbottin, G. Trapping and Detrapping Kinetics Impact on CV and I(V) Curves, from *Instabilities in Silicon Devices*, edited by Barbottin, G., and Vapaille, A., North-Holland: Elsevier Science Publishers B.V. 1989.

Barker, D., Sharif, I., Dasgupta, A., and Pecht, M. Effect of SMC Lead Dimensional Variabilities on Lead Compliance and Solder Joint Fatigue Life. *Mechanics of Surface Mount Assemblies Symposium, ASME Winter Annual Meeting*, 1-8, 1991.

Barlow, R.E., et al. Statistical Estimation Procedures for the 'Burn-In' Process. *Technometrics*, 10(1), 51-62, 1968.

Barrett, C.R. and Smith, R.C. Failure Modes and Reliability of Dynamic RAMs. *International Electronic Device Meeting*, 1978.

Bart, J.J. Analysis of Chemical and Metallurgical Change in Microcircuits. *IEEE Transactions on Electronic Devices*, ED-16, 351, 1969.

Barton, D. and Maze, C. A Two Level Metal CMOS Process for VLSI Circuits. *Semiconductor International*, 98-102, 1985.

Barton, J.L. and Bockris, J.O. The Electrolytic Growth of Dendrites from Ionic Solutions. *Royal Society of London Proceedings, Series A.*, 268 A, 485-505, 1962.

Beasley, K. New Standards for Old. *Quality and Reliability International*, 6 289-294, 1990.

Beatty, B.A., Krishna, S., and Adley, M.S. Second breakdown in Power Transistors due to Avalanche Injection. *IEEE Transactions Electron Devices*, ED-23, 851-857, 1976.

Bell, J.J. Recovery Characteristics of Ionic Drift Induced Failures under Time/Temperature Stress. *18th Annual Proceedings of International Reliability Physics Symposium*, 217-219, 1980.

Berman, A. Time Zero Dielectric Reliability Test By A Ramp Method. *19th International Reliability Physics Symposium*, 204, 1981.

Berry, K.A. Corrosion Resistance of Military Microelectronics Package at the Lead-glass Interface. *Proceedings of the ASM's Third Conference on Electronic Packaging: Materials and Processes and Corrosion in Microelectronics*, ASM International, Menlo Park, OH, 55, 1987.

Berry, R.S., Rice, S.A., and Ross, J. *Physical Chemistry*, Wiley, NY, 1980.

Bhattacharya, B.K., Huffman, W.A., Jahsman, W.E., Natarajan, B., Moisture Absorption and Mechanical Performance of Surface-Mountable Plastic Packages, *Proceedings, 38th Electronic Components Conference*, 1994.

Bhattacharya, S. S., Bannerjee, S. K., Nguyen, B-Y, and Tobin, P. J. Temperature Dependence of the Anomalous Leakage Current in Polysilicon-on-Insulator MOSFET's, *IEEE Transactions on Electron Devices*, Vol. 41, No. 2, pp. 221-227, 1994.

Billinton, R. and Allen, R.N. — *Reliability Evaluation of Engineering Systems: Concepts and Techniques, Longman Scientific and Technical Publishing Co.*, Burnt Mill, Harlow, Essex CM202JE, U.K.

Black, J.R. Electromigration of Al-Si Alloy Films. *Proceedings IEEE International Reliability Physics Symposium*, 233-240, 1978.

Black, J.R. Electromigration: A Brief Survey and Some Recent Results. *IEEE Transactions Electron Devices*, ED-16, 338, 1969.

Black, J.R. Electromigration Failure Modes in Aluminum Metallization for Semiconductor Devices. *Proceedings IEEE*, 57, 1957, 1968.

Black, J.R. Current Limitation of Thin Film Conductor. *International Reliability Physics Symposium*, 300-306, 1982.

Black, J.R. Physics of Electromigration. *IEEE Proceedings of the International Reliability Physics Symposium,*, 142-149, 1983.

Blair, J.C., Ghate, P.B., and Haywood, C.T. Electromigration Induced Failures in Aluminum Film Conductors. *Applied Physics Letters,* 17, 281, 1970.

Blanks, Henry J. Arrhenius and the Temperature Dependence of Non-constant Failure Rate, *Quality and Reliability Engineering International*, 6, 259-265, 1990.

Blanks, H.S. Temperature Dependence of Component Failure Rate. *Microelectronics and Reliability*, 20, 219-246, 1980.

Blech, I.A., Meieran, E.S. Direct Transmission Electron Microscope Observations of Electrotransport in Aluminum Films. *Applied Physics Letters*, 11, 263, 1967.

Bloomer, R.L. Franz, Johnson, M.J., Kent, S., Mepham, B., Smith, S., Sonnicksen, R.M., Walker, L.E. Motorola Inc., Failure Mechanisms in Through Hole Packages. *Electronic Materials Handbook, ASTM International*, 976, 1989.

Bolger, J.C. Polyimide Adhesive to Reduce Thermal Stresses in LSI ceramic Packages. *14th National SAMPE Technical Conference*, 394-398, 1982[a].

Boudry, M.R. and Stagg, J.P. *Journal of Applied Physics*, 50, 942, 1979, 942.

Bolger, J.C. and Mooney, C.T. Die Attach in Hi-rel DIPS: Polyimides or Low Chloride Epoxies. *IEEE Transactions on Component Hybrids Manufacturing Technology*, CHMT-7, 257-266, 1984[b].

Bowden, F.P. and Throssell, W.R. Adsorption of Water Vapor on Solid Surfaces, *Royal Society of London Proceedings, Series A.*, 209, 1951.

Bowles, J. A Survey of Reliability Prediction Procedures for Microelectronic Devices, *IEEE Transactions on Reliability*, 41, No. 1, 2-12, March 1992.

Boyko, K.C. and Gelarch, D.L. Time Dependent Dielectric Breakdown of 210A Oxides. *27th Annual Reliability Physics Symposium, IEEE,* 1-8, 1989.

Brambilla, P., Fantini, F., and Malberti, P., Mattana, G. CMOS Reliability: A Useful Case History to Revise Extrapolation Effectiveness, Length and Slope of the Learning Curve. *Microelectronics and Reliability,* 21, 191-201, 1981.

Bresse, J.F. *Microelectronics and Reliability,* 25, 411, 1985.

Broadbent, E.K., Towner, J.M., Townsend, P.H., Vander Plas, H.A. Electromigration Induced Short-circuit Failure in Aluminum underlaid with Chemically Vapor Deposited Tungsten. *Journal of Applied Physics,* 63, 1917-1923, 1988.

Broek, D. *Elementary Engineering Fracture Mechanics,* Boston, Martinus Nijhoff, 1986.

Bromstead, J.R., Weir, G.B., Johnson, R.W., Jaeger, R.C., and Baumann, E.D. Performance of Power Semiconductor Devices At High Temperatures. *Transactions of the First International High Temperature Electronics Conference,* NM, 27-35, 1991.

Brooke, L. Pulsed Current Electromigration Failure Model. *Proceedings of 25th IEEE International Reliability Physics Symposium,* 136-139, 1987.

Brummet, S.L., et al. Reliability Parts Derating Guidelines, RADC-TR-82-177 (Boeing Aerospace, Seattle, WA), AD-A120367, 1982.

BT *Handbook of Reliability Data for Electronic Components Used in Telecommunications Systems,* British Telecom Handbook HRD3, Issue 3, January 1984.

Budenstein, P.P. A Survey of Second Breakdown Phenomena, Mechanisms, and Damage in Semiconductor Junction Devices, *Report No. RG-TR-70-19,* U.S. Army Missile Command, Redstone Arsenal, AL, 1970.

Budenstein, P.P., Pontius, D.H., and Wallace, B.S. Second Breakdown and Damage in Semiconductor Junction Devices, *Contract No. DAAH 01-71-C-0708,* U.S. Army Missile Command, Redstone Arsenal, AL, 1972.

Buhanan, D. *IEEE Transactions on Electron Devices,* ED-16, 117, 1969.

Bukkett, T.A. and Miller, R.L. Electromigration Evaluation: MTF Modeling and Accelerated Testing. *Proceedings of 22nd International Reliability Physics Symposium,* 264-272, 1984.

Campbell, M. Monitored Burn-in Improves VLSI IC Reliability. *Computer Design,* 24, 143-144, 1985.

Campisano, S.U., Costanzo, E., Scaccianoce, F., and Cristofollini, R. *Thin Solid Films,* 52, 97-101, 1978.

Canali, C, Fatini, F., Gaviraghi, S., and Senin, A. Reliability Problems in TTL-LS Devices. *Microelectronics and Reliability,* 21, 637-651, 1981.

Canali, C., Fantini, F., Zanoni, E., Giovannetti, A., and Brambilla, P. Failures Induced By Electromigration in ECL 100k Devices. *Microelectronics and Reliability,* 24, 77-100, 1984.

Carslaw, H.S. and Jaegar, J.C. *Conduction of Heat in Solids*, London: Oxford Press, 255-263, 1959.

Chabra, D.S. and Ainslie, N.G. Open Circuit Failures in Thin Film Conductors, Technology Report *IBM Component Div., E. Fishkill Facility*, New York, Technical Report No. 22.419, 1967.

Chandrasekaran, R. Optimal Policies for Burn-in Procedures, *Opsearch*, 149-160, 1990.

Chang, L., Vo, K., and Kerg, J. A Simplified Model to Predict the Linear Temperature Coefficient of CMOS Inverter's Delay Time. *IEEE Transactions on Electron Devices*, ED-34, 1834-1837, 1987.

Chang, P.H., Hawkins, R., Bonifield, T.D., and Melton, L.A. Aluminum Spiking At Contact Windows in Al/Ti-W/Si. *Applied Physics Letters*, 52, 272-274, 1988.

Chen, I.C., Holland, S.E., and Hu, C. Electrical Breakdown in Thin Gate and Tunneling Oxides. *IEEE Transactions on Electron Devices*, ED-32, 413, 1985.

Chen, D.Y., Lee, F.C., Blackburn, D.L., and Berning, D.W. Reverse-bias Second Breakdown of High Power Darlington Transistors. *IEEE Transaction Aerospace Electronic Systems*, AES19, 840-846, 1983.

Chiang, S.S. and Shukla, R.K. Failure Mechanism of Die Cracking due to Imperfect Die Attachment. *34th Electronics Components Conference*, 195-202, 1984.

Chiang, K.L. and Lauritzen, P.O. Thermal Instability in Very Small p-n Junctions. *IEEE Transactions in Electron Devices*, ED-18, 94-97, 1970.

Christou, A. Report on the 1982 GaAs Device Workshop. *20th Annual Proceedings International Reliability Physics Symposium*, 276-277, 1982.

Christou, A., Cohen, E., and MacPherson, A. C. Failure Modes in GaAs Power FETs: Ohmic Contact Electromigration and Formation of Refractory Oxides. *19th International Reliability Physics Symposium*, 182-187, 1980.

Chu, R.C. and Simons, R.E. Evolution of Cooling Technology in Medium and Large Scale Computers: An IBM Perspective. *Proceedings International Symposium on Heat Transfer in Electronic and Microelectronic Equipment*, International Center for Heat Transfer, Dubrovnik, Yugoslavia, Hemisphere Publishing Co. 1988.

Clark, I.D., Harrison, L.G., Kondratiev, V.N., Szabo, Z.G., Wayne, R.P. *The Theory of Kinetics*, Edited by C.H. Bamford and C.F.H. Tipper, Elsevier Scientific Amsterdam, 1979.

CNET Recueil de Donnees de Fiabilitie du CNET (Collection of Reliability DATA from CNET), Centre National d'Etudes des Telecommunications (National Center for Telecommunication Studies), 1983.

Coffin, L.F., Jr. *Transactions of the ASME*, 76, 931, 1954.

Coffin, L.F., Jr. Low Cycle Fatigue. *American Society of Metals*, 15-24, 1963.

Coffin, L.F., Jr. Fatigue at High Temperatures. *ASTM, STP 520*, PA, 1973.

Commizolli, R.B., White, L.K., Kern, W., and Schnable, G.L. Corrosion of Aluminum IC Metallization with Defective Surface Passivation Layer. *18th Annual Proceedings Reliability Physics Symposium,* 282, 1980.

Crook, D.A. Techniques for Evaluating Long Term Oxide Reliability At Wafer Level. *IEDM Technical Digest,* 444-448, 1978.

Crook, D.A. Method of Determining Reliability Screens for Time Dependent Dielectric Breakdown. *17th Annual Reliability Physics Symposium,* 1-7, 1979.

Curry, J., Fitzgibbon, G., Guan, Y., Muollo, R., Nelson, G., and Thomas, A. New Failure Mechanisms in Sputtered Aluminum-silicon Films. *International Reliability Physics Symposium,* 6-8, 1984.

Cushing , M. et al. Design Reliability Evaluation of Competing Causes of Failure in support of Test-Time Compression. *Proceedings of the Institute of Environmental Sciences,* (1994).

Cushing M., et al. Comparison of Electronics Reliability assessment Approaches. *IEEE Transactions on Reliability,* 42 (December 1993), 600-607.

Cushing, M.J., Mortin D.E., Stadterman, T.J., and Malhotra, A. "Comparison of Electronics-Reliability Assessment Approaches", *IEEE Transactions on Reliability,* Vol. 42, No.4, pp.13-17 (December 1993).

Cypress Semiconductor *BiCMOS/CMOS Databook,* San Jose, CA, 13-15, 1990.

d'Heurle, F.M., Ainslie, N.G., Gangulee, A., and Shine, M.C. Activation Energy for Electromigration Failure in Aluminum Films Containing Copper. *Journal of Vacuum Science and Technology,* 9, 289-293, 1972.

d'Heurle, F.M., Gangulee, A., Alliotta, C.F., and Ranieri, V.A. Effects of Mg Addition on the Electromigration Behavior of Al Thin Film Conductors. *Journal of Electronic Materials,* 4, 497-515, 1975.

Dais, J.L. and Howland, F.L. Fatigue Failure of Encapsulated Au-Beam Lead and TAB Devices. *IEEE Transactions on Components, Hybrids, and Manufacturing Technology,* CHMT-1, 158-166, 1978.

Dais, J.L., Erich, J.S., and Jaffe, D. Face-down TAB for Hybrids. *IEEE Transactions on Components, Hybrids, and Manufacturing Technology,* CHMT-3, 623-634.

Danso, K.A. and Tullos, T. Thin Film Metallization Studies and Device Lifetime Prediction Using Al-Si, Al-Cu-Si Conductor Test Bars. *Microelectronics Reliability,* 21, 513, 1981.

Dantowitz, A., Hirshberger, G., and Pravidlo, D. Analysis of Aeronautical Equipment Environmental Failures, Report No. AFFDL-TR-71-32, 1971.

Darveaux, R., K. Banerji, A. Mawer, G. Dody, Reliability of Plastic Ball Grid Array Assembly, *Ball Grid Array Reliability* edited by John Lau, Chapter 13, pp.379-442, John Wiley, New York, 1994

Dasgupta, Pecht, M., D. and Barker An Approach to the Development of Package Design Guidelines, *International Journal of Microcircuits and Electronic Packaging*, Vol. 16(3), pp. 217-240, 1993.

Davey, J.E. and Christou, A. Reliability and Degradation of Active III-V Semiconductor Devices, in M.J. Howes and D.V. Morgan, eds., *Reliability and Degradation*, New York, Wiley and Sons, 237-300, 1981.

Davis, T.L. *Journal of Applied Physics*, 38, 3756-3760, 1967.

Davis, J.R. et al. *Metals Handbook: Corrosion*, ninth ed., 13 ASM International, Metals Park, OH, 145, 1987.

DeChairo, L.F. Electrothermomigration in NMOS LSI Devices. *19th Annual Proceedings International Reliability Physics Symposium*, 212-277, 1981.

DELCO Data (as of 11-2-88) ETDL LABCOM from 636 nos IC's analyzed 513 nos Bipolar & 123 nos MOS; *quoted from* Dicken, H. K. *Physics of Semiconductor Failures*. DM Data Inc. 6900 E. Camelback Road Suite 1000, Scottsdale, AZ 85251, 119, 1988.

Denton, D.L. and Blythe, D.M. The Impact of Burn-in on IC Reliability. *The Journal of Environmental Sciences*, 29, 19-23, 1986.

Derbenwick, G.F. *Journal of Applied Physics*, 48, 1127, 1977.

DerMarderosian, A. and Gionet, V. Water Vapor Penetration Rate into Enclosures with Known Air Leak Rates, *16th International Reliability Physics Symposium*, 179-186, 1978.

Devanay, J.R. Failure Mechanisms in Active Devices. *Electronic Materials Handbook, ASTM International*, 1007, 1989.

DiGiacomo, G., McLaughlin, P. Current Leakage Kinetics Across Tinned Cr/Cu Lands Having Polyimide Coating, *International Society for Hybrid Microelectronics Proceedings*, 38-42, 1991.

DiGiacomo, G. Metal Migration (Ag, Cu, Pb) in Encapsulated Modules and Time to Fail Model as A Function of the Environment and Package Properties. *IEEE/ Proceedings of the IRPS*, 27-33, 1982.

DiGiacomo, G. Current Leakage Kinetics Across Tinned Cr/Cu Lands Having Epoxy Overlay, *IEEE Transactions on Components, Hybrids, and Manufacturing Technology*, CHMT-8, No.4, 440-445, 1985.

Dimitrijev, S.S., Stojadinovic, N.D., and Pruic, Z.D. Analysis of Temperature Dependence of CMOS Transistors, Threshold Voltage. *Microelectronics and Reliability*, 31, 33-37, 1991.

DiStefano, T.H. and Shatzkes, M. Dielectric Stability and Breakdown in Wide Bandgap Insulators. *Journal Vacuum Science Technology*, 12, 37, 1975.

DM Data Inc., System Design Guidelines for Reliability, 96, 1990.

Dolny, G. and Nostrand, G.E. Characteristics of Power Integrated Circuits at High Operating Temperatures. *Transactions: First International High Temperature Electronics Conference*, 43-49, 1991.

Draper, B.L. and Palmer, D.W. Extension of High Temperature Electronics. *IEEE Transactions on Components, Hybrids, and Manufacturing Technology*, CHMT-2, 399-404, 1979.

Drukier, I. and Silcox, J.F. on the Reliability of Power GaAs FETs. *IEEE International Reliability Physics Symposium*, 150-155, 1979.

Dummer, G. W. And Griffin, N. B. Galvanic Corrosion. *Environmental Testing Techniques for Electronics and Materials*, Macmillan Co., New York, 115-118, 1962.

Duvvury, R.A. et al. ESD Reliability Protection in 1 µm CMOS Technologies. *International Reliability Physics Symposium*, 1983.

Eachus, J., Klema, J., and Walker, S. Monitored Burn-in of MOS 64K Dynamic RAMs. *Semiconductor International*, 1984.

Eckel, J.F. The Influence of Frequency on the Repeated Bending Life of Acid Lead. *Proceedings ASTM 51*, 1951.

Edwards, D.G. Testing for MOS IC Failure Modes. *IEEE Transaction Reliability*, R-31, 9-17, 1982.

Ellwanger, R.C. and Towner J.M. The Deposition and Film Properties of Reactively Sputtered Titanium Nitride. *Thin Solid Films*, 161, 289-304, 1988.

Engelmaier, W. Fatigue Life of Leadless Chip Carrier Solder Joints During Power Cycling. *IEEE Transactions on CHMT*, CHMT-6, 232-237, 1983.

Engelmaier, W. Surface Mount Solder Joint Long-term Reliability: Design, Testing, Prediction. *Soldering and Surface Mount Technology*, 1, 14-22, 1989.

Engelmaier, W. Surface Mount Attachment Reliability of Clip Leaded Ceramic Chip Carriers on FR-4 Circuit Boards. *Proceedings of the International Reliability Physics Symposium*, 9, 3-11, 1988.

Engelmaier, W. Functional Cycling and Surface Mountiñg Attachment Reliability. *ISHM Technical Monograph Series 6984-002*, 87-114, 1984.

English, A.K., Tai, K.L., and Turner, P.A. Electromigration of Ti-Au Thin-film Conductors at 180°C. *Journal of Applied Physics*, 45, 3757-3767, 1974.

Eskin, D.J. et al. Reliability Derating Procedures, RADC-TR-84-254, Martin Marietta Aerospace, Orlando, FL, AD-A153268, 1984.

Estes, R.H. The Effect of Porosity on Mechanical, Electrical and Thermal Characteristics of Conductive Die-attach Adhesives. *Solid State Technology*, 191-197, 1974.

Estes, R.H. A Practical Approach to Die-attach Adhesive Selection. *Hybrid Circuit Technology*, 44-47, 1991.

Estes, R.H., Kulesza, F.W., and Banfield, C.E. Recent Advances Made in Die-attach Adhesives for Microelectronic Applications. *Proceedings of the International Symposium on Microelectronics*, ISHM, 191-198, 1985.

Evans, J., Lall, P. and Bauernschub, R. A Framework for Reliability Modeling of Electronics. *Reliability and Maintainability* Symposium, 144-151, 1995.

Evans, M.G. and Polanyi, M. *Faraday Society (London) Transactions*, 34, 11, 1938.

Eyring, H., Lin, S.H., Lin, S.M., *Basic Chemical Kinetics,* Wiley, New York, 1980

Fantini, F., Specchiulli, G. and Caprile, C. The Validity of Resistometric Technique in Electromigration Studies of Narrow Stripes. *Thin Solid Films,* 172, L85-L89, 1989.

Farahani, M.M., Turner T.E., and Barnes, J.J. Evaluation of Titanium as A Diffusion Barrier Between Aluminum and Silicon for 1.2 mm CMOS Integrated Circuits. *Journal Electrochemical Society,* 134, 2835-45, 1987.

Farrow, R.H. and Parker, G.W. The Failure Physics Approach to IC Reliability. *Microelectronics and Reliability*, 11, 151-157, 1972.

Feinstein, L.G. and Bindell, J.B. The Failure of Aged Cu-Au Thin Films by Kirkendall Porosity. *Thin Solid Films 62*, 37-47, 1979[a].

Feinstein, L.G. and Pagano, R.J. Degradation of Thermocompression Bonds to Ti-Cu-Au and Ti-Cu by Thermal Aging. *29th Proceedings Electronic Components Conference,* Cherry Hill, NJ, 346-354, 1979[b].

Felton, J.J., Horowitz, S.J., Gerry, D.J., Larry, J.R., and Rosenberg, R.M.,Recent Developments in Gold Conductor Bonding Performance and Failure *Mechanisms, Solid State Technology* 22 37-44, 1979.

Fox, L.R., Sofia, J.W., and Shine, M.C. Investigation of Solder Fatigue Acceleration Factors. *IEEE Transactions on Components, Hybrids, and Manufacturing Technology*, CHMT-8, 275-282, 1985.

Fraser, A. and Ogbounah, D. *Proceedings of GaAs Symposium*, 161, 1985.

Frear, D.R. Thermo-mechanical Fatigue of Solder Joints: A New Comprehensive Test Method. *Proceedings of 39th Electronic Component Conference*, 293-300, 1989.

Fried, L.J., Hares, J., Lecdeton, J.S., Logen, J.S., Paal, G., and Totta, P.A. *IBM Journal Research and Development*, 26, 263, 1982.

Frost, H.J., and M.F. Ashby, Deformation Mechanism Maps, Pergamon Press, 1982, Chapter 2.

Fukuzawa, I., Ishiguro, S., and Nanbu, S. Moisture Resistance Degradation of Plastic LSI's By Reflow Soldering. *Proceedings 23rd Annual International Reliability Physics Symposium*, 192-197, 1985.

Funamizu, Y. and Watanabe, K. Interdiffusion in the Aluminum-Copper System. *Transactions of Japan Institute Metals*, 12, 147-152, 1971.

Gaffeny, J. Internal Lead Fatigue Through Thermal Expansion in Semiconductor Devices. *IEEE Transactions, Electronic Devices,* ED-15, 617, 1968.

Gargini, P.A., Tseng, C., and Woods, M.H. Elimination of Silicon Electromigration in Contacts By the Use of An Interposed Barrier Metal. *Proceedings of 20th International Reliability Physics Symposium,* 66-76, 1982.

Garofalo, G. Fundamentals of Creep and Creep-Rupture in Metals, The Macmillan Company, New York, N.Y. 1965.

Garrigues, M. and Hellouin, Y. *IEEE Transactions on Electron Devices,* 28, 928, 1981.

Garrigues, M. and Balland, B. Hot Carrier Injection Into SiO_2, in *Instabilities in Silicon Devices,* edited by Barbottin, G. and Vapaille, A. North-Holland: Elsevier Science Publishers B.V. 1986.

Gerling, W. Electrical and Physical Characterization of Gold-ball Bonds on Aluminum Layers. *34th Proceedings IEEE Electronic Components Conference,* New Orleans, LA 13-20, 1984.

Gershinskii, A.E., Formin, B.I., Cherepov, E.J., and Edelman, F.L. *Thin Solid Films,* 42, 269-275, 1977.

Ghate, P.B. and Blair, J.C. Electromigration Testing of Ti:W/Al and Ti:W/Al-Cu Film Conductors. *Thin Solid Films,* 55, 113-123, 1978.

Ghate, P.B. Aluminum Alloy Metallization for Integrated Circuits. *Thin Solid Films,* 83, 195-205, 1981.

Gohn, G.R. and Ellis, W.C. The Fatigue Test as Applied to Lead Cable Sheath. *Proceedings ASTM 51,* 1951.

Gottesfeld, S. and Gibbons, L. Reliability Characterization of High Speed Logic IC's. *RCA Review,* 45, 179-193, 1984.

Grabe, E. and Schreiber, H.U. *Solid State Electronics,* 26, 1023, 1983.

Grivas, D., Frear, D., Quan, L., and Morris, J. W. Jr. The Formation of Cu_3Sn Intermetallic on the Reaction of Cu with 95Pb-5Sn Solder. *Journal of Electronic Materials,* 15, 355-359, 1986.

Gueguen, M., Boussois, J.L., Goudard, J.L., and Sauvage, D. Screening and Burn-in: Application to Optoelectronic Device Selection for High Reliability S280 Optical Submarine Apparatus, *Semiconductor Device Reliability, Proceedings of the NATO Advanced Research Workshop,* edited by Christou, A., and Unger, B.A., Netherlands, Kluwer Publishers, 43-73, 1990.

Guess, F., Walker, E., and Gallant, D. Burn-in to Improve which Measure of Reliability. *Microelectronics and Reliability,* 32, 759, 1992.

Hagge, J. Predicting Fatigue Life of Leadless Chip Carriers Using Manson-Coffin Equations. *Proceedings International Electronic Packg. Society,* 1982.

Hakim, E.B. *Microelectronic Reliability, Reliability Test and Diagnostics*, I, Artech House, 1989.

Hakim, Edward B. Reliability Prediction: Is Arrhenius Erroneous?, *Solid State Technology*, 57, August 1990.

Hall, P.M., Panousis, N.T., and Menzel, P.R. Strength of Gold Plated Copper Leads on Thin Film Circuits under Accelerated Aging. *IEEE Transactions on Parts, Hybrids, and Packaging*, PHP-11, 202-205, 1975.

Hall, P.M. Creep and Stress Relaxation in Solder Joints in Surface Mounted Chip Carriers. *Proceedings of 37th Electronic Component Conference*, 579-588, 1987.

Hall, P.M. and Morabito, J.M. Thin Film Phenomena-interfaces and Interactions. *Surface Science*, 67, 373-392, 1977.

Hall, P.M., Marobito, J.M., and Panousis, N.T. Interdiffusion in Cu-Au Thin Film System at 25°C to 250°C. *Thin Solid Films*, 41, 342-361, 1977.

Hallberg, O, and Peck, D.S., Recent Humidity Accelerations, a Base for Testing Standards, John Wiley and Sons, pp. 169-179, 1994.

Hallberg, O. Hardware Reliability Assurance and Field Experience in a Telecom Environment, *Quality and Reliability Engineering International*, 10, 195-200, 1994.

Hannemann, R. Thermal Control for Mini- and Microcomputers: the Limits of Air Cooling. *Proceedings International Symposium on Heat Transfer in Electronic and Microelectronic Equipment*, International Center for Heat Transfer, Dubrovnik, Yugoslavia: Hemisphere Publishing Co, 1988.

Hara, K., Tokura, N., Himi, H., Endoh, K., Mori, H., and Nakano, Y. A New Method of Reliability Testing for CMOS VLSI Evaluation. *International Symposium on Reliability and Maintainability 1990-Tokyo R&M ISRM '90 Tokyo*, 102-106, 1990.

Harari, E. Dielectric Breakdown In Electrically Stressed Thin Films of Thermal SiO_2. *Journal of Applied Physics*, 49, 2478, 1978.

Harman, G.G. Reliability and Yield Problems of Wire Bonding in Microelectronics, A Technical Monograph of the ISHM, 1989.

Harman, G. Acoustic-emission-mounted Tests for TAB Inner Lead Bond Quality. *IEEE Transactions on Components, Hybrids, and Manufacturing Technology*, CHMT-5, 445-453, 1982.

Harman, G.G. Metallurgical Failure Modes of Wire Bonds. *12th International Reliability Physics Symposium*, 131-141, 1974.

Hart, A., Teng, T.T., and McKenna, A. Reliability Influences from Electrical Overstress on LSI Devices. *18th Annual Proceedings International Reliability Physics Symposium*, 190-196, 1980.

Hawkins, G., Berg, H., Mahalingam, M., Lewis, G., and Lofran, J. Measurement of Silicon Strength as Affected By Wafer Back Processing. *25th Annual Proceedings Reliability Physics,* 216-223, 1987.

Heiman, F.P. and Miller, H.S. Temperature Dependence of n-type MOS Transistors. *IEEE Transactions on Electron Devices*, ED-12, 142-148, 1965.

Hemmert, R.S. Recoverable Ionic Contaminant Induced Failures on N-channel Memory Products. *Microelectronics and Reliability,* 21, 63-77, 1981.

Hemmert, R.S. Temperature Dependent Defect Level for an Ionic Failure Mechanism. *19th Annual Proceedings International Reliability Physics Symposium,* 172-174, 1981.

Hertzberg, R. W. Deformation and Fracture Mechanica of Engineering Materials. John Wiley and Sons, New York (1989).

Hickmott, T.W. *Journal Applied Physics*, 46, 2583, 1975.

Hillen, M.W. Thesis, University of Groningen, the Netherlands, (DAI-C 81-70, 066), 1981; cited from Hillen, M.W. and Verwey, J.F. Mobile Ions in SiO_2 Layers on Si, in Barbottin, G. and Vapaille, A. *Instabilities In Silicon Devices*, North Holland, Elsevier Science Publishers B.V., 1986.

Hinode, K., Owada, N., Nishida, T., and Mukai, K. Stress-Induced Grain Boundary Fractures in Al-Si Interconnects, *Journal of Vacuum Science and Technology*, B5(2), 518-522, 1987.

Hinode, K., Asano, I., and Homma, Y. Void Formation Mechanism in VLSI Aluminum Metallization, *IEEE Transaction on Electron Devices*, ED-36(6), 1050-1055, 1989.

Ho, P.S. and Howard, J.K. *Journal of Applied Physics*, 45 1974, 3229, quoted from Ho, P. S. and Kwok, T. Electromigration on Metals, *Reports on Programs in Physics*, 52, 301-348, 1989[a].

Ho, P.S. and Kwok, T. Electromigration on Metals. *Reports on Programs in Physics,* 52, 301-348, 1989[b].

Hoang, H. Effects of Annealing Temperature on Electromigration Performance of Multilayered Metallizations Systems, *26th Annual Symposium IRPS*, 1988.

Hodges, D. A. and Horace G. J. *Analysis and Design of Digital Integrated Circuits*, New York, McGraw-Hill Publishing Company, 1988.

Hokari, Y., Baba, T., and Kawamura, N. Reliability of Thin SiO_2 Films Showing Intrinsic Dielectric Integrity. *IEDM Technical Digest*, 46, 1982.

Holland, S., Chen, I.C., Ma, T.P., and Hu, C. Physical Models for Gate Oxide Breakdown. *IEEE Electron Device Letters*, 5, 302, 1984.

Honda, K., Ohsawa, A., and Toyokura, N. Breakdown in Silicon Oxides: Correlation with Cu Precipitates. *Applied Physics Letters*, 45, 270, 1984.

Honda, K., Ohsawa, A., and Toyokura, N. Breakdown in Silicon Oxides(II): Correlation with Fe Precipitates. *Applied Physics Letters*, 46, 582, 1985.

Hosoda, T., Niwa, H., Yagi, H., and Tsuchikawa, H. Effects of Line Size on Thermal Stress in Aluminum Conductors. *29th Annual Proceedings of the International Reliability Physics Symposium*, 77-83, 1991.

Hosticka, B.J., Dasab, K.G., Krey, D., and Zimmer, G. Behavior of Analog MOS Integrated Circuits at High Temperatures. *IEEE Journal of Solid State Circuits*, SC 20, 871-874, 1985.

Howard, R.T. (IBM Corp, Armonk, NY, USA), Semiconductor Device Package Having a Substrate with a Coefficient of Expansion Matching Silicon, *IBM Tech Disclosure Bull.* (USA) Vol 20, No. 7, pp. 2849-2850.

Howard, R.T. Edge Seal for Multilevel Integrated Circuit with Organic Interlevel Dielectric, *IBM Tech Disclosure Bull*, Vol 20, No. 8,, pp. 3002-3003, Jan 1978.

Howard, R.T. Packaging Reliability: How to Define and Measure It. *IEEE Transactions on Components, Hybrids, and Manufacturing Technology*, 376-384, 1982.

Hower, P.L. and Reddi, V.G.K. Avalanche Injection and Second Breakdown in Transistors. *IEEE Transactions Electron Devices*, ED-17, 320-335, 1970.

Howes, M.J. and Morgan, D.V. *Reliability and Degradation : Semiconductor Device and Circuits*, Wiley-Interscience Publication, John Wiley and Sons, Bath, Great Britain, 1981.

HRD5 *Reliability Prediction* British Book. British Telecom, March 1995.

Hsu, F.C. and Chiu, K.Y. Temperature Dependence of Hot Electron Induced Degradation in MOSFET's. *IEEE Electron Devices Letters*, EDL-5, 1984.

Hu, C. and Chi, M. Second Breakdown of Vertical MOSFET's. *IEEE Transactions Elec. Devices*, ED-29, 1287-1293, 1982.

Hu, J.M., Pecht, M., and Dasgupta, A. Design of Reliable Die Attach, *The International Journal of Microcircuits and Electronic Packaging*, 16 (1), 1993.

Hu, J., Pecht, M., and Dasgupta, A. A Probabilistic Approach for Predicting Thermal Fatigue Life of Wirebonding in Microelectronics, *ASME Journal of Electronic Packaging*, 113, 275, 1991.

Hu, C. Hot Electron Effects in MOSFET'S. *IEDM Technical Digest*, 1983, 176.

Huang, C. J., Grotjohn, T. A., Sun, C. J., Reinhard, D. K., and Yu, C-C. W. Temperature Dependence of Hot-Electron Degradation in Bipolar Transistors, *IEEE Transactions on Electron Devices*, Vol. 40, No. 9, pp. 1669-1674, 1993 .

Hull, R. and Jackson, R. Analysis of High-Voltage ESD Pulse Testing on CMOS Gate Array Technology. *Electrical Overstress/Electrostatic Discharge Symposium Proceedings*, 1988.

Hummel, R.E. and Geier, H.J. Activation Energy for Electrotransport in Thin Silver and Gold Films. *Thin Solid Films*, 25, 335-342, 1975.

Huntington, H.B. and Grone, A.R. Current-induced Marker Motion in Gold Wires, *Journal Physics Chemistry Solids*, 20 1961, 76, quoted from Ho, P.S. and Kwok, T. Electromigration on Metals, *Reports on Programs in Physics*, 52, 301-348, 1989.

Huston, H.H., Wood, M.H., DePalma, V.M. Burn-in Effectiveness: Theory and Measurement. *29th Annual Proceedings of the Reliability Physics Symposium*, 271-276, 1991.

IITRI/Honeywell, *VHSIC/VHSIC-Like Reliability Prediction Modeling,* Contract No F30602-86-C-0261, 15 January (1988).

Inayoski, H., Nishi, K., Okikawa, S., and Wakashima, Y. Moisture Induced Aluminum Corrosion and Stress on the Chip in Plastic Encapsulated LSI's. *Proceedings 17th Reliability Physics Symposium,* 113-117, 1979.

Institute of Environmental Sciences Integrated Circuit Screening Report, 1988.

Integrated Device Technology High Performance CMOS Databook Supplement, Santa Clara, CA, S3-8, 1989.

Intel Corporation Components Quality and Reliability Handbook, Santa Clara, CA, 9-18, 1989.

Intel Corporation Military Volume I: Military Quality Assurance:, Santa Clara, CA, 1-3, 1990.

IPC Guidelines for Accelerated Reliability Testing of Surface Mount Solder Attachments, *IPC-SM-785*, March 1992.

Irvin, J.C. and Loya, A. Failure Mechanisms and Reliability of Low Noise GaAs FETs, *Bell System Technology Journal,* 57 1978, 2823; quoted from Ricco, B., Fantini, F., Magistralli, F., Brambilla, P., Reliability of GaAs MESFET's, in Christou, A. and Unger B. A., *Semiconductor Device Reliability*, Netherlands: Kluwer Academic Publishers, 455-461, 1978.

Irwin, G.R. The Crack Extension Force for a Part-through Crack in a Plate. *Journal of Allied Mechanics*, 651-654, 1962.

James, K. Reliability Study of Wire Bonds to Silver Plated Surfaces. *IEEE Transactions on Parts, Hybrids, and Packaging,* PHP-13, 419-425, 1977.

Jaspal, J.S. and Dalal, H.M. A Three Fold Increase in Current Carrying Capability of Al-Cu Metallurgy By Predepositing of A Suitable Underlay Material. *Proceedings of 19th International Reliability Physics Symposium*, 238-242, 1981.

Jellison, J.L. Effect of Surface Contamination on Thermocompression Bondability of Gold. *IEEE Transactions on Parts, Hybrids, and Packaging,* PHP-11, 206-211, 1975.

Jensen, F. and Peterson, N.E. *Burn-in: An Engineering Approach to the Design and Analysis of Burn-in Procedures*, New York, John Wiley and Sons, 1982.

Jensen, F. Component Burn-in: the Changing Attitude. *Semiconductor Device Reliability, Proceedings of the NATO Advanced Research Workshop*, edited by Christou, A. and Unger, B.A. Dorchrecht, Netherlands, Kluwer Publishers, 97-106, 1990.

Jensen, F. Component Failures Based on Flaw Distributions. *Proceedings of the Annual Reliability and Maintainability Symposium*, 21-25, 1989.

Johnson, G.M. Accelerated Testing Highlights CMOS Failure Modes, EAS-CON-76 Record, 142-A-142-I, 1976.

Johnson, J.E. Die Bond Failure Modes. *IEEE/Proceedings of the International Reliability Physics Symposium,* 150-153, 1983.

Jones, E. and Sheppe, R. Alternate Approach to Traditional Burn-in, *Evaluation Engineering*, 30, 16-26, 1991.

Kakar, A.S. Electromigration Studies on Aluminum-copper Stripes, *Solid State Technology*, 47-50, 1973.

Karam, D. Burn-In: Which Environmental Stress Screens Should Be Used? RADC-TR-81-87, In-House Report, Rome Air Development Center, Griffis Air Force Base, 1991.

Kashiwagi, S., Takase, S., Usui, T., and Ohono, T. Reliability of High Frequency High Power GaAs MESFETs. *25th Annual Proceedings of International Reliability Physics Symposium*, 97-98, 1987.

Kasley, K. L., Oleszek, G. M., and Anderson, R. L. Investigation of the Kink and Hysteresis from 13 K to 30 K in n-Channel CMOS Transistors with a Floating Well, **Solid State electronics**, Vol. 36, No. 7, pp. 945-948, 1993.

Kato, M., Niwa, H., Yagi, H., and Tsuchikawa, H. Diffusional Relaxation and Void Growth in an Aluminum Interconnect of Very Large Scale Integration. *Journal of Applied Physics*, 68, 334-338, 1990.

Kauffmann, N.L. and Bergh, A.A. *IEEE Transactions on Electron Devices*, ED-15, 732, 1968.

Kejzlar, M. and Slunsky, L. Monitoring and Optimization of Burn-in Process of Electronic and Microelectronic Components, *Relectronic* 85, 365-374, 1985.

Kelly, G., et al., Investigation of thermo-mechanically induced stress in a PQFP 160 using finite element techniques, *IEEE 42nd ECTC Proc.*, pp. 467-472, 1993.

Kenichi, M. and Mori, T. Thermal Stability of Various Ball Limited Metal Systems under Solder Bumps. *IEEE Transactions on Components, Hybrids, and Manufacturing Technology*, 11, 481-484, 1988.

Kessel, C.G., Gee, S.A., and Murphy, J.J. The Quality of Die Attachment and Its Relationship to Stresses and Vertical Die-cracking. *Proceedings of the 33rd Electronic Components Conference*, IEEE, 237-244, 1983.

Khajezadah, H. and Rose, A.S. Reliability Evaluation of Trimetal Integrated Circuits in Plastic Packages. *15th Annual Reliability Physics Symposium*, 244-249, 1977.

Khan, M.M. and Fatini, H. Gold-Aluminum Bond Failure Induced By Halogenated Additives in Epoxy Molding Compounds. *Proceedings of ISHM*, 420-428, 1986.

Kidson, G.V. Some Aspects of the Growth of Diffusion Layers in Binary Systems. *Journal of Nuclear Materials 3*, 1, 21-29, 1961.

Kim, K., Kniffin, M., Sinclair, R., and Helms, C.R. Interfacial Reactions in the Ti/GaAs System. *Journal of Vacuum Science and Technology*, A6, 1473-1477, 1988.

Kinsborn, E., Blech, I. A., and Komen, Y. Threshold Current Density and Incubation Time to Electromigration in Gold Films. *Thin Solid Films*, 46, 136-150, 1977.

Kinsborn, E., Melliar-Smith, C.M., English, A.T., and Chynoweth, T. Failure of Small Thin Film Conductors due to High Current Density Pulses. *16th Annual Proceedings International Reliability Physics Symposium*, 248-254, 1978.

Kitano, M., Nishimura, A., Kawai, S., and Nishi, K. Analysis of Package Cracking During Reflow Solder Process. *Proceedings 26th Annual International Reliability Physics Symposium*, 90-95, 1988.

Klema, J., Pyle, R., and Domangue, E. Reliability Implications of Nitrogen Contamination During deposition of Sputtered Aluminum/Silicon Metal Films, *Proceedings of the 1984 International Reliability Physics Symposium*, 1-5, 1984.

Klinger, D. J. On the Notion of Activation Energy in Reliability: Arrhenius, Eyring, and Thermodynamics, *1991 Proceedings of Annual Reliability and Maintainability Symposium*, 295-300, 1991.

Knott, J. Comments at the Electronic Industries Association Meeting of G-12 Committee on Solid State Devices, Baltimore. *Quoted from M.S. Thesis (A. Arora, University of Maryland at College Park)*, 1992.

Ko, P.K. Hot Electron Effects in MOSFET'S, Ph.D Dissertation, EECS Dept., U.C. Berkeley, 1980.

Kobayashi, A.S., Zi, M., and Hall, L.R. Approximate Stress Intensity Factor for an Embedded Elliptical Crack Near Two Parallel Free Surfaces. *International Journal of Fracture Mechanics*, 1, 81-95, 1965.

Kopanski, J., Blackburn, D.L., Harman, G.G., Berning, D.W. Assessment of Reliability Concerns for Wide-Temperature Operation of Semiconductor Device Circuits, *Transactions of the First International High Temperature Electronics Conference*, Albuquerque, New Mexico, 137-142, 1991.

Kossowsky, R., Pearson, R.C., and Christovich, L.C. Corrosion of In-based Solders. *Proceedings of the 16th Reliability Physics Symposium*, 200-206, 1978.

Kotlowitz, R.W. Comparative Compliance of Representative Lead Designs for Surface Mounted Components. *Proceedings IEEE Electronic Components Conference*, 791-831, 1989.

Kotlowitz, R.W. and Taylor, L. Compliance Metrics for the Inclined Gull-Wing, Spider J-Bend and Spider Gull-Wing Lead Designs for Surface Mount Components. *Proceedings of the IEEE Electronic Components Conference*, 299-312, 1991.

Kucera, V. And Mattson, E. Atmospheric Corrosion of Bimetallic structures. *Atmospheric Corrosion*, New York, John Wiley and sons, 183-192, 1982.

Kuo, W. and Kuo, Y. Facing the Headache of Early Failures: State of the Art Review of Burn-in Decisions. *Proceedings IEEE*, 71, 1257-1266, 1983.

Kwok, T., Ho, P.S., Yip, S., Ballufi, R. W., Bristow, P. D., and Brokman, A. Evidence for Vacancy Mechanism in Grain Boundary Diffusion in bcc Iron: A Molecular-Dynamics Study. *Physical Review Letters*, 47, 1148-1151, 1981.

Kwok, T. Effects of Grain Growth and Grain Structure on Electromigration Lifetime in Al-Cu Submicron Interconnects. *Proceedings of the 4th International Multilevel Interconnection Conference*, 456, 1987.

Kwok, T., Ho, P.S., and Yip, S. Computer Simulation of Vacancy Migration in a fcc Tilt Boundary. *Surface Science*, 144, 44, 1984.

LaCombe, D. J. and Parks, E. L. The Distribution of Electromigration Failures. *24th Annual Reliability Physics Symposium, IEEE*, 1-6, 1986.

LaCombe, D.J., Dening, D.C., and Christou, A. A New Failure Mechanism in Thin Gold Films at Elevated Temperature. *20th Proceedings of International Reliability Physics Symposium*, 81-87, 1982.

Lane, C.H. Aluminum Metallization and Contacts for Integrated Circuits. *Metallurgical Transactions*, 713, 1970.

Lau, J.H., Rice, D.W., and Harkins, C.G. Thermal Stress Analysis of Tape Automated Bonding and Interconnections. *Proceedings of the 39th IEEE Electronic Components Conference*, 456-463, 1989.

Lau, J.H., Erasmus, S.J., and Rice, D.W. Overview of Tape Automated Bonding Technology. *Electronic Materials Handbook*, Materials Park, ASM International, 1, 274-295, 1991.

Lau, J. *Handbook of Tape Automated Bonding*, New York, Van Nostrand Reinhold, 1992.

Learn, A.J. Electromigration Effects in Aluminum Alloy Metallization. *Journal of Electronic Materials*, 3, 531-552, 1974.

Lee, J., Chen, I.C., and Hu, C. Modeling and Characterization of Gate Oxide Reliability. *IEEE Transactions on Electron Devices*, 35, 2268, 1988.

Lee, J., Chen, I.C., and Hu, C. Statistical Modeling of Silicon Dioxide Reliability. *26th Annual International Reliability Physics Symposium*, 131-138, 1988.

Leonard, C. T., and Pecht, M. How Failure Prediction Methodology Affects Electronic Equipment Design, *Quality and Reliability Engineering International*, Vol. 6 (4) pp. 243-250. October, 1990.

Leonard, C.T. Mechanical Engineering Issues and Electronic Equipment Reliability: Incurred Costs without Compensating Benefits, 113, 1-7, 1991.

Leonard, C.T. Passive cooling for Avionics Can Improve Airplane Efficiency and Reliability. *Proceedings of the National Aerospace and Electronics Conference NAECON*, Vol.4, pp. 1248-1253, 1989.

Lerner, I. and Eldridge, J.M. Effects of Several Parameters on Corrosion Rates of Al Conductors in Integrated Circuits. Journal of Electrochemical Society, *Solid State Science and Technology*, 129, 2270-2273, 1982.

Levi, R.A., Parrillo, L.C., Lecheler, L.J., and Knoell, R.V. In-source Al-0.5%Cu Metallization for CMOS Devices. *Journal of Electrochemical Society*, 132, 159-168, 1985.

Levine, E. and Kitcher, J. Proceedings of the 22nd International Reliability Physics Symposium, 242, 1984.

Lewis, R. and Awkward, K. Methods for Determining a Maximum Operating Frequency for TTL Gates. *Proceedings 1990 Reliability and Maintainability Symposium*, 372-377, 1990.

Liehr, M., Bronner, G.B., and Lewis, J.E. Stacking-fault-induced Defect Creation in SiO_2 on Si(100). *Applied Physics Letters*, 52, 1892, 1988.

Liew, B.K., Cheung, N.W., and Hu, C. Projecting Interconnect Electromigration Lifetime for Arbitrary Current Waveforms. *IEEE Transactions on Electron Devices*, ED-37, 1343-1351, 1990.

Liew, B.K., Cheung, N.W., and Hu, C. Electromigration Interconnect Lifetime under AC and Pulsed DC Stress. *Proceedings of 27th IEEE International Reliability Symposium*, 215-219, 1989.

Liljestrand, L.G. Bond Strength of Inner and Outer Leads on TAB Devices. Hybrid Circuits, 10, 42-48, 1986.

Lim, T.B. The Impact of Wafer Back Surface Finish on Chip Strength. Proceedings of the 27th. *Annual Reliability Physics Symposium, IEEE*, 131-136, 1989.

Lin, P.S.D., Marcus, R.B., Sheng, T.T. Leakage and Breakdown in Thin Oxide Capacitors: Correlation with Decorated Stacking Faults. *Journal of Electrochemical Society*, 130, 1878, 1983.

Lin, P., Lee, J., and Im, S. Design Considerations for a Flip-chip Joining Technique. *Solid State Technology*, 48-54, 1970.

Lloyd, J.R. and Smith, P.M. *Journal Vacuum Science Technology*, A1, 455, 1983.

Lloyd, J.R., Koch, R.H. Study of Electromigration-induced Resistance and Resistance Decay in Al Thin Film Conductors. *Proceedings of 25th IEEE International Reliability Physics Symposium*, 161-168, 1987.

Lloyd, J.R., and Shatzkes, M., and Challaner, D.C. Kinetic Study of Electromigration Failure in Cr/Al-Cu Thin Film Conductors Covered with Polyimide and the Problem of the Stress Dependent Activation Energy. *IEEE International Reliability Physics Symposium*, 216-225, 1988.

Lloyd, J.R. and Koch, R.H. Study of Electromigration-induced Resistance and Resistance Decay in Al Thin Film Conductors. *Applied Physics Letters*, 52, 194-196, 1988.

LSI Logic Reliability Manual and Data Summary, Milpitas, CA, 12-18, 1990.

Lycoudes, N. and Childers, C.G. Semiconductor Instability Failure Mechanism Review. *IEEE Transactions Reliability*, 29, 237-247, 1980.

Lyman, J. Special Report: Film Carriers Star in High-volume IC Production. *Packaging and Production Editor*, 175-182, 1975.

Lyman, J. Fairchild Slashes DIP Burn-in Rejects. *Electronics*, 59, 38, 1986.

Machiels, F., Lijbers, G., Allaire, R., and Poiblaud, G. Derating of Results According to Various Models and Prediction of Lifetimes of Plastic SMD's under Humidity Stress Conditions, *5th International Conference Quality in Electronic Components, Failure Prevention, Detection and Analysis*, Bordeaux, France, 869-874, 1991.

Mahalingam, M., Nagarkar, M., and Lofgran, L. Thermal Effects of Die Bond Voids in Metal, Ceramic and Plastic Packages. *Proceedings of the 34th Electronic Components Conference, IEEE*, 469-474, 1984.

Mahalingam, M. Design Considerations. *Electronic Material Handbook: Packaging, Materials Park: ASM International, 1*, 409-421, 1989.

Maiz, J.A. and Segura, I. A Resistance Change Methodology for The Study of Electromigration in Al-Si Interconnects. *Proceedings of the 26th IEEE International Reliability Physics Symposium*, 209-215, 1988.

Maiz, J.A. Characterization of Electromigration under Bidirectional (BC) and Pulsed Unidirectional (PDC) Currents. *Proceedings of the 27th IEEE International Reliability Physics Symposium*, 220-228, 1989.

Majni, G. and Ottavini, G. Au-Al Compound Formation by Thin Film Interactions. *Journal Crystal Growth*, 47, 583-588, 1979.

Manchester, K.E. and Bird, D.W. Thermal Resistance: A Reliability Consideration. *IEEE Transactions on Components, Hybrids and Manufacturing Technology*, CHMT, 580-587, 1980.

Manno, P.T. RADC Failure Rate Prediction Methodology: Today and Tomorrow. *Electronic Systems Effectiveness and Life Cycle Costing*, edited by Skwirzynski, J.K. Springer-Verlag Berlin-Heidelberg, NATO ASI Series, F3, 177-200, 1983.

Manson, S.S. and Hirschberg, M.H. Fatigue, *An Interdisciplinary Approach*, Syracuse: Syracuse University Press, 133, 1964.

Manzione, L.T. *Plastic Packaging of Microelectronic Devices*, New York, Van Nostrand Reinhold, 1990.

Marazas, B. M. Observations of the Behavior of Bipolar and MOS Transistors and Bipolar and MOS Inverters at High Temperatures, M.S. Thesis directed by Electrical Engineering Department and Mechanical Engineering Department, University of Maryland at College Park, 1992.

Marcus, R. and Blumenthal, S. A Sequential Screening Procedure, *Technometrics*, 16(2), 229-234, 1974.

Martin, C.A., Onduresk, J.C., and McPherson, J.W. Electromigration Performance of CVD-W/Al-Alloy Multilayered Metallization. *Proceedings of 28th International Reliability Physics Symposium*, 31-36, 1990.

Matsumoto, H., Sawada, K., Asai, S., Hirayama, M., and Nagasawa, K. Effect of Long Term Stress on IGFET Degradations Due to Hot Electron Trapping. *IEEE Transactions on Electron Devices,* ED-28, 923-928, 1981.

Mayerfeld, P. Flexible Burn-in Architecture Maximizes System Utilization. *Evaluation Engineering*, 29, 74-83, 1990.

Mayerfeld, P. ESS/Burn-in: The Challenges of ASIC Burn-in. *Evaluation Engineering*, 30, 21-23, 1991.

Mayumi, S., Umemoto, T., Shishino, M., Nantsue, N., Useda, S., and Inoue, M. The Effect of Cu Addition to Al-Si Interconnects on Stress Induced Open-circuit Failures, *Proceedings of the International Reliability Physics Symposium,* 15-21, 1987.

McAteer, O.J. *Electrostatic Discharge Control,* New York, McGraw-Hill Publishing Company, 1989.

McDonald, N.C., and Riach, G.E. Thin Film Analysis for Processes Evaluation, *Electronic Packaging and Production,* 50-56, 1973.

McDonald, N.C., and Palmberg, P.W. Application of Auger Electron Spectroscopy for Semiconductor Technology, *International Electron Device Meeting*, Washington, D.C., 42, October 11-13, 1971.

McKitterick, J. Very Thin Silicon on Insulators Devices for CMOS at 500°C. *Transactions: First International High Temperature Electronics Conference,* 37-41, 1991.

McLeod, P.S. and Hartsough, L.D. High-rate Sputtering of Aluminum for Metallization of Integrated Circuits. *Journal Vacuum Science Technology,* 14, 263-265, 1977.

McLinn, J.A. Constant Failure Rate: A Paradigm Transition. *Quality and Reliability International,* 6, 237-241, 1990.

McPherson, J.W. and Baglee, D.A. Acceleration Factors for Thin Gate Oxide Stressing. *23rd Annual Reliability Physics Symposium, IEEE,* 1-5, 1985.

McPherson, J.W. and Dunn, C.F. A Model for Stress-Induced Metal Notching and Voiding in Very Large-scale-integrated Al-Si (1%) Metallization, *Journal of Vacuum Science and Technology,* B5(5), 1321-1325, 1987.

Mentor Graphics Modeling Air Flow in Electronic Packages. *Mechanical Engineering*, 56-58, 1989.

Meyer, D.E. Transactions TMS-AIME, 245, 593-599, 1969, quoted from Pasco, R.W. and Schwarz, J.A. Application of A Dynamic Technique to the Study of Electromigration Kinetics, 21st *Proceedings of Reliability Physics Symposium*, 10-23, 1983.

MIL-HDBK 217F *Reliability Prediction of Electronic Equipment*, MIL-HDBK 217F, U.S. Department of Defense, Washington D.C., 1991.

MIL-STD-883C Electrostatic Discharge Sensitivity Classification. *Technical Report Notice* 8, DOD March, 1989.

Milnes, A.G. *Deep Impurities in Semiconductors*, New York: John Wiley and Sons, 1973.

Mittal, S., Zaragoza, M., and Doi, J. Failure of Gold-silicon Die-attach by the Presence of Nickel Silicide in the Bonding of Larger Size Dice to Ceramic Chip-carriers. *Proceedings of the 34th Electronic Components Conference, IEEE*, 463-471, 1984.

Mizugashira, S. and Sakaguchi, E. 15th Symposium on Reliability and Maintainability, 1985, 53; Quoted from Ricco, B., Fantini, F., Magistralli, F., Brambilla, P., Reliability of GaAs MESFET's, in Christou, A. and Unger B.A., *Semiconductor Device Reliability*, Netherlands: Kluwer Academic Publishers, 455-461, 1985.

Moazzami, R., Lee, J., Chen, I.C., and Hu, C. Projecting the Minimum Acceptable Oxide Thickness for the Time-dependent Dielectric Breakdown. IEDM Technical Digest, 710, 1988.

Moazzami, R., Lee, J., and Hu, C. Temperature Acceleration of Time-dependent Dielectric Breakdown. *IEEE Transactions on Electron Devices*, 36, 1989.

Moazzami, R. and Hu, C. Projecting Gate Oxide Reliability and Optimizing Reliability Screens. *IEEE Transactions on Electron Devices*, 37, 1990.

Morgan, D.V. Interdiffusion of Metal Films on Gallium Arsenide and Indium Phosphide, Chapter 3, in M.J. Howes and D.V. Morgan (eds), *Reliability and Degradation*, New York, Wiley & Sons, 151-190, 1981.

Moss, R.Y. Caution: Electrostatic Discharge at Work, *IEEE Transactions Component Hybrids and Manufacturing Technology*, 5, 512-515, 1982.

Motorola Inc. Semiconductor Products Sector, Discrete Materials and Technologies Group, *Reliability Audit Report*, (1st Quarter 1991).

Motorola TMOS Power MOSFET *Reliability Report*, Phoenix, Motorola Literature Distribution, 1988.

Motorola Semiconductor Products Sector Reliability and Quality Handbook, Phoenix, AZ, 1990.

Muller, R.S. and Kamins, T.I. *Device Electronics for Integrated Circuits*, New York, John Wiley & Sons, 2nd Edition, 1986.

Murthi, A.K. and Shewchen, J. The Effect of Hydrogen Ambients on Failure Mechanisms in CMOS Metallization. *IEEE/Proceedings* IRPS, 55-65, 1982.

Murty, K.L., Clevinger, G.Z. and Papazoglou, T.P. Thermal Creep of Zircaloy-4 Cladding, *Proceedings of 4th International Conference on Structural Mechanics in Reactor Technology*, C 3/4, 1977.

Murty, K.L., and Turlik, I., Deformation Mechanisms in Tin-lead Alloys, Application to Solder Reliability in Electronic Packages, *Proceedings of the 1st Joint Conference on Electronic Packaging*, ASME/JSME, pp.309-318, 1992

Nagasawa, E. and Okabayashi, H. Electromigration and Ohmic Contacts Properties of the Magnetron-sputtered Al-2%Si Alloy Films. *NEC Research and Development Journal*, 59, 1-11, 1980.

Nagesh, V.K. Reliability of Flip Chip Solder Bump Joints. *Proceedings of 20th International Reliability Physics Symposium*, 6-15, 1982.

Nakayama, W. and Bergles, A.E. Cooling Electronic Equipment: Past, Present, and Future. *Proceedings International Symposium on Heat Transfer in Electronic and Microelectronic Equipment*, International Center for Heat Transfer, Dubrovnik, Yugoslavia, Hemisphere Publishing Co, 1988.

Nanda, V. and Black, J.R. Electromigration of Al-Si Alloy Films. *International Reliability Physics Symposium*, 1978.

NASA Requirements to Preclude the Growth of Tin Whiskers, information received from Jack Shaw, Manager, NASA Parts, Projects Office, NASA Goddard Space Flight Center. Quoted from *M.S. Thesis (A. Arora, University of Maryland at College Park)*, 1992.

Naumchik, P. Burn-in of Integrated Circuits....Required or Not?, Signetics Report, 1988.

Nauta, P.K. and Hillen, M.W. *Journal of Applied Physics*, 49, 2862, 1978.

Naval Air Systems Command Department of the Navy, Application and Derating Requirements for Electronic Components, *AS-4613*, 1976.

Neff, G. R. Hybrid Hermeticity and Failure Analysis. *Hybrid Circuit Technology*, 3, 19-24, 1986.

Neri, B., Diligenti, A., Aloe, P., and Fine, V.A. Electromigration in Thin Metal Films, Activation Energy Evaluation By Means of Noise Technique. Results and Open Problems for Indium and Gold. *Vuoto (Scienza a Tecnologia)*, 4, 219-222, 1989.

Neudeck, Gerald W. *Modular Series on Solid State Devices*; *The Bipolar Junction Transistor*, vol. III, Addison-Wesley Publishing Company, 1989.

Newsome, J.L. and Oswald, R.G. Metallurgical Aspects of Aluminum Wire Bonds to Gold Metallization, *14th IRPS*, 63-74, 1976.

Nguyen, T.H. and Foley, R.T. The Chemical Nature of Aluminum Corrosion, Part III. *Journal of Electrochemical Society*, 127, 2563-2566, 1980.

Nicollian, E.H. Interface Instabilities. *12th Annual Proceedings International Reliability Physics Symposium*, 267-272, 1974.

Nield, B.J. and Quarrel, A.G. Intercrystalline Cracking in Creep of Some Aluminum Alloys. *Journal Inst. Metals*, 85, 480-488, 1956.

Ning, T.H., Osburn, C.M., and Yu, H.N. *Applied Physics Letters*, 48, 286, 1977.

Ning, T.H., Cook, P.W., Dennard, R.H., Osburn, C.M., Schuster, S.E., and Yu, H.W. 1mm MOSFET VLSI Technology Part IV: Hot Electrons Design Constraints. *IEEE Transactions on Electron Devices*, ED-26, 346-353, 1979.

Ninomaya, T. and Harada, K. Multilayer Debugging Process (A Method of Screening), *IEEE Transactions on Reliability*, R-24, 230-238, 1975.

Nippon Telegraph and Telephone Corporation (NTT) *Standard Reliability Table for Semiconductor Devices*, March 1985.

Nischizawa, J.I. et al. Field Effect Transistor Versus Analog Transistor Static Induction Transistor, *IEEE Transactions*, 22, 185-197, 1975.

Nishimura, A., Tatemichi, A., Muira, H., and Sakamoto, T. Life Estimation for IC Plastic Packages under Temperature Cycling Based on Fracture Mechanics. *IEEE Transactions on Components, Hybrids, Manufacturing Technology*, CHMT-12, 637-642, 1987.

Niwa, H., Yagi, H., and Tsuchikawa, H. Stress Distribution in an Aluminum Interconnect of Very Large Scale Integration. *Journal of Applied Physics*, 68(1), 328-333, 1990.

Noble, W.P. and Ellenberger, A.R. Temperature Effects on Device Functionality. *Thermal Management Concepts in Microelectronic Packaging*, ISHM Technical Monograph, 45-66, 1984.

Noon, D.W. Corrosion and Reliability Industrial Process Control Electronics. *Electronic Packaging and Corrosion in Microelectronics*, 49-53, 1987.

Nordquist, S.E., Haslett, J.W., and Trifimenkoff, F.N. High Temperature Leakage Current Suppression in CMOS Integrated Circuits. *Electronics Letters*, 25, 1133-1135, 1989.

Norris, K.C. and Landzberg, A.H. Reliability of Controlled Collapse Interconnections. *IBM Journal Res. Devices (USA)*, 13, 1969.

NTT *Standard Reliability Table for Semiconductor Devices*, Nippon Telegraph and Telephone Corporation, March 1985.

O'Connor, P.D.T. Reliability Prediction: Help or Hoax?, *Solid State Technology*, 59-61, August 1990.

O'Donnell, S.J., Bartling, J.W., and Hill, G. Silicon Inclusions in Aluminum Interconnects, *International Reliability Physics Symposium*, 9-16, 1984.

Okabayashi, H. An Analytical Open-circuit Model for Stress Driven Diffusive Voiding in Al Lines. *5th International Conference Quality in Electronic Components; Failure Prevention, Detection and Analysis*, 171-175, 1991.

Okikawa, S., Suzuki, H., and Mikino, H. Stress Analysis of Poor Gold-silicon Die-attachment for LSIs. *Proceedings of the International Symposium for Testing and Failure Analysis, International Society for Testing and Failure Analysis*, 180-189, 1984.

Oliver, C.B. and Bower, D.E. Theory of the Failure of Semiconductor Contacts by Electromigration. *8th Annual Proceedings Reliability Physics Symposium*, 116-120, 1970.

Olsen, D.R. and James, K.L. Evaluation of The Potential Reliability Effects of Ambient Atmosphere on Aluminum-copper Bonding in Semiconductor Products, *IEEE Transactions on Components Hybrids, and Manufacturing Technology*, CHMT-7, 357-362, 1984.

Onduresk, J.C., Nishimura, A., Hoang, H.H., Sugiura, T., Blumenthal, R., Kitagawa, H., McPherson, J.W. Effective Kinetic Variations with Duration for Multilayered Metallizations. *26th Annual IRPS*, 179-184, 1988.

Ost, G. The Practice and Economy of Burn-in, *Electronic Engineering*, 58, 37-40, 1986.

Oswald, R.G. and Miranda, R. Application of Tape Automated Bonding Technology to Hybrid Microcircuits. *Solid State Physics*, 33-38, 1977.

Owada, N., Hinode, K., Horiuchi, M., Nishada, T., Nakata, K., and Mukai, K. Stress Induced Slit-like Void Formation in a Fine Pattern Al-Si Interconnect During Aging Test, *Proceedings of the 1985 IEEE VLSI Multi-level Interconnect Conference*, 173-179, 1987.

Padmanabhan, R. Metal Corrosion in an Experimental Tape Automated Bonded Die. *Proceedings of IEEE Electronic Components Conference*, 309-313, 1985.

Palkuti, L.J., Prince, J.L., and Glista, A.S. Jr. Integrated Circuit Characteristics at 260°C for Aircraft Engine Control Application. *IEEE Transactions on Component Hybrid, and Manufacturing Technology*, CHMT-2, 405-412, 1979.

Palmer, D.W. and Gaynard, F.P. Aluminum Wire to Thick Film Connections for High Temperature Operation. *IEEE Transactions on Components, Hybrids and Manufacturing Technology*, CHMT-1, 219-222, 1978.

Palmour, J.W., Kong, H.S., and Davis, R.F. High Temperature Depletion Mode MOSFET's in Beta-SiC Thin Films. *Applied Physics Letters*, 51, 2028-2030, 1987.

Palumbo, W. and Patrick, Dugan M. Design and Characterization of Input Protection Networks for CMOS/SOS Application. *Electrical Overstress/Electrostatic Discharge Symposium Proceedings*, 1986.

Pancholy, R.K. and Jhnoki, T. CMOS/SOS Gate Protection Networks. *IEEE Transactions on Electronic Devices*, ED-25, 917-925, 1978.

Panousis, N. and Hall, P. The Effects of Gold and Nickel Plating Thicknesses on the Strength and Reliability of Thermocompression-bonded External Leads. *IEEE Electronic Components Conference*, 74-79, 1976.

Paris, P.C., Gomez, M.P., and Anderson, W.E. A Rational Analytical Theory of Fatigue. *The Trend in Engineering, University of Washington*, 13, 9-14, 1961.

Parsons, R. Semiconductor Device Burn-in: Is There A *Future?Quality and Reliability Engineering International*, 2, 255-258, 1986.

Pasco, R.W. and Schwarz, J.A. Application of A Dynamic Technique to the Study of Electromigration Kinetics. *21st Proceedings of Reliability Physics Symposium*, 10-23, 1983.

Patridge, J., Marques, A., and Camp, R. Electromigration, Thermal Analysis and Die Attach: A Case History. *20th Annual Proceedings Reliability Physics Symposium, IEEE*, 34-44, 1982.

Pearne, N.K. Future Trends in IC Packaging. *19th Annual Connectors and Interconnection Technology Symposium Proceedings*, 321, 1986.

Pecht, M., Ramappan, V. Are Components Still the Major Problem: A Review of Electronic System and Device Field Failure Returns, *IEEE Transactions on Components, Hybrids, and Manufacturing Technology*, 15, No.6, 1-5, December 1992.

Pecht, M. and Ko, W. A Corrosion Rate Equation for Microelectronic Die Metallization, *International. Journal for Hybrid Microelectronics*, 13, 41-52, 1990.

Pecht, M., Lall, P., Leonard, C., and Hakim, E. Temperature Dependence on Integrated Circuit Failure Mechanisms, *Advances in Thermal Modeling III, edited by Avaram Bar-Cohen, Prentice Hall*, 1991.

Pecht, M. *Handbook of Electronic Package Design*, New York, Marcel Dekker, 1991.

Pecht, M., Lall, P., and Dasgupta, A. A Failure Prediction Model for Wire Bonds. *Proceedings of the 1989 International Symposium on Hybrids Microelectronics*, 607-613, 1989.

Pecht, M., Lall, P., Leonard, C., and Hakim, E. Temperature Dependence on Integrated Circuit Failure Mechanisms, *Advances In Thermal Modeling III*, edited by Avaram Bar-Cohen and Allan D. Kraus, ASME Press, New York/IEEE Press, New York, 61-152, 1993.

Peck, Stewart D. Comprehensive Model for Humidity Testing Correlation, *Proceedings of the 1986 International Reliability Physics Symposium*, 44-50, 1986.

Peeters, J. Burn-in for Increased Automotive Reliability Test, 13(3), 24-25, 1991.

Pentic, D. Questioning the Benefits of Burn-in. *Electronic Engineering*, 56, 45-47, 1984.

Phillips, W.E. Microelectronic Ultrasonic Bonding, G.G. Harman, ed., National Bureau of Standards (U.S.), Spec. Publ. 400-2, 80-86, 1974.

Philosky, E.M. and Ravi, K.V. Measuring the Mechanical Properties of Aluminum Metallization and Their Relationship to Reliability Problems. *11th Annual Proceedings of Reliability Physics Symposium*, 33-40, 1973.

Philosky, E. Intermetallic Formation In Gold-aluminum Systems. Solid State *Electronics*, 13, 1391-1399, 1970.

Philosky, E. Purple Plague Revisited. *8th Annual Proceedings of the International Reliability Physics Symposium*, 177-185, 1970.

Philosky, E. Design Limits When Using Gold Aluminum Bonds. *9th Annual Proceedings IEEE Reliability Physics Symposium*, 11-16, 1972.

Pierce, J.M. and Thomas, M.E. Applied Physics Letters, 39(2), 165, 1981.

Pierret, R.F. *Field Effect Devices*, 2nd. ed., Reading: Addison-Wesley Publishing Company, IV, 1990.

Pierret, R.F. *Volume IV: Modular Series on Solid State Devices; Advanced Semiconductor Fundamentals*, Addison-Wesley Publishing Company, 1989.

Pierret, R.F. *The PN Junction Diode*, 2nd. ed., Reading: Addison-Wesley Publishing Company, II, 1989.

Pinnel, M.R. and Bennett, J.E. Mass Diffusion in Polycrystalline Copper/ Electrodeposited Gold Planar Couples. *Metallurgical Transactions 3*, 1989-1997, 1972.

Pitt, V.A., Needles, C.R.S., and Johnson, R.W. Ultrasonic Aluminum Wire Bonding to Copper Conductors. *Electronics Components Conference*, 18-23, 1981.

Pitt, V.A. and Needles, C.R.S. Thermosonic Gold Wire Bonding to Copper Conductors. *IEEE Transactions on Components, Hybrids, and Manufacturing Technology*, CHMT-5, 435-440, 1982.

Pollino, E. Microelectronic Reliability, *Integrity Assessment and Assurance* 2, Boston: Artech House, 1987, 364.

Price, P.B., Vermilyea, D.A., and Webb, M.B. Growth and Properties of Electrolytic Whiskers. *Acta Metallurgica*, 6, 524-531, 1958.

Prince, J.L., Draper, B.L., Rapp, E.A., Kronberg, J.N., and Fitch, L.T. Performance of Digital Integrated Circuit Technologies at Very High Temperatures. *IEEE Transactions on Components, Hybrids, and Manufacturing Technology*, CHMT-3, 571-579, 1980.

Pugacz-Muraszkiewicz, I. A Methodical Approach to Accelerated Screening of Card Assemblies, Static Burn-in Case, *The Journal of Environmental Sciences*, 28, 42-47, 1985.

Punches, K. Burn-in and Strife Testing, *Quality Progress*, 19, 93-94, 1986.

Puttlitz J.K. Corrosion of Pb-50In Flip-chip Interconnections Exposed to Harsh Environment. *Proceedings of 39th Electronic Component Conference*, 438-444, 1989.

Puttlitz, J.K. Preparation, Structure, and Fracture Modes of Pb-Sn and Pb-In. Terminated Flip-chips Attached to Gold Capped Microsockets. *IEEE Transactions on Components, Hybrids, and Manufacturing Technology*, 13, 647-655, 1990.

Rashid, M. H. *Spice for Circuits and Electronics Using PSpice*, New Jersey, Prentice Hall, 1990.

Ravi, K.V. and Philosky, E.M. Reliability Improvement of Wire Bonds Subjected to Fatigue Stresses. *10th Annual Proceedings IEEE Reliability Physics, Symposium*, 143-149, 1972.

RCA Solid State Division, RCA Power Transistors, Data Book SSD-220B, Somerville, New Jersey, 1978; quoted from: Blicher, A., *Field-Effect and Bipolar Power Transistors Physics*, New York, Academic Press, 1981.

Reich, B. and Hakim, E.B. Secondary Breakdown Thermal Characteristics and Improvement of Semiconductor Devices. *IEEE Transactions in Electron Devices*, ED-13, 734-737, 1966.

Reif, F. *Fundamentals of Statistical and Thermal Physics*, McGraw Hill, New York, 1965.

Reimer, J.D. The Effect of Contaminants on Aluminum Film Properties. *Journal Vacuum Science Technology*, 2, 242-243, 1984.

Reiss, H. *Methods of Thermodynamics*, Blaisdell, New York, 1965.

Rhoden, W.E., Banton, D.W., Kitchen, D.R., and Maskowitz, J.V. Observation of Electromigration At Low Temperatures. *IEEE Transactions on Reliability*, 40, 524-530, 1991.

Ricco, B., Azbel, M. Y., and Brodsky, M.H. Novel Mechanism for Tunneling and Breakdown of Thin SiO_2 Films. *Physics Review Letters*, 51, 1795, 1983.

Robinson, P.H. and Heiman, F.P. Use of HCl Gettering In Silicon Device Processing. *Journal Electrochemical Society*, 118, 141-143, 1971.

Rodbell, K.P. and Shatynski, S.R. Electromigration in Sputtered Al-Cu Thin Films. *Thin Solid Films*, 108, 95-102, 1983.

Rofail, S. S. and Elmasry, M. I. Temperature-Dependent Characteristics of BiCMOS Digital Circuits, *IEEE Transactions on Electron Devices*, Vol. 41, No. 1, pp. 169-177, 1993.

Rome Air Development Center RAC (Reliability Analysis Center), IC Quality Grades: Impact on System Reliability and Life Cycle Cost, *SOAR-3*, 1985.

Rome Air Development Center Reliability Prediction of Electronic Equipment, Military Handbook MIL-HDBK 217F, Griffis AFB, New York, 1991.

Rosenberg, R. and Ohring, M. Void Formation and Growth During Electromigration In Thin Films. *Journal of Applied Physics*, 42, 5671, 1971.

Rosenburg, R. and Berenbaum, L. Resistance Monitoring and Effects of Non-adhesion During Electromigration in Aluminum Films. *Applied Physics Letters*, 12, 201, 1968.

Roulston, D.J. Bipolar Semiconductor Devices, New York, McGraw-Hill Company, 1988.

Rountree, R.N., et al. A Process-tolerant Input Protection Circuit for Advanced CMOS Processes. *Electrical Overstress/Electrostatic Discharge Symposium Proceedings*, 1988.

RPP Reliability Prediction Procedure for Electronic Equipment, TR-TSY-000332, Issue 2, Bellcore, July 1988.

Runayan, W.R. *Silicon Semiconductor Technology*, New York, McGraw-Hill Company, 1965.

Saito, M. and Hirota, S. Effect of Grain Size on the Lifetime of Aluminum Interconnections. *Review of the Electrical Communication Laboratories*, 22, 678-694, 1974.

Sangiorgi, E., Johnston, R.L., Pinto, M.R., Bechtold, P.F., and Fichtner, W. Temperature Dependence of Latch-up Phenomenon in Scaled CMOS Structures. *IEEE Electron Device Letters*, EDL-7, 28-31, 1986.

Sardo, G.M. Masters Thesis, Solid State Sci. and Technology, Dept. Chem. Eng.& Mat'l Sci., Syracuse University, Syracuse, NY, 1981; quoted from Pasco, R.W. and Schwarz, J.A., Application of A Dynamic Technique to the Study of Electromigration Kinetics, 21st *Proceedings of Reliability Physics Symposium*, 10-23, 1983.

Satake, T., Yokoyama, K., Shirakawa, S., and Sawaguchi, K. Electromigration in Aluminum Film Stripes Gated with Anodic Aluminum Oxide Films. *Journal of Applied Physics*, 12, 518-522, 1973.

Schafft, H.A. and French, J.C. Second Breakdown in Transistors. *IRE Transactions in Electron Devices*, ED-9, 129-136, 1962.

Schafft, H.A. and French, J.C. A Survey of Second Breakdown. IEEE Transactions in Electron Devices, ED-13, 613-618, 1966.

Schafft, H.A. and French, J.C. Second Breakdown and Current Distribution in Transistors. *Solid State Electronics*, 9, 681-688, 1966.

Schafft, H.A., Grant, T.C., Saxena, A.N., and Kao, C.Y. Electromigration and the Current Density Dependence. *Proceedings of 23rd IEEE International Reliability Physics Symposium*, 93-99, 1985.

Scherier, L.A. Electrostatic Damage Susceptibility of Semiconductor Devices. *16th Annual Proceedings Reliability Physics Symposium*, 151-153, 1978.

Schnable, G.L. Failure Mechanisms in Microelectronic Devices. *Microelectronics and Reliability*, Artech House Inc., 1, 25-87, 1988.

Schnable, G.L. and Keen, R.S. Aluminum Metallization-advantages and Limitations for Integrated Circuit Applications. *Proceedings of IEEE*, 57, 1570, 1969.

Schreiber, H.U. Activation Energies for the Different Electromigration Mechanisms in Aluminum. *Solid State Electronics*, 24, 583-589, 1981.

Schroeder, D.K. Advanced MOS Devices, Volume VII, Reading: Addison-Wesley Publishing Company, 1987.

Schwarzenberger, A.P., Ross, C.A., Evetts, J.E., and Greer, A.L. Electromigration in Presence of A Temperature Gradient: Experimental Study and Modeling. *Journal of Electronic Materials*, 17, 473-478, 1988.

Sedra, A.S. and Smith, K. C. Microelectronic Circuits. Holt, Rinehart and Winston, Inc., 1987.

Setliff, J. E. A Review of Commercial Microcircuit Qualification and Reliability Methodology, Proceedings of the 1991 Advanced Microelectronics Technology, Qualification, Reliability and Logistics Workshop, August 13-15, Seattle, WA, 325-335, 1991.

Severns, R. and Armijos, J. MOSPOWER Applications Handbook, Santa Clara: Siliconix Corporation.

Shah, H.J. and Kelly, J.H. Effect of Dwell Time on Thermal Cycling of the Flip-chip Joint. *Proceedings of the International Hybrid Microelectronics Symposium*, 341-346, 1970.

Shatzkes, M. and Lloyd, J.R. A Model for Conductor Failure Considering Diffusion Concurrently with Electromigration Resulting in A Current Exponent of 2. *Journal of Applied Physics*, 59(11), 3890-3893, 1986.

Shih, D.Y. and Ficalora, P.J. The Effect of Hydrogen on the Rate of Formation of Intermetallics in the Cu-Sn, Ag-Sn, and Ni-Sn Systems. *Proceedings IEEE International Reliability Physics Symposium*, 87-90, 1979.

Shih, D.Y. and Ficalora, P.J. The Reduction of Au-Al Intermetallics Formation and Electromigration in Hydrogen Environments. *Proceedings IEEE International Reliability Physics Symposium*, 268-272, 1978.

Shih, D.Y. and Ficalora, P.J. Effect of Oxygen and Argon on the Interdiffusion of Au-Al Thin Film Couples. *Proceedings IEEE International Reliability Physics Symposium*, 253-256, 1981.

Shine, M.C., Fox, L.R., and Sofia, L.W. A Strain Range Partitioning Procedure for Solder Joint Fatigue. *Brazing and Soldering*, 9, 11-20, 1985.

Shirley, G.C. and Hong, C.E. Optimal Acceleration of Cyclic THB Tests for Plastic Packages Devices. *29th International Reliability Physics Symposium*, 12-21, 1991.

Shoucair, F., Hwang, W., and Jain, P. Electrical Characteristics of Large Scale Integration MOSFETs at Very High Temperatures Part II: Experiment. *Microelectronics and Reliability*, 24, 487-510, 1984.

Shoucair, F. and Early, J.M. High Temperature Diffusion Leakage Current Dependent MOSFET Small Signal Conductance. *IEEE Transactions on Electron Devices*, ed. 31, 1866-1872, 1984.

Shoucair, F. Small Signal Drain Conductance of MOSFET in Saturation Region: A Simple Model. *Electronics Letters*, 22, 239-241, 1986.

Shoucair, F. Design Considerations in High Temperature Analog CMOS Integrated Circuits. *IEEE Transactions on Components, Hybrids, and Manufacturing Technology, CHMT-9*, 243-251, 1986.

Shoucair, F. CMOS Logic Cell Switching Speed Thermal Characterization. *Electronics Letters*, 23, 458-460, 1987.

Shoucair, F. High Temperature Latchup Characteristics in VLSI CMOS Circuits. *IEEE Transactions on Electron Devices*, 35, 2424-2426, 1988.

Shoucair, F. Potential and Problems of High Temperature Electronics and CMOS Integrated Circuits (25-150°C): An Overview. *Microelectronics Journal*, 22, 39-54, 1991.

Shoucair, F. Analytical and Experimental Methods for Zero Temperature Coefficient Biasing of MOS Transistors. *Electronics Letters*, 25, 1196-1198, 1989.

Shoucair, F. Scaling, Subthreshold, and Leakage Current Matching Characteristics in High Temperature (25-150°C) VLSI CMOS Devices. *IEEE Transactions on Components, Hybrids, and Manufacturing Technology*, 12, 780-788, 1989.

Siemens Standard, SN29500 Reliability and Quality Specification Failure Rates of Components, 1986.

Sigsbee, R.A. Electromigration and Metallization Lifetimes. *Journal Applied Physics*, 44, 1973.

Sim, S.P. Procurement Specifications Requirements for Protection Against Electromigration Failures in Al Metallizations. *Microelectronics and Reliability*, 19, 207-218, 1979.

Sim, S.P. and Lawson, R.W. Influence of Plastic Encapsulants and Passivation Layers on Corrosion of Thin Al Films Subjected to Humidity Stress. *Proceedings 17th Annual Reliability Physics Symposium*, 103-112, 1979.

Slunsky, L. Monitored Burn-In as the Method of Quality Control in Electronic Components and Devices Production. *Microelectronics and Reliability*, 29(3), 447, 1989.

Smith, R.W., Hirschberg, M.H., and Manson, S. S. NASA TN D-1574, NASA, 1963.

Smith, J.S. Overstress Failure Analysis in Microcircuits. *16th Annual Proceedings International Reliability Physics Symposium*, 1978, 41-46.

Sneddon, I.N. The Distribution of Stress in the Neighborhood of a Crack in a Elastic Solid. *Proceedings of Royal Society London*, A187, 229-260, 1946.

Speakman, T.S. A Model for the Failure of Bipolar Silicon Integrated Circuit Subjected to Electrostatic Discharge. *Proceedings of the International Reliability Physics Symposium*, 60-70, 1974.

Specchiulli, G., Vanzi, M., Caprile, C., Fantini, F., and Sala, D. Effect of The TiN/Ti Diffusion Barrier on the Electromigration. *Proceedings of International Symposium for Testing and Failure Analysis (ISTFA)*, 343-353, 1989.

Speerschneider, C. and Lee, J. Solder Bump Reflow Tape Automated Bonding. Microelectronic Packaging Technology: *Materials and Processes*, ASM, 7-12, 1989.

Stagg, J.P. and Boudry, M.R. *Journal of Applied Physics*, 52, 885, 1977.

Steppan, J.J., Roth, J.A., Hall, L.C., Jeanotte, D.A., and Carbone, S.P. A Review of Corrosion Failure Mechanisms During Accelerated Tests. *Journal Electrochemical Society*, 134, 175-190, 1987.

Stojadinovic, N.D. Failure Physics of Integrated Circuits: A Review. *Microelectronics and Reliability*, 23, 609-707, 1983.

Stone, D., Wilson, H., Subrahmanyan, R., and Li, C.Y. Mechanisms of Damage Accumulation in Solder During Thermal Fatigue. *Proceedings of 36th Electronic Components Conference*, 630-635, 1986.

Streetman, B.G. Solid State Electronic Devices, New Jersey, Prentice Hall, 1990.

Suehle, J.S. and Schafft, H.A. The Electromigration Damage Response Time and Implications for DC and Pulsed Characterizations. *Proceedings of 27th IEEE International Reliability Physics Symposium*, 229-233, 1989.

Suhir, E. and Poborets, B. Solder-glass Attachment in Cerdip/Cerquad Packages: Thermally Induced Stresses and Mechanical Reliability. *Transactions of the ASME*, 112, 204-209, 1990.

Suhir, E. Calculated Thermally Induced Stress in Adhesively Bonded and Soldered Assemblies. Presented at *International Symposium on Microelectronics*, ISHM, 1986.

Suhir, E. Stress in Adhesively Bonded Bi-material Assemblies Used in Electronic Packaging. *Proceedings of Material Research Society Symposium*, 72, 133-138, 1986.

Suhir, E. and Lee, Y. Thermal, Mechanical, and Environmental Durability Design Methodologies, Electronic Material Handbook. Package, *ASM International*, 1, 45-75, 1989.

Suhir, E. Die Attachment Design and Its Influence on Thermal Stresses in the Die and the Attachment. *Proceedings of 37th Electronic Components Conference*, 508-517, 1987.

Sunshine, R.A. Avalanching and Second Breakdown in Silicon-on-sapphire Diodes. *Technical Report PRRL-70-TR-245, RCA Laboratories*, Princeton, NJ., 1970.

Suyko, A. and Sy, S. Development of a Burn-in Time Reduction Algorithm Using the Principles of Acceleration Factors. *29th Annual Proceedings of the Reliability Physics Symposium*, 264-290, 1991.

Swanson, D.W. and Licari, J.J. Effect of Screen Tests and Burn-in on Moisture Content of Hybrid Microcircuits. *Solid State Technology*, 29, 125-130, 1986.

Swartz, G.A. Gate Oxide Integrity of NMOS Transistor Arrays. *IEEE Transactions on Electron Devices*, ED-33, 1826-1829, 1986.

Sze, S.M. Physics of Semiconductor Devices, 2nd edition, N.Y., John Wiley and Sons, 30-47, 1981.

Tai, K.L. and Ohring, M. Grain Boundary Electromigration in Thin Films. Part II: Tracer Measurements In Pure Au. *Journal of Applied Physics*, 48, 36-45, 1977.

Takeda, E., Nakagome, Y., Kume, H., Suzuki, N., and Asai, S. Comparison of Characteristic of N-Channel and P-Channel MOSFET's for VLSI. *IEEE Transactions on Electron. Devices*, ED-30, 675-680, 1983.

Tam, S. and Hsu, F.C. A Physical Model for Hot Electron Induced Degradation in MOSFET'S. *IEEE Semiconductor Interface Specialist Conference*, 1983.

Tan, S.L., Chang, K.W., Hu, S.J., and Fu, K.K.S. Failure Analysis of Die Attachment on Static Random Access Memory (SRAM) Semiconductor Devices. *Journal of Electronic Materials*, 16, 7-11, 1987.

Tatsuzawa, T., Madokoro, S., and Hagiwara, S. Si Nodule Formation in Al-Si Metallization. *Proceedings of 23rd International Reliability Physics Symposium*, 138-141, 1985.

Tavernelli, J.F. and Coffin, L.F., Jr. *Transactions ASM 51*, 438, 1959.

Texas Instruments Military Products: Products Spectrum Nomenclature and Cross Reference, Designer's Reference Guide, Dallas, TX, 4-10 to 4-11; 8-9; 12-29, 1988.

Texas Instruments Advanced CMOS Logic Applications Engineering and Logic Marketing Dept., Texas Instruments General Purpose Logic Reliability Bulletin, SDLT053, 1987.

Tezaki, Atsumu, Mineta, T., Egawa, H., and Noguchi, T. Measurement of Three Dimensional Stress and Modeling of Stress Induced Migration Failure in Aluminum Interconnects, *Proceedings of the International Reliability Physics Symposium*, 221-229, 1990.

Thomas, R.W., and Calabrese, D.W. Phenomenological Observations on Electromigration. *21st Proceedings of International Reliability Physics Symposium*, 1-9, 1983.

Thornton, C.G. and Simmons, C.D. A New High Current Mode of Transistor Operation. *IRE Transactions in Electron Devices*, ED-5, 6-10, 1958.

Towner, J. Electromigration Testing of Thin Films At the Wafer Level. *Solid State Technology*, 197-200, 1984.

Towner, J.M. and Van de Ven, E.P. Aluminum Electromigration under Pulsed D.C. Conditions. *Proceedings 21st IEEE International Reliability Physics Symposium*, 36-39, 1983.

Towner, J.M., Dirks, A.G., and Tein, T. Electromigration in Titanium Doped Aluminum Alloys. *Proceedings of 24th International Reliability Physics Symposium*, 7-11, 1986.

Trindale, D.C. Can Burn-in Screen Wear out Mechanisms? Reliability Modeling of defective Sub-populations: A Case Study. *29th Annual Proceedings of the Reliability Physics Symposium*, 260-263, 1991.

Tsividis, Y.P. and Antognetti, P. Design of MOS VLSI Circuits for Telecommunications, Englewood Cliffs, NJ, Prentice Hall, 1985.

Tu, K.N. and Berry, B.S. *Journal of Applied Physics*, 43, 3283, 1972.

Tummala, R.R. and Rymaszewski, E.J. Microelectronics Packaging Handbook, New York, Van Nostrand and Rienhold, 1989.

Tummala, R.R. Stress Corrosion of Low Temperature Solder Glass. Noncrystalline Solids, 19(1), 263-272, 1975.

Turner, T. and Wendel, K. The Influence of Stress on Aluminum Conductor Life. *International Reliability Physics Symposium*, 142-147, 1985.

Unger, B.A. Early Life Failures. *Quality and Reliability Engineering International*, 4, 27-34, 1988.

Usell, R.J. and Rao, B.S. Designing-in and Building-in Reliability: An Operational Definition with Examples from Electronic Packaging Products. *Proceedings of the International Electronics Packaging Society Conference*, 540-547, 1984.

Uyemura, J. P. Fundamentals of MOS Digital Integrated Circuits, Addison-Wesley Publishing Company, Reading, Massachusetts, 1988.

Vadasz, L. and Grove, A.S. Temperature Dependence of MOS Transistor Characteristics Below Saturation. *IEEE Transactions on Electron Devices*, ED-12, 863-866, 1966.

Vaidya, S., Fraser, D.B., and Lindenberger, W.S. Electromigration In Fine-line Sputter Gun Al. *Journal of Applied Physics*, 51, 4475-4482, 1980.

Vaidya, S., Sheng, T.T., and Sinha, A.K. *Applied Physics Letters*, 36(6), 464, 1980.

Vaidya, S. Electromigration In Aluminum/Polysilicon Composites. *Applied Physics Letters*, 39, 900-902, 1981.

Valeri, S.J. and Robinson, A.L. A High Temperature, High Voltage Silicon on Insulator Composite Device. *Transactions: First International High Temperature Electronics Conference*, 50-58, 1991.

Van Gurp, G.J. Electromigration in Cobalt Films. *Thin Solid Films*, 38, 295-311, 1976.

Van Gurp, G.J. Electromigration in Al Films Containing Si. *Applied Physics Letters*, 19, 476-478, 1971.

Vassilev, V.Z. and Nenkova, B.G. Cost Effectiveness of Burn-in Procedures of Semiconductor Devices and Integrated Circuits. *Microelectronics and Reliability*, 29(3), 453, 1989.

Vassilev, V. Effectiveness from Penalization Operations Burn-in for Semiconductor Devices and Integrated Circuits. *Standards and Quality*, 10, 1986.

Venables, J.R. and Lye, R.G. A Statistical Model for Electromigration-induced Failures in Thin Film Conductors. *Proceedings 10th Annual Reliability Physics Symposium IEEE*, 159-164, 1972.

Villela, F. and Nowakowski, M. F. Investigation of Fatigue Problems in 1-mil Diameter Thermocompression and Ultrasonic Bonding of Aluminum Wire. NASA Technical Memorandum, NASA, TM-X-64566, 1970.

Villela, F. and Nowakowski, M.F. Thermal Excursion Can Cause Bond Problems. *9th Annual Proceedings IEEE Reliability Physics, Symposium*, 172-177, 1971.

Wada, T., Higuchi, M., and Asiky, T. Electromigration in Double-level Metallization. *IEEE Transactions on Reliability*, R-34, 2-7, 1985.

Wager, A.J. and Cook, H.C. Modeling the Temperature Dependence of Integrated Circuit Failures. Thermal Management Concepts in Microelectronic Packaging, *International Society of Hybrid Microelectronics*, 1-45, 1984.

Wakabayashi, Y. Nagai, Y., and Suzuki, T. (Fujitsu Labs. Ltd., Kawasaki, Japan), Development and Evaluation of Epoxy Molding Compounds for Encapsulation Use, Fujitsu Sci. Technol. J. 11, (2) 115-133 June 1975.

Wang, S.Y. and Smith, D.M. (Los Alamos Scientific Lab, NM, USA), In vitro Mutagenicity Testing, (1) Kermide 601resin, Sylgard 184 encapsulating resin and Sylgard 184 curing agent, Aug 1978. Accession No LA-7437-MS.

Wang, T.C. (Harbour Branch Foundation Lab, Fort Pierce, Fla., USA), Prouch Method for Measuring Permeability of Organic Vapors through Plastic Film, *Am. Chem. Soc. Divn Polymer Chem.* Prepr 16, (1) pp. 758-764 (1975) for Meet, Philadelphia, Pa, 6-11 1971.

Wendt, H., Cerva, H., Lehmann, V., and Pamler, W. Impact of Copper Contamination on the Quality of Silicon Oxides. *Journal of Applied Physics*, 65, 2402, 1989.

Westergaard, H.M. Bearing Pressure and Cracks. *Journal of Applied Mechanics*, A-49-A-53, 1939.

Westinghouse Electric Corp. Summary Chart of 1984/1987 Failure Analysis Memos, 1989.

Westinghouse Electric Co. Application and Derating Requirements for Electronic and Electromechanical Components. Appendix to Defense Unit Guide DU 1.4.2.4, 1986.

What Is the Latest Ratio of Failure Causes? Semiconductor Reliability News, 4, 1990.

Whitbeck, C.W. and Leemis, L.M. Component Versus System Burn-in Techniques for Electronic Equipment. *IEEE Transactions on Reliability*, 38(2), 193, 1989.

White, P.M. *Proceedings of European Microwave Conference*, 405, 1978.

White, P.M., Hewett, B.L., and Turner, J.A. *Proceedings of European Microwave Conference*, 1978, 405. Quoted from Ricco, B., Fantini, F., Magistralli, F., Brambilla, P. Reliability of GaAs MESFET's, pp.455-461, (in Christou, A. and Unger B.A. Semiconductor Device Reliability), Kluwer Academic Publishers, Netherlands, 1990.

White, M.L. Encapsulation of Integrated Circuits. *Proceedings of the IEEE*, 57, 1610, 1969.

Wigner, E.P. Faraday Society (London) Transactions, vol. 34, 29, 1938.

Wild, A. and Triantafyllou, M. Electromigration on Oxide Steps. Microelectronics and Reliability, 28, 248-255, 1988.

Wild, R.N. Some Fatigue Properties of Solders and Solder Joints. Nat. *Electronic Packaging Conference*, 409-414, 1974.

Witzman, S., Smith, K., and Metelski, G. Silicon Interconnect: A Critical Factor in Device Thermal Management, *IEEE Components, Hybrids, and Manufacturing Technology*, 13, 946-952, 1990.

Witzmann, S. and Giroux, Y. Mechanical Integrity of the IC Device Package: A Key Factor in Achieving Failure Free Product Performance, *Transactions of the First International High Temperature Electronics Conference*, Albuquerque, 137-142, June 1991.

Wong, Kam L. What Is Wrong With the Existing Reliability Prediction Methods?, *Quality and Reliability Engineering International*, 6, 251-257, 1990.

Wong, K.L. The Bathtub Does Not Hold Water Any More. *Quality and Reliability International*, 6, 279-282, 1990.

Wood, J. Reliability and Degradation of Silicon Devices and Integrated Circuits, Chapter 4, Reliability and Degradation: Semiconductor Devices and Circuits, edited by M.J. Howes and D.V. Morgan, John Wiley & Sons, Wiley Interscience 1981.

Woods, M.H. and Rossenburg, S. EPROM Reliability; Parts I and II: Electronics, *Microelectronics and Reliability*, 53, 133-141, 1980.

Workman, W. Failure Modes of Integrated Circuits and Their Relationship to Reliability. *Microelectronics and Reliability*, 7, 257-264, 1968.

Wu, C.J. and McNutt, M.J. Effects of Substrate Thermal Characteristics on the Electromigration Behavior of Al Thin Films Conductors. *Proceedings of 21st IEEE International Reliability Physics Symposium*, 24-31, 1983.

Wunsch, D.C. Applications of Electronic Overstress Models to Gate Protective Networks. *IEEE Proceedings - Reliability Physics*, 47-55, 1978.

Wunsch, D.C. and Bell, R.R. Determination of Threshold Failure Levels of Semiconductor Diodes and Transistors Due to Pulse Voltages. *IEEE Transactions on Electron Devices*, NS-15, 244-259, 1968.

Wurnik, F. Integrated Circuits with or without Burn-in? World Quality Congress in Brighton, *Proceedings 1*, 201-209, 1984.

Wurnik, F. and Pelloth, W. Functional Burn-in for Integrated Circuits. *Microelectronics and Reliability*, 30 (2), 265, 1990.

Yeu, L., Hong, C., and, Crook, D. Passivation Material and Thickness Effects on the MTTF of Al-Si metallization. *Proceedings of the 23rd International Reliability Physics Symposium*, 115, 1985.

Yost, F.G., Romig, A.D., Jr., and Bourcier, R.J. Stress Driven Diffusive Voiding of Aluminum Conductor Lines: A Model for Time Dependent Failure, Sandia National Laboratories, *Report SAND 88-09 46, NTIS DE 89-0010507*, 1988.

Yost, F.G., Amos, D.E., and Romig, A.D. Stress-Driven Diffusive Voiding of the Aluminum Conductor Lines. *International Reliability Physics Symposium*, 193-201, 1989.

Yue, J.T., Funsten, W.P., and Taylor, R.V. Stress Induced Voids in Aluminum Interconnects During IC Processing. *International Reliability Physics Symposium*, 126-137, 1985.

Zakraysek, L. Metallic Finish Systems for Microelectronic Components. *IEEE Transactions on Components, Hybrids and Manufacturing Technology*, CHMT-4, 462, 1981.

Zhu, Lily H. High Temperature Modeling and Thermal Characteristics of GaAs MESFETs on Diamond Heat Sinks, M.S. Thesis directed by Electrical Engineering Department and Mechanical Engineering Department, University of Maryland at College Park, 1992.

Index

Permissions

Chapter 1

Figure 1.1: Valley high output versus free air temperature (74AC11373 compared to end-pin product) Texas Instruments - Advanced CMOS Logic Applications Engineering and Logic Marketing Dept., Texas Instruments General purpose Logic Reliability Bulletin, SDLTO53, 1987 (*Reprinted by Permission of Texas Instruments*)

Figure 1.2: Evaluation of temperature effects (peak low-level output voltage versus free-air temperature) Texas Instruments - Advanced CMOS Logic Applications Engineering and Logic Marketing Dept., Texas Instruments General purpose Logic Reliability Bulletin, SDLTO53, 1987 (*Reprinted by permission of Texas Instruments*)

Figure 1.3: The Arrhenius plot depicts the inverse of Junction temperature versus time to failure.

Table 1: Dominant VLSI failure mechanisms based on the survey response [*IITRI 15 January 1988, RADC*].

Figure 1.4: The sensitivity of change in activation energy on the mean time between failure as a function of temperature

Figure 1.5: Scatter diagram showing the lack of correlation between observed failure rate and junction temperature of different types of bipolar logic ICs. [Hallberg 1994] (© *1994 John Wiley and Sons*)

Figure 1.6: The reliability of electronic devices has often been represented by an idealized graphic called the bathtub curve

Figure 1.7: Plot of mean time between failure (MTBF) as a function of package case temperature for a Boeing E-3A multiplexer hybrid

Figure 1.8: Guidance or reliability allocations of a system

Table 3: Historical perspective of dominant failures in microelectronic devices. [Pecht 1992] (© *1992 IEEE*)

Table 4: Temperature-acceleration factors [Bowles 1992] (© *1992 IEEE*)

Chapter 2

Figure 2.1: Failure mechanisms in microelectronic devices have been classified according to failure sites at die (or chip) level and first level package.

Figure 2.2: SEM Micrograph - Aluminum Wirebond corrosion. (*Courtesy of Motorola*)

Figure 2.2: SEM Micrograph - Aluminum Wirebond Corrosion Magnified. *(Courtesy of Motorola)*

Table 1: Valid combinations of corrosion exponent (n_{corr}) and Activation Energy ($E_{a,corr}$) [Pecht and Ko 1989]

Figure 2.3: MTF is shorter for longer on times, if device power dissipation is large [Ajiki 1979] MTF is not much effected by ON:OFF time ratio when device power dissipation is small. (© *1979 IEEE*)

Figure 2.4: The amount of moisture absorbed increases with the increase in OFF time, i.e., the more the time spent at lower temperature, the larger is the amount of moisture absorbed. [Ajiki 1979] (© *1979 IEEE*)

Figure 2.5: The average amount of moisture remains the same as ON:OFF ratio is the same [Ajiki 1979] (©*1979 IEEE*)

Figure 2.6: Higher power dissipation devices evaporate moisture due to junction temperature rise, irrespective of environmental conditions. [Ajiki 1979] (© *1979 IEEE*)

Figure 2.7: Time to failure due to corrosion versus duty cycle [Pecht 1990]

Figure 2.8: The lifetime due to electromigration is a complex function of temperature and can be represented by an apparent activation energy which changes with operating temperature (© 1972 IEEE)

Figure 2.9: Temperature dependence of vacancy supersaturation distribution [Rosenberg 1971] (reprinted with permission from *ROSENBERG, R. AND OHRING, M. VOID FORMATION AND GROWTH DURING ELECTROMIGRATION IN THIN FILMS. JOURNAL OF APPLIED PHYSICS, 42, 5671,1971, Copyright 1971 American Institute of Physics)*

Figure 2.10: Combination of current density and temperature which will produce electromigration damage in Ti-Pt-Au metallization. [English 1974] (*reprinted with permission from ENGLISH, A.K., TAI, K.L., AND TURNER, P.A. ELECTROMIGRATION OF TI-AU THIN-FILM CONDUCTORS AT 180C. JOURNAL OF APPLIED PHYSICS, 45, 3757-3767, 1974, Copyright 1974 American Institute of Physics*)

Figure 2.11: The mean time to failure vary with current density and temperature in a complex manner. Simple power laws can represent experimental observations over a small range of current densities [Venables 1972] (© *1972 IEEE*)

Figure 2.12: Temperature dependence has been represented by an apparent activation energy which changes with test conditions [Venables 1972] (© *1972 IEEE*)

Figure 2.13: Influence of baseline temperature on mean time to failure [Venables 1972] (© *1972 IEEE*)

Table 3: Estimated MTTF values for electromigration at 85C, as a function of the exponent used in Black's Equation (© *IEEE*)

Figure 2.14: Location of failure versus temperature distribution along a stripe for high current condition [Lloyd 1988] (©*1988 IEEE*)

Figure 2.15: Location of failure sites versus temperature distribution along a stripe length for low current condition [Lloyd 1988] (© 1988 IEEE)

Figure 2.16: Effect of temperature on the MTF versus grain size [Attardo 1970] (*reprinted with permission from ATTARDO, M.J. AND ROSENBERG, R. ELECTROMIGRATION DAMAGE IN ALUMINUM FILM CONDUCTORS. JOURNAL OF APPLIED PHYSICS, 41, 2381, 1970, Copyright 1970 American Institute of Physics*)

Figure 2.17: MTF due to electromigration versus aluminum line width @ 182°C [Kwok 1989] (© *1989 IEEE*)

Figure 2.18: MTF due to electromigration versus Al-Cu line width @ 182°C Kwok 1987] (© *1987 IEEE*)

Figure 2.19: Growth kinetics of wedge voids at various aging temperatures [Yost 1989] (© *1989 IEEE*)

Figure 2.20: Far field stress in aluminum as a function of temperature [Yost 1989] (*Courtesy of Sandia Laboratories*)

Figure 2.21a: Time to failure for crack-like growth [Yost 1989]. The results presented here are very conservative, because the models do not take into account stress relaxation effects at high temperature. (*Courtesy of Sandia Laboratories*)

Figure 2.21b: Time-to-failure for wedge like growth [Yost 1989]; The results presented here are very conservative, because the models do not account for stress relaxation effects at high temperatures (*Courtesy of Sandia Laboratories*)

Figure 2.22a: Temperature dependence of lifetime under stress driven diffusive voiding for various notch angles [Okabayashi 1991] (*Courtesy of ESREF*)

Figure 2.22b: Lifetime under stress driven diffusive voiding versus temperature for different line widths [Okabayashi 1991] (*Courtesy of ESREF*)

Table 4: Effects of H_2, He, Ar and N_2 ambients versus temperature. (© *1978, 1979, 1982 IEEE*)

Table 5: Summary of kinetic parameters for electromigration and extent of damage sustained by AL-2% Cu thin stripes (current stressed at $3x10^6$ A/cm^2). An almost constant activation energy at different heating rates suggests that the phenomena is temperature - independent in hydrogen and helium ambients. (© *1988 IEEE*)

Figure 2.23: SEM micrograph shows electrostatic discharge damage to a metallized runner which caused failure indicated by a latched signal. The ESD event originated in the silicon substrate of an integrated circuit and erupted outward along one side of the runner causing two distinct craters along with the disturbed metal. (*Courtesy of Motorola*)

Figure 2.24: SEM micrograph (upper) shows the site of an ESD artifact on an IC indicated by an arrow. SEM micrograph (lower) details the ESD damage evidenced by a large crater in one edge of the metallized runner. (*Courtesy of Motorola*)

Figure 2.25: Photomicrograph (upper) shows the site of an ESD subsurface artifact on a digital switch transistor marked with an arrow. Photomicrograph (lower) provides a magnified view of the polyp-like pattern of discoloration caused by ESD-induced breakdown between the emitter and collector. (*Courtesy of Motorola*)

Figure 2.26: Temperature dependence of TDDB for electric field of 8MV/cm [McPherson 1985] (© *1985 IEEE*)

Figure 2.27: The electric field acceleration parameter is inversely dependent on temperature [McPherson and Baglee 1985] (© *1985 IEEE*)

Table 6: Comparison of various temperature dependent dielectric breakdown models. (© *1985 IEEE*)

Figure 2.28a: Photomicrograph locates an area of electrical overstress an IC which was probably initiated by an ESD event. (*Courtesy of Motorola*)

Figure 2.28b: Photomicrograph shows the EOS-charred metallization with a probable ESD crater indicated by an arrow. (*Courtesy of Motorola*)

Figure 2.29: Photomicrograph shows the site of an electrical overstress artifact on a load switching integrated circuit marked with an arrow. The circular area of EOS-charred metallization surrounds a circular surface crack in the IC. the size and shape of the crack strongly suggests damage due to disturbance of a wirebonding tool, which probably predisposed the failure. (*Courtesy of Motorola*)

Figure 2.30: Photomicrograph shows effects of electrical/thermal overstress on parallel segments of a radio frequency power transistor. Charred, melted or premolten emitter and base metallization resulted when the device was unable to dissipate heat generated by the transistor. This was due to a conchoidal fracture of the ceramic attachment between the transistor and heat sink. (*Courtesy of Motorola*)

Figure 2.31: Photomicrograph shows the electrical/thermal overstress of a radio frequency power transistor evidenced by all emitter fusing links (Arrow) being blown. Some melted emitter and base metallization is also apparent. Fusing links may sometime limit the effects of transient electrical overstress and enable the transistor to function at reduced power. (*Courtesy of Motorola*)

Figure 2.32: Silicon resistivity versus temperature [Runayan, 1965] (© *1965 McGraw Hill*)

Figure 2.33: Photomicrograph shows the effect of massive electrical/thermal overstress on a PNP transistor. This evidenced by a large crater and cracks in the phenolic plastic encapsulant. A short circuit of a related printed circuit board runner caused this primary failure. (*Courtesy of Motorola*)

Figure 2.34: Photomicrograph shows the electrical overstress indicated by fusing of three parallel wirebonds that carried unfused B+ to the collector pedestal of a power transistor, part of a radio frequency power amplifier. (*Courtesy of Motorola*)

Figure 2.35: Photomicrograph shows the effects of massive overstress on Zenner diode chip. The anode metallization was melted and discolored in a semicircle around the remnant wirebond, suggesting the failure might have been predisposed by wirebonding tool damage. Black spots on the anode are remnant bits of plastic irreversibly polymerized by localized heating due to the overstress. (*Courtesy of Motorola*)

Figure 2.36: SEM micrograph (upper) shows decapsulated avalanche diode that failed in a transient suppressed application due to massive electrical overstress. SEM micrograph (lower) details the large crater extending from the anode metallization and well into the silicon substrate. Failure was indicated by an anode to cathode short circuit. (*Courtesy of Motorola*)

Figure 2.37: Influence of temperature on ionic current due to sodium and potassium ions [Hillen 1986] (*Reprinted from Barbottin, G, and Vapaille, A./ Instabilities in Silicon devices, ©1986, page 413, with kind permission from Elsevier Science - NL, Sara Burgerhartstraat 25, 1055 KV Amsterdam, the Netherlands*)

Figure 2.38: Change in threshold voltage due to hot electrons is much greater at lower temperature than at higher temperature [Matsumoto 1981] (© *1981 IEEE*)

Figure 2.39: Substrate characteristics for a 5μm device at 20°C and 100°C, V_g=3V and 7V, V_T=9V [Hsu 1984]. The variation in device characteristics between 20°C and 100°C indicates that degradation due to hot electrons is almost temperature independent in this range. (© *1984 IEEE*)

Figure 2.40: Drain characteristics for a 5μm device at 20°C and 100°C, V_g=3V and 7V, V_T=0.9V [Hsu 1984]. (© *1984 IEEE*)

Figure 2.41: substrate current dependence of device lifetime at 20°C and 100°C [Hsu 1984]. The small variation in device lifetime for temperature variation between 20°C and 100°C clearly indicates that the degradation due to hot electrons is temperature independent in this range. (© *1984 IEEE*)

Chapter 3

Figure 3.1: Chip cracking from the surface flows in a chip seal package (*Courtesy of Motorola*)

Figure 3.2: Die cracks caused by crack initiation on the backside of the die, caused by ejector pins. (*Courtesy of Motorola*)

Figure 3.3: Vertical depth of indentation measured using the Laser Scanning Micrograph (*Courtesy of Motorola*)

Figure 3.4a: Scanning Electron Microscope Image of a Wirebond - overbonded to a thick film. (*Courtesy of Motorola*)

Figure 3.4b: Scanning Electron Micrograph Image of a Wirebond - with heel cracks induced by overbonding. (*Courtesy of Motorola*)

Figure 3.5a: Scanning Electron Micrograph Image of a Wirebond - normal configuration. (*Courtesy of Motorola*)

Figure 3.5b: Scanning Electron Micrograph Image of a Wirebond - lifted wirebond due to underbonding. (*Courtesy of Motorola*)

Figure 3.6: Scanning Electron Micrograph Image of a Wirebond - with neck down induced by overbonding of a gold ball bond. (*Courtesy of Motorola*)

Figure 3.7a: Scanning electron Microscope Image of an overbonded gold ball bond. (*Courtesy of Motorola*)

Figure 3.7b: Magnified Scanning Electron Microscope Image of an overbonded gold ball bond. (*Courtesy of Motorola*)

Figure 3.8: De-adhesion of a copper pad from a Neodymium-Titanate substrate. (*Courtesy of Motorola*)

Figure 3.9: Phase identification versus temperature for LMCA 212B, Owen-Illinois 99+ and Honeywell thin film systems [Newsome 1976] (© *1976 IEEE*)

Figure 3.10: Change in wirebond resistance versus temperature [Newsome 1976] (© *1976 IEEE*)

Table 1: Summary of optical and SEM findings for three gold systems [Newsome 1976] (© *1976 IEEE*)

Figure 3.11: The formation of significant amounts of purple plague can take years in the equipment operating range of -55 to 125°C [DM Data, 1990]

Figure 3.12: Temperature dependence of time to decrease copper-gold bond strength 40% below the as bonded strength [Hall, 1975] (©*1975 IEEE*)

Table 2: Temperature dependence of interdiffusion in various metal systems

Figure 3.13a: CSAM image of the top of the die, showing that greater than 75% of the interface is delaminated (shown in red) - (*Courtesy of Motorola*)

Figure 3.13b: CSAM Image of the leadframe, showing that all of the tips of the bonding fingers are delaminated (shown in yellow red) - (*Courtesy of Motorola*)

Figure 3.14: Temperature dependencies of plastic package cracking [Intel 1989]

Figure 3.15: Drawing of a 48 pin PLCC showing areas of delamination. (*Courtesy of Motorola*)

Figure 3.16a: CSAM Image of the top of the die, showing 0% delamination is detected at this interface. (*Courtesy of Motorola*)

Figure 3.16b: CSAM image of the leadframe, showing minor delamination on six bonding fingers (shown in yellow) - (*Courtesy of Motorola*)

Figure 3.17: CSAM Image of the die paddle from an "New Virgin" part. 100% delamination is detected at this interface surrounding the die (yellow-red). Leadframe and die surfaces are not in focus. Only paddle is in focus and can be interpreted. (*Courtesy of Motorola*)

Figure 3.18:Differential scanning calorimetry plot of a typical epoxy resin for plastic packages. The two degradation exotherms - in the neighborhood of 320°C and 350°C indicate degradation of bulk molding compound. [Tummala 1989] (*Reprinted from Tummala, et. Al., Microelectronics packaging Handbook, © 1997, Figure 5 - 31, Page I-457, with permission from Chapman and Hall*)

Figure 3.19a: Raw nickel-silver shield solder joint. (*Courtesy of Motorola*)

Figure 3.19b: Tin plated nickel-silver shield solder joint. (*Courtesy of Motorola*)

Figure 3.20a: Raw nickel-silver shield solder joint. *(Courtesy of Motorola)*

Figure 3.20b: Tin plated nickel-silver shield solder interface with copper pad and shield. (Note: similar intermetallics at the solder interface with the copper pad and the shield) *(Courtesy of Motorola)*

Table 3: Major causes of lead seal failure. [Neff 1986] *(©1986 IHS Publishing Group)*

Figure 3.21: Fatigue cracks in a ball grid array package after liquid-liquid thermal shock. *(Courtesy of Motorola)*

Figure 3.22: Solder joint fatigue cracks in a TSOP package subjected to liquid-liquid thermal shock. *(Courtesy of Motorola)*

Figure 3.23: Temperature shock profile versus time. *(Courtesy of Motorola)*

Figure 3.24: Model prediction of inelastic shear strain in solder joints versus time in 7 cycles of temperature shock from -40 to 125°C. *(Courtesy of Motorola)*

Figure 3.25: Model prediction of inelastic shear stress in solder joints versus time in 7 cycles of temperature shock from -40 to 125°C. *(Courtesy of Motorola)*

Figure 3.26: Hysteresis loop for solder joint fatigue in 7 cycles of temperature shock from -40 to 125°C. *(Courtesy of Motorola)*

Figure 3.27: Plastic work done per unit volume in the solder joints in 7 cycles of temperature shock from -40 to 125°C. *(Courtesy of Motorola)*

Chapter 4

Figure 4.1 Temperature dependence of current gain for a bipolar transistor. The temperature sensitivity of current gain is a contributing factor to hot spot formation and effects the second breakdown energy limit. It is therefore advantageous to reduce the current gain temperature by lightly doping the transistor base and limiting the phosphorus doped emitter surface concentration to about 7e19/cc. [Kaufmann and Bergh 1968, Buhanan 1969, RCA 1978] *(© 1968 IEEE)*

Table 1: Variation in intrinsic carrier concentration, epitaxial resistivity, and collector-emitter saturation voltage for temperatures between 20 and 200°C. [Bromstead 1991] *(Courtesy of HiTec Organizing Committee)*

Chapter 5

Figure 5.1: Threshold voltage variation of n-and p-channel MOSFETs (a). with bulk concentration 10^{15} cm^{-3} and (b). with bulk concentration $3x10^{16}$ cm^{-3} [Wang 1971]. The change in threshold voltage produces a very significant change in circuit performance, because even a 200 mV change in V_T does not cause a larger percentage change in V_{GS} - V_T [Hodges and Jackson 1988] (© *1971 IEEE*)

Figure 5.2: Effective channel mobility characteristics for n and p channel MOSFETs. Mobility in an inverse function of absolute temperature [Shoucair 1981] (© *1981 IEEE*)

Figure 5.3: The variation of in drain-to-source ON resistance. The ON resistance increases by a factor of 3 over a temperature range of -40 to 120 °C [Blicher 1981] *(Courtesy of International Rectifier)*

Figure 5.4: The variation in transconductance versus temperature in a temperature range of -55 to 120 °C [Blicher 1981] *(Courtesy of International Rectifier)*

Figure 5.5: The variation in drain current in the equipment operating range of -55 to 125 °C is not very significant [Blicher 1981] *(Courtesy of International Rectifier)*

Figure 5.6: Subtreshold parameter versus temperature. At a temperature greater than 150 - 200°C, diffusion leakage currents completely dominate the weak inversion drain characteristics, causing the subthreshold parameter n(T) to exponentially increase with temperature. Depending on the device design, the temperature at which the subthreshold currents state to decrease exponentially, typically represents the practical upper operation junction temperature above which no reasonable device turn-off can be achieved [Shoucair 1989] (*©1989 IEEE*)

Figure 5.7: DC transfer characteristics of a typical CMOS inverter with temperature as a parameter. At temperature > 270 °C curves degenerate into a flat line due to onset of pn-pn latchup phenomenon. [Shoucair 1984] (*©1984 IEEE*)

Figure 5.8: The variation in the peak-low-level output voltage versus steady state temperature. The reduction in output voltage indicates an increase in the noise margins at higher temperatures. [Texas Instruments 1987] *(Reprinted with permission of Texas Instruments)*

Figure 5.9: The variation in the peak high-level output voltage versus steady state temperature. The increase in the output indicates an increase in noise margins at high temperatures. [Texas Instruments 1987] *(Reprinted with permission of Texas Instruments)*

Figure 5.10: Typical safe operating area (SOA) of a power MOSFET with device protected from second breakdown. [Blicher 1981] *(Courtesy of International Rectifier)*

Chapter 7

Figure 7.1: Derating methodology for temperature tolerant design

Figure 7.2: Derating curves of life under corrosion versus temperature and duty cycle. The curves identify the various combinations of temperature and duty cycle which will result in desired life. The paradigm of higher reliability associated with lower temperature is misleading, since the same mission life of 29 years can be obtained for any temperature from 40 to 160 °C depending on the duty cycle.

Figure 7.3: Derating curves for electromigration stresses - the curves identify various combinations of current density and temperature which will result in desired life.

Figure 7.4: The variation in time to failure versus aspect ratio and temperature shows that SDDV changes its temperature dependence from steady state to inverse temperature dependence, at a temperature threshold which is a function of the passivation temperature. [Kato 1990, Niwa 1990] *(reprinted with permission from KATO, M., NIWA, H., YAGI, H. AND TSUCHIKAWA, H. DIFFUSIONAL RELAXATION AND VOID GROWTH IN AN ALUMINUM INTERCONNECT OF VERY LARGE SCALE INTEGRATION. JOURNAL OF APPLIED PHYSICS, 68, 334-338, 1990, Copyright 1990 American Institute of Physics)*

Figure 7.5a: Due to non-linear temperature dependence of life under TDDB on temperature, worst case manufacturing defect magnitude, or electric field, derating the stress below a particular value may not result in noticeable benefit in terms of increased life because the time to failure is much beyond wear-out life of the device. Derating curve for TDDB versus steady-state temperature and effective oxide thickness. The curves identify the various combinations of temperature, T, and effective oxide thickness x_{eff} which will result in desired life.

Figure 7.5b: Due to non-linear temperature dependence of life under TDDB on temperature, manufacturing defect magnitude, or electric field, derating the stress below a particular value may not result in noticeable benefit in terms of increased life because the time to failure is much beyond wear-out life of the device. Derating curve for TDDB versus steady state temperature and electric field, E_{ox}. The curves identify the various combinations of temperature, T, and electric field, E_{ox} which will result in desired life.

Table 2: Physical and chemical properties of metallization materials [Pecht and Ko 1991]

Table 3: Coating integrity index [Pecht 1991]

Table 4: Black's constants for various metallization materials [Black 1982] (*©1982 IEEE*)

Chapter 8

Figure 8.1a: For gold wire bonded to aluminum metallization. Time for heel of the wire to transform to intermetallic versus temperature. The temperature at which intermetallic formation has a dominant dependence on steady state temperature is a function of the bond geometry. [Philosky 1970, 1972] (*© 1970, 1972 IEEE*)

Figure 8.1b: For aluminum wire bonded to gold metallization - time for heel to transform to intermetallic versus temperature. The temperature at which intermetallic formation has dominant dependence on steady state temperature is a function of bond geometry [Philosky 1970, 1971] (*©1970, 1971 IEEE*)

Figure 8.1c: Time for heel to transform to intermetallic versus temperature. The temperature at which intermetallic formation has a dominant dependence on temperature is a function of bond geometry [Philosky 1970, 1971]. Time to failure versus temperature and bond pad thickness. The time to failure at any temperature may be much greater than the mission life depending on bond pad thickness. (*© 1970, 1971 IEEE*)

Figure 8.2 Various time-temperature products which will result in gold-aluminum intermetallic growth [Philosky 1970]. A horizontal line on the graph at a thickness equal to the wire thickness at the bond pad gives the time-temperature product which will result in failure after time equal to the abscissa. The slope of each line gives the rate constants at various temperatures. (*© 1970 IEEE*)

Chapter 9

Figure 9.1: Derating curves for mission life versus the operational parameters for dominant failure mechanisms at constant operating temperature of 125 ° C. The derating plot can be used to derate non-temperature operational stresses for cost-effective designs which are reliable at high temperature.

Figure 9.2: Derating curves for mission life versus steady state temperature for dominant failure mechanisms at a specified value of non-temperature operational stresses. The derating plot can be used to derate non-temperature operational stresses for cost-effective designs which are reliable at high temperature.

Milton Keynes UK
Ingram Content Group UK Ltd.
UKHW052019071024
449327UK00027B/2342

9 780367 400972